ENVIRONMENT AND GLOBAL MODERNITY

SAGE STUDIES IN INTERNATIONAL SOCIOLOGY

Editor
Julia Evetts, *University of Nottingham, UK*

ENVIRONMENT AND GLOBAL MODERNITY

edited by

Gert Spaargaren, Arthur P.J.Mol
and Frederick H. Buttel

SAGE Studies in International Sociology 50
Sponsored by the International Sociological Association/ISA

First published 2000

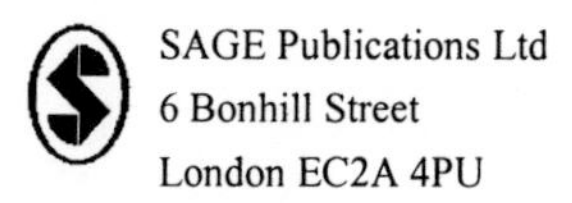

SAGE Publications Ltd
6 Bonhill Street
London EC2A 4PU

SAGE Publications Inc
2455 Teller Road
Thousand Oaks, California 91320

SAGE Publications India Pvt Ltd
32, M-Block Market
Greater Kailash – I
New Delhi 110 048

British Library Cataloguing in Publication data

A catalogue record for this book is available from the British Library

ISBN 0 7619 6766 4

ISBN 0 7619 6767 2 (pbk)

Library of Congress catalog record available

Printed and bound in Great Britain by Athenaeum Press, Gateshead

Contents

Preface vii
Abbreviations ix
List of Contributors xi

1 Introduction: Globalization, Modernity and the Environment 1
Gert Spaargaren, Arthur P.J. Mol and Frederick H. Buttel

2 Classical Theory and Contemporary Environmental Sociology: some reflections on the antecedents and prospects for reflexive modernization theories in the study of environment and society 17
Frederick H. Buttel

3 Ecological Modernization Theory and the Changing Discourse on Environment and Modernity 41
Gert Spaargaren

4 Modern Theories of Society and the Environment: the Risk Society 73
Eugene A. Rosa

5 Social Constructions and Social Constrictions: toward analyzing the social construction of 'the naturalized' as well as 'the natural' 103
William R. Freudenburg

6 Globalization and Environment: between apocalypse-blindness and ecological modernization 121
Arthur P.J. Mol

7 Environmental Social Theory for a Globalizing World Economy 151
Michael Redclift

8 The Ideology of Ecological Modernization in 'Double-Risk' Societies: a case study of Lithuanian environmental policy 163
Leonardas Rinkevicius

9 Political Modernization Theory and Environmental Politics 187
Pieter Leroy and Jan van Tatenhove

10 Ecological Modernization and Post-Ecologist Politics 209
Ingolfur Blühdorn

11 Self-organizing Complexity, Conscious Purpose and 'Sustainable Development' 229
Ernest Garcia

Index 255

Preface

This volume originated out of a conference on 'Social Theory and the Environment' organized under the auspices of the International Sociological Association (ISA) by the Research Group on 'Environment and Society'. It was a so-called 'regional conference' organized in between two ISA-world congresses which have been aimed at presenting the whole spectrum of sociological activity throughout the world. This regional conference had a more specific focus, concentrating itself on theoretical or conceptual issues within the field of environmental social sciences.

There are several reasons for having both a conference and a book which give a certain primacy or priority to theoretical issues within the field of environmental social sciences. We will shortly discuss three reasons for doing so.

First, the relative lack of a common conceptual ground can be said to be one of the key factors negatively influencing the future development of the environmental social sciences. This lack of common ground is rooted of course in the disciplinary boundaries that exist also within the environmental social sciences. Leaving the economists aside, we still are left with a great variety of disciplines which all have modest or more substantive records in the environmental field: philosophers, political and administrative scientists, sociologists, (social) psychologists and historians. Although these disciplines may in principle or in theory share a methodological foundation, in practice they sometimes seem more eager to stress the differences that come along with the specific sets of societal issues they traditionally address. The fragmentation that results from these processes of distinction and competition among the different environmental social sciences seems to weaken the position of the social sciences vis-à-vis the natural sciences. The natural sciences are known for their still dominant position in the environmental field, both with respect to the research funds and facilities they have access to as well as regarding the definition of the environmental problem they put forward. While the call for a really interdisciplinary approach seems to be nowhere stronger than within the environmental field, we think the social sciences are in some respects not yet ready for the kind of collaboration with the natural sciences that policy makers are asking for. In our view, reflecting on the theoretical and conceptual issues that the environmental social sciences have in common could strengthen their position vis-à-vis the natural sciences and highlight the specific contributions that can and cannot be expected from social scientists when it comes to doing interdisciplinary research in the future.

Second, theoretical issues are not so well developed within the environmental social sciences because a significant share of the practitioners are simply not interested in the kind of research that is regarded as 'abstract theoretical' or 'highly formal' in character. In their endeavor to please policy makers with results that are 'relevant' in terms of being applicable in the short run as well as fitting smoothly within the existing policy frames, they keep conceptual exercises on a level that makes them easy to understand for non-scientists as well. Although there seems to exist some differences in this respect between the environmental social science tradition in the USA on the one hand and some European countries on the other, the overall conclusion – that the mainstream environmental social research can be said to be predominantly empirical in character – seems to be valid to a considerable degree. Though we consider empirical research as an indispensable ingredient of environmental social sciences, we think that one cannot and should not stick to the most recent tables, figures, and data even when the main objective is to do policy-relevant research. In view of the incredibly high pace of change that characterizes modern policies, one runs the risk of figures being outdated the very moment they are published. Moreover, the definition of policy-relevant research might, against this background of accelerating change, soon become adjusted in the direction of medium- and long-term research which can stand on its own and which is theoretically well informed.

Third and finally, theoretical research in the environmental social sciences has been frustrated or at least been handicapped by the fact that the founding fathers, the classical thinkers who delineated the field of social sciences so far, paid little attention to environmental problems at all. This is as much true for Marx, Kant and Hegel as it is for Weber, Hobbes, Durkheim, Simmel and Mead. When leading contemporaries argue that the legacy of the classics needs to be rethought and taken away from its 19th century footing, they should have the immediate consent of environmental social scientists too. When this challenge of reinventing and redefining the social science classical tradition is taken up in a serious way, environmental issues inevitably belong to the core themes to be discussed. We hope that this volume will contribute to strengthening the disciplinary identity of environmental sociology as well as to the greening of sociology.

This book would not exist if Neil Guppy, as editor of the Sage Studies in International Sociology, had not asked us to start this project. The book would not be as attractive as it is without the efforts made by Corry Rothuizen. We also would like to thank the other members of the organizing board, Riley Dunlap and Peter Dickens, and especially Guus Gijswijt from SISWO in the Netherlands because of his decisive role in organizing the Woudschoten conference from which this book resulted.

Wageningen/Madison, April 1999.

Gert Spaargaren
Arthur P.J. Mol
Frederick H. Buttel

Abbreviations

BSE	Bovine Spongiform Encephalopathy (mad cow disease)
CEE	Central and Eastern Europe
CFCs	Chloro Fluoro Carbons
ECE	Economic Commission for Europe
EMAS	Environmental Management and Audit Scheme
EU	European Union
FOEI	Friends Of the Earth International
G7	Group of Seven (richer countries of the world)
GATT	General Agreement on Tariffs and Trade
GEC	Global Environmental Change
GEF	Global Environmental Facility
GNP/GDP	Gross National/ Domestic Product
GPP	Greatest Permissible Pollution
HCRs	High Consequence Risks
HEP	Human Exemptionalist Paradigm
HIID	Harvard Institute of International Development
IMF	International Monetary Fund
IPCC	Intergovernmental Panel on Climate Change
ISO	International Standard Organization
LEIF	Lithuanian Environmental Investment Fund
MEMO	MEns en Milieu-vriendelijk Ondernemen (Man and environmental friendly enterprises)
MNE	Multi National Enterprises
NAFTA	North American Free Trade Agreement
NEP	New Ecological Paradigm
NEPP	(Dutch) National Environmental Policy Plan
NGO	Non-Governmental Organization
NICs	New Industrializing Countries
NIDL	New International Division of Labor
NIMBY	Not In My BackYard
OECD	Organization for Economic Cooperation and Development
PESTO	Public Participation and Environmental Science and Technology Policy Options
PHARE	Pologne/Hongrie: Assistance á la Restruction des Economies (EU aid programme for former CEE countries)
PPP	Pollution Prevention Pays/Polluter Pays Principle
PRA	Probabilistic Risk Assessment
RAP	Rational Actor Paradigm

SMCS	Socio-Material-Collective-Systems
SSK	Sociology of Scientific Knowledge
TPP	Temporarily Permissible Pollution
TRI	Toxic Release Inventory
UN	United Nations
UNCED	United Nation Commission on Environment and Development
USAID	United States Agency for International Development
WCED	World Commission on Environment and Development (Brundtland-commission)
WHO	World Health Organization
WTO	World Trade Organization
WWF	World Wide Fund for Nature

List of Contributors

Ingolfur Blühdorn teaches politics and social theory in the Department of European Studies at the University of Bath, UK. He obtained his PhD from the University of Keele. In recent years his research has focused on the relationship between ecological issues and contemporary social theory.

Frederick H. Buttel is Professor of Rural Sociology and Environmental Studies at the University of Wisconsin, Madison. He is currently the President of the Environment and Society Research Committee (RC 24) of the International Sociological Association and Co-Editor of *Society and Natural Resources*.

William R. Freudenburg is a Professor of Rural Sociology and Environmental Studies at the University of Wisconsin, Madison, who has written well over 100 articles and book chapters on society-environment relationships. His books include *Nuclear Power and the Public: are there critical masses?* (with Eugene Rosa) and *Oil in Troubled Waters: perceptions, politics, and the battle over offshore oil drilling* (with Bob Gramling).

Ernest Garcia is Professor of Environmental Sociology and Social Change at the University of Valencia, Spain. He has carried out research on regional sustainable development, social conflict and the environment, and household consumption and sustainability.

Pieter Leroy is Professor of Political Sciences of the Environment and chair of the Department of Environmental Policy Sciences at Nijmegen University, the Netherlands. He chairs the four year education program on Social and Political Sciences of the Environment. His research and publications focus on new developments and arrangements in environmental politics and policies.

Arthur P.J. Mol is an environmental sociologist at Wageningen University. He has published widely in the field of environmental sociology and environmental policy. He is the central coordinator of several major research projects on environmental transformations in Europe and in South-east and East Asia. He is connected to the new *Journal for Environmental Policy and Planning* and also serves as co-secretary for the ISA Research Group on Environment and Society.

Professor Michael Redclift is a member of the Department of Geography, King's College, University of London, where he is leader of the Environment research group. Formerly he was Professor of International Environmental Policy at Keele University, and Professor of Environmental Sociology, Wye College, University of London. He was the first Director of the ESRC's Global Environmental Change research programme (1990-1995).

Leonardas Rinkevicius is Associate Professor in environmental sociology and policy at the Faculty of Social Sciences, Kaunas University of Technology, Lithuania. He is a member of the board of the Environment and Society Research Committee (RC 24) of the International Sociological Association, and he is also in the action planning committee of the International Greening of Industry Network.

Eugene A. Rosa is Professor and Chair of Sociology and the Edward R. Meyer Distinguished Professor of Natural Resource and Environmental Policy at Washington State University. He has published widely in the fields of environmental sociology, risk, and environmental policy. Among his recent published works is the co-authored *Risk, Uncertainty, and Rational Action* (Earthscan).

Gert Spaargaren is an environmental sociologist working at Wageningen University. He has published several articles on ecological modernization theory and environmental sociology. His dissertation is titled *'The ecological modernization of production and consumption'*. He serves as co-secretary for the ISA Research Group on Environment and Society.

Jan van Tatenhove is lecturer in Political Sciences of the Environment at the Department of Environmental Policy Sciences at Nijmegen University, the Netherlands. His research focuses on the institutionalization of environmental policy and new policy arrangements in environmental politics and physical planning.

1

Introduction: Globalization, Modernity and the Environment

Gert Spaargaren, Arthur P.J. Mol
and Frederick H. Buttel

Scattered landscapes: globalization and the changing nature of borders

The process of globalization did not just alter the character of modern societies themselves, it also influenced sociological theorizing on modernity. The post hoc, seemingly well-ordered nation-state system was transformed into a global system which surrounded the nation-state in a tight web of socio-economic, political and cultural relations. Although perhaps suggested by the term itself, globalization is not to be investigated only or even primarily at the highest analytical level possible. The globalization process also implies the reshaping of social relations at the local and regional levels.

The net result of globalization as an historical process is a wide range of new emerging local-global relationships. Because of these new types of interrelations, concepts like democracy, sovereignty and even the very notions of 'society' and 'state' have to be rethought and redefined (cf. Held, 1995). The nation-state is no longer the 'natural unit' or logical starting point for theorizing social systems. It is widely recognized by contemporary sociologists that major parts of 19th century sociological thinking need to be refined and reformulated in this respect. Because this task is dealt with from a variety of theoretical perspectives, one may conclude that the scattered world-landscape of our globe – consisting of a variety of socio-political and economic units and therefore sometimes referred to in terms of a new mediavialism – is more or less mirrored by the scattered landscape of sociological theorizing. For starters, sociology must be a puzzling terrain, now that the post hoc, seemingly well-ordered system of the 'orthodox consensus' and of distinguishable streams of thinking developed out of a fixed set of classical founders and of intuitive knowing about left-wing and right-wing social theorizing is gone.

One of the obvious themes that gained new relevance throughout the process of globalization is the notion of borders. Today we witness profound changes both between socio-political and geographical units as well as between disciplines and theoretical streams of thought. Hence the importance of rethinking the notion of borders both in terms of its empirical content and with regard to its theoretical meaning. The city-wall of traditional societies is very different from the fixed borders of the nation-state system, which in turn are very different from the sometimes permeable and multi-faceted borders that come along with the inter-societal systems of the global order.

There exists, however, one additional border-transformation that has to be distinguished, one that has special relevance for environmental sociology. When the notion of society or social system is redefined, at the same time its relationship with the social and natural environment is reformulated. In what follows it will be argued that in the era of global modernity we should not only reconsider the implicit identification of the concept of 'social system' with that of the 'nation-state', but also and at the same time we have to rephrase the notion of borders between social systems and their social and natural 'environments'.

Borders between environment and society

The notion of scattered landscapes is also relevant when investigating only certain aspects of these landscapes and some more or less well circumscribed streams of thought within sociology. Environmental sociology can be regarded as a by now reasonably circumscribed sub-discipline which has society-environment dialectics as its main object of theorizing and research and which is indeed not left untouched with respect to the dynamics behind the scattered landscapes. It will be argued throughout this book that the theoretical landscape within environmental sociology has become more diverse – and according to some more confusing – in recent years and that this diversity has something to do with transformations in its object, in the society-environment relations themselves. The changing character of society-environment interdependence is related to the shifting borders between social systems and their social and natural environments, as will be shortly illustrated in what follows.

In dealing with society-environment relationships, it was most often implicitly suggested by (environmental) sociologists that society meant the 'nation-state'. Now that in the era of global modernity the notion of nation-state is no longer taken for granted, the concept of society-environment interrelationships stands in need for discussion and more precise definition too. Characteristic of societies in the context of the nation-state system was the obvious significance of borders. Like the city-wall in earlier times, these borders closed off the social system at hand from its 'outside' environment. Borders were not only relevant for protecting inhabitants of the city or citizens of the nation-state; they also made clear for everyone the meaning of 'inside' and 'outside' and distinctions such as 'endogenous' and 'exogenous' factors and developments. In explaining the development of the city or the state as a social system, internal or endogenous factors and processes tend to be privileged in sociology, as is argued by Anthony Giddens in his critique of 'unfolding models of change' or models of 'endogenous development' (Giddens, 1984: 164).

The above mentioned idea of a social system which is neatly closed off from its 'outside environment' by fixed borders also seems to have been very influential when it comes to conceptualizing society-nature relationships. The literal meaning of the concept of 'environment' already suggests a basic scheme of 'something to be surrounded by something else' – that something else being located outside the social system itself. In most of the basic books on environmental sciences, the first figure that one comes across does visually tear apart and counterpose the social and the natural system, society and its environment. This tearing apart can be – and is – done in a variety of ways. Sometimes, the environmental system is depicted as the outside unit delivering inputs to the social system, and also handling outputs that are originated by the social system. The idea of environmental 'functions' or similar concepts is derived from this kind of model. In other models, the environment is depicted as the sustenance-base or the material substratum on which society is based, on which it is 'footed'. This model always bears with it some notion of 'limits' to social development set by physical factors. Some other pictures are more inspired by the idea of a biosphere surrounding the globe, referring to climatic problems in particular.

Nature and environment pulled into society

Whatever their concrete shape, all these models rest on the assumption of a strict – more or less physical or symbolic – wall or border closing off the social system from its natural environment. Especially as illustrated by Ulrich Beck (1986) in his influential book on the risk-society, this counteropposing of society and nature as two separate bodies or realities is very much a characteristic of 19th century social thinking on society and its natural environment. Only after Chernobyl have we come to fully recognize that our border mentality no longer suffices. There is no 'inside' and 'outside' when it comes to dealing with environmental risks in the context of a globalized modernity. The concept of the risk-society literally captures this idea of society and (environmental) risks being inseparable. Society and nature are not only interconnected or intertwined: nature/the environment is 'pulled into society'. When you buy the new kitchen, the environment comes along with it in terms of the risk of formaldehyde. When you eat your red meat, drink your tap-water or simply relax in the sun, environmental risks are always with you and you know it. Nature has become an integral part of societal reproduction both in its positive guise as a provider of the material assets of social life and in its negative dimensions as a risk to our health, safety and the possibilities for future development. Some of the major environmental problems are 'democratic' (as Beck calls them) or borderless (as we prefer). Climate change and the hole in the ozone layer are prototypes of the High Consequence Risks that cannot be isolated from society or from certain fractions or social classes within society. The fact that these problems cannot be bordered in terms of 'contained' within a definite time-space zone gives them their very threatening character: they don't bother about human made borders. As Buttel and Freudenburg point out in their contribution, this does not mean that these High Consequence Risks – and let alone the majority of 'normal' environmental risks – affect every group, class or fraction in a similar way.

Ecological modernization in between human ecology and postmodernity

The loss of borders or bordering mechanisms separating 'societies' and their 'environments' seems to us to be one of the major challenges for environmental sociology in the present day. From the macro-ecology of climate change to the micro-ecology of domestic waste, we must come to grips with the idea that our relationship with nature or the environment cannot be taken for granted any longer but has to be reflexively organized. This can be done in different ways, and within environmental sociology we think at least three major schools of thought can be distinguished in this respect: the human ecology tradition, the ecological modernization school of thought and postmodern views of the environment. These three environmental sociological perspectives are directly related to the wider sociological debates on the character of modernity. As will become clear below, the human ecology tradition can be understood as a reaction to the long time neglect by 'mainstream' sociology of the materialist dimension of social practices and institutional developments. Ecological modernization, together with risk-society theories and some moderate versions of constructivism, are the environmental pendants of reflexive modernization perspectives, while the last tradition, including relativist constructivism, can be labeled as green versions of postmodernity perspectives. From these three streams of thought, the environmental pendants of reflexive modernization theories, and especially ecological modernization theory, are of most central concern in this volume. It is for this reason that the human ecology tradition and the postmodern perspective will not be dealt with at length in this introduction but will only be sketched in order to position the ecological modernization approach – and to a lesser extent risk-society and moderate constructivism – within the broader field of environmental social science.

Human ecology

The human ecology tradition stood at the birth of environmental sociology. The 'New Environmental Paradigm' that Riley Dunlap and others advocate was, in fact, one great effort to redefine the relationships between human societies and their natural environments. The notion of the 'Web of Life' was taken to illustrate the interconnectedness of social and natural processes and factors, and at the same time the concept served as an instrument for criticizing the borderline notions that figured in mainstream sociology at that time. NEP, the New Ecological Paradigm, was opposed against HEP, the Human Exemptionalist Paradigm that dominated the social sciences. The HEP not only defined the environment as something 'out there' but also did not seem to concern itself about its possible impacts on human societies.

The HEP-NEP debate had a high mobilizing potential for environmental social sciences and scientists, and it did contribute significantly to the establishment of a new field of study within the social sciences. Nevertheless, we think the basic theoretical notions and the main general direction that came along with this tradition, are somewhat flawed: the 'winding road towards human ecology' in the end could perhaps turn out to be a dead-end. The main reason for this is the way in which the issue of dealing with the boundaries between the social and the natural world is settled in the human ecology tradition.

What made human ecology from the Chicago School onward into a distinguishable school of thought was the general idea that there is a realm within societies that is shaped primarily by non-social or sub-social factors: the biotic community (Nelissen, 1972). Within this biotic sub-sphere of society there are facts and phenomena that can and must be explained with reference to the realm of the biological and natural sciences. The biotic community is counterposed to 'society' or 'culture', and both realms are supposed to have their own, different logics or dynamics. The non-social sphere of the biotic community is thought to be governed by ecological laws or dynamics which must be explained with the help of concepts such as survival of the fittest, adaptation, invasion, overshoot, and so on. Should the social system be left untouched by human made policies and politics, one could witness some patterns to occur which result from these bio-natural mechanisms.

To the degree that this picture of the basics of human ecology is valid, it can be concluded that crossing the society-nature border within the human ecology tradition means taking us 'beyond the social' in a very specific way. From the Chicago School up to Peter Dickens' (1992) influential book *Society and Nature*, it is argued that some social facts or behavioral patterns should be understood or explained by making use of 'laws' or regularities that exist across different – human and non-human – species. Behavioral practices in the field of child raising, dating and mating or the practices of defending a territory or hierarchical order are thought to be governed by – and thus should be partly explained by referring to – 'basic' laws that in fact take us beyond the social realm and logics.

Ecological modernization theory: looking beyond the social without lapsing into ecologism

The value of human ecology in all its different forms – and its contribution to the emergence of environmental sociology in the 1970s – is the fact that nature and the environment are no longer simply disregarded or done without. The border, or sometimes the iron wall, between the social and the natural as it was created and sustained by most of the classical sociological thinking was criticized, and a more reflexive mode of relating the social and the natural was plead for. The social should not be treated in isolation from the natural, as modern societies are an inherently 'materialistic' affair. The environment is not the passive realm of risks and opportunities that exists somewhere 'out there', waiting to be used one day and one way for serving mankind. We cannot explain or understand the human project by referring to 'endogenous' or 'internal' social facts only. History is an inherently natural affair, and human ecology in general and the HEP-NEP debate within environmental sociology in particular have contributed to the better understanding of this naturalness of history.

The unsatisfactory element of human ecology – from the classical Chicago School on to present forms of so called 'deep ecology' – has to do with the tendency to try and restore the interrelationship between the social and the natural world in such a way that they seem to underscore the fact that all fact, events, goals, outcomes, patterns etc. *as we know of them*, are socially mediated. There is no such thing as the 'biotic community' when this should mean sub-social or non-social. The naturalness of history is mirrored by the historicity of nature (Harmsen, 1992).

Ecological modernization theory takes up the task of redefining the borders between modern societies and their social and natural environments. The need for such a redefinition is fully recognized, and in this respect there is general agreement with the proponents of the HEP-NEP approach and other human ecologists. Ecological modernization theorists also argue that the notion of 'environment' should be taken seriously and not be left un- or under-theorized by social scientists by first constructing a city-wall as a border between social systems and their 'outside environments' and then argue that 'social facts should be explained by using social facts and factors alone'. What is conceived of as 'social' – e.g. that what happens inside the city-wall – cannot be explained without reference to the natural, without taking into account the relationships with the outer-city. This has become one of the central notions in all contributions to the reflexive modernization perspective.

Within ecological modernization theory it is agreed that we must go beyond the social by taking into account naturalness, substance flows, energy flows, materials circulating throughout human societies etc. However, in restoring the analytical priority of the environment we should not throw away the baby with the bath water. The crucial difference between ecological modernization theory and the human ecologies of different kinds is the contention that we must not replace the former disregard of nature with some form of present-day biologism or ecologism. The former disregard of nature from the side of most of the classical and post-war sociological theories is linked to the crucial design-fault in some of the major institutional clusters of modern societies (Giddens, 1990). When analyzing the industrial mode of production and consumption, the attention of most sociologists used to be focused exclusively on factors such as capital, technology and labor. Environmental factors were regarded as 'external factors' in the sense not only of being 'available for free' but also in terms of being of secondary importance when it comes to explaining the dynamics of industrial production. When ecological modernization theorists talk about 'repairing' this design fault of modern industrial production, they request that environmental factors not only be taken into account, but also that they are structurally 'anchored' in the reproduction of these institutional clusters of production and consumption. To illustrate the fact that something more serious is at hand than only 'pricing' things that used to be regarded as 'external costs' – the solution as it is pursued by most of the economists working in the neo-classical tradition – ecological modernization theorists use the more encompassing vocabulary of 'rationalizing production and consumption'. This notion refers to ecological rationalities (such as the closing of substance cycles and extensifying energy-use) that have a meaning 'of their own', implying that they are independent vis-à-vis other – for example, economic – rationalities that are involved in the reproduction of production-consumption cycles (cf. Spaargaren, this volume).

Once established as independent criteria that have relevance sui generis, one can start comparing and interlinking ecological rationalities with other types of rationalities. Important concepts here are the Polluter Pays Principle (the intersection of ecological and political rationalities), the Pollution Prevention Pays Principle (intersecting ecological and economic criteria), the idea of 'Doppelnutzung' etc. It is these kinds of concepts that establish a link between ecological modernization as a general theory of societal change on the one hand and ecological modernization as a political program or policy discourse on the other. In his chapter, Rinkevicius explores these interlinking rationalities against the background of ecological modernization.

The recognition of the need to compare, link and sometimes mate with other types of rationalities involved in the industrial mode of production distinguishes the ecological modernization approach from more 'principled approaches' which lend ecological criteria an almost absolute priority above other rationalities.

Postmodern critiques of (green) grand narratives

Some will conclude from this short outline of ecological modernization theory that we are dealing here with nothing less than a new grand narrative in the making. Isn't the idea of the materiality of social systems, and the accompanying notion of ecological criteria involved in their reproduction, in principle a trans-historical and trans-cultural concept? Can one reasonably argue that the imperative of the 'sustainability' of social systems is in fact an universal one?

When understood in this way, it makes the fact of postmodern authors being among the most fierce critics of this approach understandable and predictable. It will become clear from this volume that, indeed, postmodern critiques of ecological modernization theory are as fierce as those of the more traditional critics working from a de-industrializing perspective used to be. The focus of these postmodernists is no longer on the need for 'dismantling' the institutions of modern societies instead of just 'repairing' them, as the debate on the 'technological fix' character of ecological modernization would have it. This original kind of criticism – developed for instance in the school of counter-productivity theory – does in fact underscore the need for sustainability criteria to be used in a very strict and regular way. The crucial difference of opinion between counter-productivity theorists and ecological modernization thinkers is the contention that a routinized and strict application of sustainability criteria would result in the situation in which most of the basic institutions of modernity governing contemporary production and consumption in modern industrial societies will anyhow fail (counter-productivity theory), versus the idea that these institutions could also in principle pass the test of sustainability or ecological rationality (ecological modernization theory).

Postmodernist critiques are in some respects even more radical in their consequences than those of counter-productivity theorists because they contrive the very fact that sustainability criteria could or should be developed in a feasible way whatsoever. The contribution of Blühdorn to this volume can be seen as part of this tradition. The main target of these postmodernists seems to be to show that all borders are time- and space-bound 'social constructions' which can be 'played upon' now that we have become aware of this fact in our postmodern times. So also the ways in which the borders between societies and their environments are created and sustained – from the Club of Rome in the early 1970s on to the IPCC[1]-

experts of the late 1990s – can and must be criticized in order to 'liberate' us from the grand narratives of which the ecological crisis is only the latest plot.

When trying to evaluate the relevance of postmodern theories for environmental sociology in general and in relation to ecological modernization theory in particular, it is important to distinguish between different brands of postmodernism and between the different meanings of the term itself. However, distinguishing different brands of postmodern theory or schools of thought within the postmodern tradition hardly seems to be possible due to the complicating fact that the denial of borders is one of the constituting features of postmodern thinking. Some authors judged influential in postmodern circles are themselves fiercely refusing the postmodern label. Consequently, it seems necessary to be rather precise when dealing with certain ideas of certain authors referred to as postmodern. We will first describe in what respects the ideas of postmodern thinkers are important for ecological modernization theory and then go on to discuss the issue of green narratives as something on which both approaches seem to be in serious disagreement.

Postmodernism and the sociology of consumption

Baudrillard has been among the first in sociological theory to point to the relevance of consumption. As early as 1970 he pointed to the need to consider the 'mirror of production' (Baudrillard, 1998). So at a time when mainstream thinking about the industrial mode of production in social sciences was definitely productivist in its outlook, Baudrillard made us aware of the fact that the dynamics of production-consumption-cycles can no longer be properly understood when neglecting consumption. In certain respects Baudrillard can be regarded as one of the initiators of what Allen Warde, Pete Saunders, Mike Featherstone and others came to refer to later on as a 'sociology of consumption'.

In some respects this sociology of consumption can be said to stand in the tradition of urban sociology, with its emphases on 'collective consumption'. The sociology of consumption does, however, address certain questions that were not dealt with effectively in the urban sociology tradition. Questions concerning the 'meaning of consumption', the 'motives for purchasing' certain goods and services, and the different 'modes of provision' of goods and services were left un- or under-theorized. The sociology of consumption as it has been developing over the last ten years or so is also different from earlier (for example, Frankfurt School) thinking about consumer society because the former does not treat consumer-society (only) as a more or less estranging dreamworld through which people try to escape from the hard realities of the workplace. Instead, the present day study of consumption and consumer behavior is regarded as one of the vital keys to the proper understanding of the dynamics of production-consumption cycles.

Ecological modernization theory must profit from the new emphasis on consumption as it is propagated within sociology by some postmodern thinkers. Because ecological modernization was originally developed mainly in relation to the production sphere, focusing on government agencies, companies, branch-associations, social movements and other 'institutional actors', the role of citizen-consumers was not adequately dealt with in its initial formulations. In correcting the productivist bias in ecological modernization, the citizen-consumer is no longer regarded as the 'end-user' or 'final stage in the chain' but instead treated as a decisive factor for explaining the dynamics of production-consumption dynamics – dynamics that are thoroughly social in character indeed, and for that reason cannot properly be understood using natural science or ecology based models describing the flows of energy and material moving up and down the chain as, for instance, is the case in most contributions to 'industrial ecology' and 'industrial metabolism'. Consuming services and products is more than just 'converting energy and materials', and postmodern thinking cannot be provocative enough to make this clear to environmental scientists.

The social dynamics of production and consumption must be studied on different analytical levels, ranging from the ways in which people make use of products and services to express their lifestyles and identities on up to the question of how post-Fordist regimes of production and consumption organization in general effect the relationship between producers and consumers. As Kumar (1995) has shown in his well written book *From Post-Industrial to Post-Modern Society*, there sometimes seems to be a very thin line between 'modernist' debates on post-industrialism and disorganized capitalism on the one hand and theories of the postmodern social order on the other. However, as Bauman (1987: 117) puts it: '(...) the frequent confusion notwithstanding, the two debates do not share their respective subject-matters'.

The contribution that environmental sociologists can make to the debate on the changing character of production and consumption in late-modern societies is the fact that sustainability issues point to the materiality of services and products, lifestyles and daily lives. This plain fact again seems to be neglected or simply forgotten by most of the postmodern contributors to the sociology of consumption.[2] Postmodern consumption analyses should take into account the fact that material product-qualities *do* matter, and that even the most 'virtual' consumption practices can and should be evaluated in terms of their environmental impacts from cradle to grave.

Sustainability: postmodern construct or universal value?

According to postmodern thinking, every grand narrative can and should be deconstructed and shown to be arbitrary to a great extent. Since the need for sustainable development is one of the few problems that are recognized and accepted as a challenge to society all around the world, this seems to be a privileged objective for some postmodern critics.

Within environmental sociology the debate that postmodern authors triggered is reflected in the frequently cited dispute on 'realism' versus 'constructivism'. Several authors have contributed to this debate, thereby referring to postmodern issues and ideas in an implicit or explicit way (Yearley, 1991; Dunlap and Catton, 1994; Hannigan, 1995). In this volume Fred Buttel, William Freudenburg and Ingolfur Blühdorn take different positions in this debate. Standpoints vary from 'hard' or radical to 'soft' or moderate constructivism. Especially the radical or relativist variant of constructivism seems to have as a goal in itself to deconstruct or dismantle the naive beliefs that come along with environmental stories about global change, nuclear waste or soil erosion. From the fact that the environmental discourse has been changing from the early 1970s to the late 1990s with regard to priorities and approaches, it is concluded that environmental problems do not have a 'real', 'objective' existence but are instead the result of a process of framing of certain social problems by certain social actors in a very specific, sometimes arbitrary way. As these relativist constructivists would have it, sustainability as grand narrative, dominant discourse or 'story line' stands in need for a deconstruction, showing that the story could have been framed otherwise, leading to different kind of conclusions and priorities.

Ecological modernization theorists are not immune to the kind of epistemological issues touched upon by the relativist constructivists. In his book on ecological modernization Hajer (1995) seems to end up taking a position which is not too far away from where postmodernists would feel comfortable. In a more or less similar way Peter Wehling (1992) evaluates the position taken by Huber, Jänicke and other ecological modernists as being too little aware of the limitations of modernization theory in general and ecological modernization theory in particular. A more 'reflexive' approach is requested, especially when dealing with the role of science and technology in promoting sustainable production and consumption.

Mol (1996) has addressed the challenge to confront ecological modernization theory with the debate on late- or reflexive-modernity as it has been developed by Beck, Giddens, Lash and others. Although it is doubtful whether it has ever been the case, under the condition of reflexive modernity the ecological modernization of production and consumption can no longer be thought of or designed in terms of undisputed facts, values and futures. The ecological risks of reflexive modernity are no longer simply accepted on the authority of (natural) scientists, even more so if they at the same time also claim to have a privileged position in pointing out the best or most promising route towards a sustainable future. Science and technology are indeed disenchanted, and this has some potentially far reaching consequences for the ways in which environmental problems are perceived by lay-actors as well as policy makers.

The fact of science and technology being no longer undisputed and bereft of that special kind of authority bestowed on them in earlier times should not be confused with epistemological issues that explain the crucial differences that exist between the natural and the social sciences. When environmental problems are discussed, these two major – but, in principle, separate – issues are very often intertwined or dealt with simultaneously. This can be said to be the case when for example the 'social' (e.g., 'constructed') character of the climate change narrative – explained in terms of different interest groups, media and environmental movements all contributing to a specific mix of policies – would be presented in a way as to serve as proof for the more encompassing (postmodern) statement that the environmental crisis is something that is 'invented' by social actors and groups whose interests are served best by making a lot of noise about this or that particular social problem. What tends to be denied then is the fact that environmental problems do have a 'real' existence in that they belong to the types of problems that need to be analyzed and understood also in terms of the language of the natural and biological sciences to a certain extent. When ignoring this fact, we would end up where we started in environmental sociology, namely with the HEP-NEP distinction, with constructivist environmental sociology as the latest variant of exemptionalist thinking.

Local-global diversity in environmental arrangements

In a certain respect, then, ecological modernization theorists do indeed claim that there is a new grand narrative in the making. This narrative in its most fundamental form boils down to the need for all social systems to reflexively take into account and (re)organize their relationship with the environment in an era that the classical borders are dissolving to a considerable extent. From the global to the local level, specific kinds of social arrangements should be developed governing our intercourse with nature. One does not necessarily have to advance an a-historical or universal notion of 'limits' to recognize the fact that social life should be permanently monitored and reorganized with regard to its consequences for the 'environmental utilization space' that is available for our and coming generations. The need for such new socio-environmental arrangements to be developed is as 'universal' as the modern industrial system of production and consumption itself. With this system obtaining a global character these days, one can and must conclude that the need to take on board issues of sustainability is a 'universal' one indeed. That this 'universalistic' claim does not imply that the socio-environmental arrangements must take a similar shape on every spot of the earth is something that goes without saying. One just has to look at the increasing diversity of local and regional arrangements that, for example, are developed by households, neighborhoods, villages and mega-cities with respect to their handling of energy, water and waste to grasp the fact that the 'universal need' for sustainability does not imply a 'uniform' solution to result from this. Globalization means that the diversity of local-global arrangements is, in fact, increasing, also with respect to the socio-environmental arrangements that are developed all over the world. It is the task for environmental sociologists to contribute to the understanding and the future development of these types of arrangements.

About this volume

The sociological and theoretical aspects of globalization, ecological modernization and the social construction of environmental risks are among the central themes addressed by the authors contributing to this volume.

In the opening contribution Fred Buttel strongly places environmental sociology within the sociological tradition. He explores the historical roots of theories in environmental sociology by analyzing what classical sociological theory (in short, sociological contributions in line with the works of Marx, Weber, Durkheim, Simmel and others) has to offer. In doing so, he makes us aware that these 'classics' in sociology were not ignorant of the biological-material dimensions of social life and still have a major influence in contemporary environmental sociology. At the same time he also explains both the logical steps of sociological theory towards what often has been called 'exemptionalism', and the developments in American environmental sociology towards neglecting environmental improvements and overestimating the role of environmental movements in environmental reform. This forms the background of the introduction of three relatively new environmental-sociological perspectives into American environmental sociology: social constructivism, ecological modernization and risk society. While all having their drawbacks, Buttel makes a strong point in emphasizing the value of these reflexive modernization perspectives as a creative response to the traditional North American environmental-sociological literature.

These perspectives are dealt with at length in this volume, especially against the background of a globalizing world order. Ecological modernization theory as one of the environmental contributions to reflexive modernization forms the central object in the contribution of Gert Spaargaren. Against the historical background of environmental sociology, he explains the emergence of ecological modernization theory, especially in Western Europe. Spaargaren shows how from the mid-1980s onwards the central characteristics of the idea of ecological modernization have been developed and reformed against the background of, on the one hand, empirical developments in environmental policies and environmental movements in Europe, and developments and debates in (environmental) sociology on the other hand. By elaborating the idea of an ecological rationality and an ecological sphere, he seeks to strengthen the sociological foundation of ecological modernization theory. From a different perspective, Eugene Rosa introduces risk-society theory by relating it to other sociological theories and perspectives on environmental risks that have been developed, especially in constant discussion with the rational actor paradigm. Against the same background of reflexive modernization he classifies various contributions in understanding environmental risks and the way modern society deals with them, in the end concluding that it is neither possible nor desirable to integrate them in a single grand theory on risks. The third and final perspective that Buttel relates to reflexive modernization, social constructivism, is put at central stage by William Freudenburg. In describing the often fierce debates between realists and social constructivists of different kinds, he not so much chooses sides but surpasses this controversy in a specific way. He blames constructivists for their preoccupation with the social construction of environmental problems, while neglecting the social construction of what he labels social privileges: the social construction of the claims that environmental disruptions are not problematic. He provides various lines along which the social constructivists' project might be redirected.

While the three contributions above touch upon issues of globalization, the next two papers of Arthur Mol and Michael Redclift take the globalizing world order as their central focus of attention. Both authors concentrate on the interrelation of processes of globalization on the one hand, and environmental deterioration, environmental struggles and environmental reform on the other. Noting that globalization theories have tended to take account of the environment and that environmental sociology has not taken into account the processes of globalization, Arthur Mol provides insight into how globalization processes might both endanger environmental quality and construct mechanisms for triggering environmental reform. This latter process, he argues, is often misunderstood or omitted, and he fills that gap in using ecological modernization theory as a useful perspective. While Arthur Mol provides a more overarching analytical framework anchored in globalization theories, Michael Redclift focuses on the consequences of a globalizing world economy – and culture – for environmental quality, but even more for reshaping social theory. In line with the observations of Fred Buttel and others in this volume, Redclift argues convincingly that most of the classical and contemporary contributions in (environmental) sociology offer us little insight into global environmental deterioration and the failures of global environmental management. In the end, he concludes that globalization in its present forms is still destructive for both the South and the environment.

Against the background of globalization Leonardas Rinkevicius shows us in his contribution how the environmental perspectives within the framework of reflexive modernization prove valuable for analyzing developments in what he calls 'double risk societies': those Central and East European countries in transition that are confronted with not only environmental risks but economic risks as well. Using, and partly transforming, the perspective of ecological modernization, he analyses the developments in belief systems of industrialists and authorities when confronted with the need for a radical environmental transformation of Lithuanian society. He notices growing convergence in their ideas on the best strategies for environmental reforms, closely related to some core ideas of ecological modernization.

The contributions of Pieter Leroy and Jan van Tatenhove, and Ingolfur Blühdorn both assess the value of ecological modernization theory for understanding environmental change and reform in Western industrial societies. While Leroy and Van Tatenhove focus primarily on the political domain and elaborate on the theory of political modernization, they share with Blühdorn most points of discussion regarding the first generation contributions to this theoretical framework: its focus on especially the production dimension, its strong normative connotations, its Eurocentrist character, its poor attention to social struggles and its danger of evolutionism. While Leroy and Van Tatenhove acknowledge that some of these points have been addressed more recently (and even in this volume) and they themselves contribute to 'repairing' these omissions especially with regard to the political dimension, Blühdorn considers these drawbacks as too fundamental to see any future for ecological modernization theory. In writing from a postmodernist and strong social constructivist perspective, Blühdorn opts for what he calls a post-ecologist politics: the radical devaluation of the ecological critique of modern society, on the grounds that there are no longer any (ecological) grounds for an environmental redesign of the institutional order. To some extent he indeed 'falls back' to the 'exemptionalist' position, as we have indicated above.

In the last contribution Ernest Garcia compares economic, biological (system-theory) and sociological models of sustainable development. He takes up the issue of defining sustainable development from a sociological and philosophical perspective, thereby reflecting on the thoroughly social character of sustainability.

Conclusion

In the present volume we have aimed to contribute to environmental sociology by both advocating for and duly recognizing some of the shortcomings of ecological modernization theory. The co-editors are by no means of one mind regarding the attractiveness of ecological modernization theory, but they are all agreed that the rise of ecological modernization theory in the 1990s will have been one of the most creative episodes in the history of environmental sociology. Ecological modernization theory has served to highlight the importance of theorizing the processes of environmental improvement, and it suggests important new ways of understanding the political, legal, and socio-cultural roles of ecological movements. Perhaps most significantly, ecological modernization opens up new ways for environmental sociology to become linked to debates over and empirical research on modernity and postmodernity as well as counter-modernity. The ability of ecological modernization to bring new perspectives on consumption into environmental sociology is an exciting development.

As we have noted several times in the preceding, the most problematic aspect of ecological modernization is its rooting in the institutions and experiences of the Northern European countries. However, several of the contributions to this volume, particularly that of Mol, have suggested some ways in which ecological modernization can be enhanced as social theory. One such strategy is to conceptualize ecological modernization as a global process, albeit a highly uneven one, and to undertake research on the ways in which the structures and practices of globalization facilitate or undermine ecological modernization processes. A related strategy is to use the Dutch and German cases as the starting points for a comparative sociology of ecological modernization.

A good share of environmental sociology today suffers from the same weaknesses – the lack of a comparative approach, and a lack of attention to the contradictory processes of globalization – that ecological modernization has had. Ecological modernization theory, however, might well have some advantages in rectifying these shortcomings over other forms of environmental-sociological theory. Even if not, ecological modernization theory will have been useful if it can help catalyze the need to give priority to comparative research and to environmental phenomena in the globalization perspective.

Notes

1. In 1988, the United Nations Environment Programme (UNEP) and the World Meteorological Organization (WMO) set up the Intergovernmental Panel on Climate Change (IPCC). This is an intergovernmental scientific and technical body with a small secretariat and a worldwide network of scientists.
2. One example of the neglect of environmental issues in postmodern views of consumer society is the TCS-special issue edited by Featherstone (1991), which does not consider the challenges posed by the need for a 'dematerialization' of postmodern consumption patterns at all. When consumption is discussed by Baudrillard with respect to the 'environmental nuisance' that comes with it, he only points to the fact that waste-behavior can be analyzed as a kind of 'celebration of affluence' without investigating in any detail the kind of perspective that could result from this (Baudrillard, 1998, p. 5).

Bibliography

Baudrillard, J. (1998), *The Consumer Society: Myths and structures*. London: Sage.

Bauman, Z. (1987), *Legislators and Interpreters: On modernity, post-modernity and intellectuals*. Cambridge: Polity Press.

Beck, U. (1986), *Risikogesellschaft. Auf dem Weg in eine andere Moderne*. Frankfurt: Suhrkamp.

Beck, U., A. Giddens, S. Lash (1994), *Reflexive Modernization: politics, tradition and aesthetics in the modern social order*. Cambridge: Polity Press.

Buttel, F.H. (1986), Sociology and the environment: the winding road toward human ecology. *International Social Science Journal*, vol. 38, no. 3, pp. 337-356.

Dickens, P. (1992), *Society and Nature. Towards a Green Social Theory*. New York: Harvester Wheatsheaf.

Dunlap, R.E. and W.R. Catton (1979), Environmental sociology. *Annual Review Sociology*, V, pp. 243-273.

Dunlap, R.E. and W.R. Catton, Jr. (1992/93), Toward an ecological sociology: the development, current status, and probable future of environmental sociology. *Annales de l'institut international de sociologie*, vol. 3, pp. 263-284.

Dunlap, R.E. and W.R. Catton, Jr. (1994), Struggling with human exemptionalism: the rise, decline and revitalization of environmental sociology. *The American Sociologist,* vol. 25, pp. 5-29.

Featherstone, M. (1991), *Consumer Culture and Postmodernism*. London: Sage.

Giddens, A. (1984), *The Constitution of Society*. Cambridge: Polity Press.

Giddens, A. (1990), *The Consequences of Modernity*. Cambridge: Polity Press.

Hajer, M.A. (1995), *The Politics of Environmental Discourse: Ecological modernization and the regulation of acid rain*. Oxford: Oxford University Press.

Hannigan, J.A. (1995), *Environmental Sociology: A social constructionist perspective*. London: Routledge.

Harmsen, G. (1974), *Natuur, Geschiedenis, filosofie*. Sun-schrift 89. Nijmegen: Sun.

Harmsen, G. (1992), Natuurbeleving en arbeidersbeweging. Amsterdam: NIVON.

Held, D. (1995), *Democracy and the Global Order: From the modern state to cosmopolitan governance*. Cambridge: Polity Press.

Kumar, K. (1995), *From Post-Industrial to Post-Modern Society*. Oxford: Basil Blackwell.

Mol, A.P.J. (1996), Ecological modernisation and institutional reflexivity: environmental reform in the late modern age. *Environmental Politics*, vol. 5, no. 2, pp. 302-323.

Nelissen, N.J.M. (1972), *Sociale ecologie*. Utrecht/Antwerpen: Het Spectrum.

Otnes, P. (ed.) (1988), *The Sociology of Consumption*. Atlantic Highlands, New Jersey: Humanities Press Int.

Saunders, P. (1988), The sociology of consumption: A new research agenda, in: P. Otnes (ed.), *The Sociology of Consumption*. Atlantic Highlands, New Jersey: Humanities Press Int., pp. 139-156.

Warde, A. (1990), Introduction to the sociology of consumption. *Sociology*, 24, (1), pp. 1-4.

Wehling, P. (1992), *Die Moderne als Sozialmythos; Zur Kritik sozialwissenschaftlicher Modernisierungstheorien*. Frankfurt: Campus Verlag.

Yearley, S. (1991), *The Green Case: A sociology of environmental issues, arguments and politics*. London: Harper Collins Academic.

2

Classical Theory and Contemporary Environmental Sociology: some reflections on the antecedents and prospects for reflexive modernization theories in the study of environment and society

Frederick H. Buttel

Introduction

Sociology has long had something of an ambivalent relationship to the classical tradition, even if this is seldom explicitly acknowledged. On the one hand, the continued importance of the classical tradition in formal terms can be gauged by the fact that classical theory remains a core requirement in most undergraduate and graduate sociology programs throughout the Western world. A PhD-holding sociologist would be considered incompetent if s/he did not know the basics of the sociologies of Marx, Weber, and Durkheim, and the meaning of concepts such as false consciousness, traditional domination, or organic solidarity. Essentially every text on sociological theory makes the claim that classical sociology has very strongly influenced the development of the sociological discipline, and that classical theory is still highly relevant to the sociological enterprise (a typical recent example being Morrison, 1995). One frequently encounters claims that virtually all important sociological notions – even the ideas of contemporary postmodern theorists – were 'anticipated' by the classical theorists (Hughes et al., 1995).

At the same time, the typical sociologist in the world today would be very unlikely to consult, much less devote serious study to, the *Grundrisse*, *Capital*, *Suicide*, *Division of Labor in Society*, *Economy and Society*, or *The Protestant Ethic and the Spirit of Capitalism* during a typical workday (or a typical work-year, for that matter). Pieces of classical sociological scholarship are rarely cited in the contemporary sociological research literature. Many influential sociological theorists such as Anthony Giddens (1987: Chapter 2) have claimed that the contexts and assumptions of nineteenth century classical theory are no longer relevant to the late twentieth century, and that progress in social theory will require jettisoning the classical sociological tradition. Some sociologists now go so far as to say that by the 1940s, the 'classical project' – the development of sociological abstraction aimed at addressing moral and political problems – had undergone 'dissolution' (Wardell and Turner, 1986). And it becomes harder every year to convince our students they need to study classical theory when *Grundrisse*, *Capital*, *Suicide*, *Division of Labor in Society*, *Economy and Society*, and *The Protestant Ethic and the Spirit of Capitalism* are not in prominent places on our own bookshelves.

The ambivalence about the relevance and influence of the classical tradition has been of particular importance in environmental sociology. It is useful to recall that the works of the initial exemplars of environmental sociology were anchored almost entirely in mid-century normal science – e.g., the neo-Ogburnianism of Fred Cottrell (1955), the functionalism of Walter Firey (1960), the unexceptional attitude studies of outdoor recreation groups and general environmental public opinion studies (Burch et al., 1972), the rural-community sociology of natural resources (Burch and Field, 1988), and so on. Whether or not one feels that this relatively ordinary proto-environmental sociology was strongly conditioned by the classical tradition, it is clear that what separated it from mainstream sociology in 1970 was not its theoretical or methodological commitments per se. Rather, the mere fact that Cottrell, Firey, and others found environmental-ecological matters to be interesting independent and/or dependent variables was the crucial difference between 1950s and 1960s proto-environmental sociology and so-called mainstream sociology.[1] In every other major respect proto-environmental sociology from the 1950s through the early 1970s was indistinguishable from 'mainstream' sociology.

Indeed, even the most central works of the first stage (from Earth Day until approximately 1975) of modern environmental sociology in North America (e.g., Burch, 1971; Burch et al., 1972; Klausner, 1971) were not particularly rebellious about either mainstream sociology or the classical tradition. In these early days of environmental social science the principal voices of theoretical defiance on behalf of the environment were *not* those of environmental sociologists. Instead, non-sociologists such as Walter Prescott Webb (1952), Richard G. Wilkinson (1973), and Murray Bookchin (1971) were the most visible or influential social science scholars who criticized mainstream inquiry for its lack of attention to the environment.

But when environmental sociology began to gather momentum during the mid – to late – 1970s, its major preoccupations came to include explaining why Western sociology had ignored the biophysical environment and what kinds of modifications of sociological theory would be required to reflect the importance of the biosphere and biophysical forces in social structure and social life. This second stage of the development of (North American) environmental sociology consisted of the elaboration of what I have earlier (Buttel, 1986) referred to as the 'new human ecology', and what Dunlap and Catton have referred to as the 'new ecological paradigm'. Integral to the new human ecology was a certain hostility toward mainstream sociology, as well as the desire to build a new sociological theory in which biophysical variables would play a definite and central role. The most central figures in the new human ecology were, of course, Dunlap and Catton. Dunlap and Catton were not the first sociologists to criticize the classical tradition for its neglect of the biophysical environment or to see mainstream and environmental sociology/social science in 'paradigmatic' terms. But in their late 1970s and early 1980s writings Dunlap and Catton articulated the notion of 'paradigm' with the ecological critique of classical sociological theory with exceptional clarity and persuasiveness (e.g., Catton, 1976; Catton and Dunlap, 1978; Dunlap and Catton, 1979; Dunlap and Catton, 1983). There is scarcely a significant text in environmental sociology today that fails to cite one or more of these early works by Dunlap and Catton as having provided the template for modern environmental sociology.

Put most succinctly, Catton and Dunlap and several other pioneers of 1970s and early 1980s environmental sociology (e.g., Cotgrove, 1982) made three key arguments about environmental sociology and sociological theory. First, it was argued that modern sociology was strongly influenced by the 'social facts' injunction and the 'exceptionalism/exemptionalism' of the classical theorists. Second, it was held that mainstream classical and contemporary sociology, despite its apparent diversity, essentially constitutes a unitary (exemptionalist) paradigm, while environmental sociology is a distinct – and a competing or alternative – paradigm. Finally, it was argued that one of the key tasks of environmental sociology is to demonstrate to mainstream sociologists that a sounder sociological theory must be built from some of the core premises of environmental sociology (e.g., the principle that societies and biophysical environments exist in reciprocal interaction with each other).

It is arguably the case that while the second and third aspects of the Dunlap-Catton agenda are not so central anymore to the field of environmental sociology, the first posture – critique of the fundamental exemptionalism of the classical tradition, and the assertion of its irrelevance to environmental sociology – remains an article of faith within North American (and to a considerable, but lesser, extent in European) environmental-sociological circles (Dickens, 1992; Martell, 1994). Put somewhat differently, the core of North American environmental sociology – or, in other words, the new human ecology – has tended to be formed in *opposition or in response* to mainstream sociology and/or the classical tradition.

In this chapter I will employ the issue of the ir/relevance of the classical tradition to a meaningful environmental sociology as a vehicle for making some observations on the historical development of environmental sociology, and on the role that two versions of 'reflexive modernization' theory are playing, and could play, in this field. I will begin by exploring some of the insights and blinders that the classical tradition has bequeathed to the contemporary cadre of environmental sociologists. I will conclude that this is a very mixed legacy – that environmental sociology needs some of the tools that were initially developed by the classical theorists, but that the overall thrust of the classical tradition was to downplay ecological questions and biophysical forces. I will then make some comments on the historical development of environmental sociology and the role of the classical tradition and its mid-twentieth century legacies in that development. Finally, I will focus on three relatively new theoretical perspectives in environmental sociology: social constructivism and two versions of reflexive modernizationist thought – namely, 'risk society' and 'ecological modernization' theories. I will identify some strengths and weaknesses of each. As I will suggest later, ecological modernizationism is unlikely to become the dominant theory in the field, but it raises particularly critical issues for the field, and with modifications has significant synthetic potential for making environmental sociological theory both comprehensive and empirically useful.

Classical sociology and the environment

Each of the major classical theorists can be seen to have developed or elaborated one or more theoretical premises or methodological injunctions that can be regarded as exemptionalist or unconducive to considering biophysical factors as independent or dependent variables. Durkheim, for example, is well known for his 'social facts' injunction – the notion that the role of sociology is to explain or account for 'social facts', while employing (exclusively) other social facts as explanatory factors. Durkheim's social facts injunction was propounded to distinguish sociology from psychology and biology, and to distance it from utilitarianism. Marx's relentless critiques of Malthus were intended to distinguish historical materialism from the oversimplifications of Malthusian population/food arithmetic. Weber could not have been more emphatic about the fact that sociology is a discipline in which evolutionary reasoning, and any other styles of reasoning from biology and the natural sciences, must be rejected because societies and natural systems have qualitatively different dynamics.

Craig Humphrey and I (Buttel and Humphrey, 1987) have argued elsewhere, however, that there is, in a certain sense, *a classical environmental sociology*. Elements of environmental sociology have roots deep in nineteenth century social thought. Not only did Marx, Durkheim, and Weber incorporate what we might regard as ecological components in their works, they did so from a variety of standpoints. Among the multiple ecologically relevant components of their works are materialist ontologies (in the case of Marx and Engels), biological analogies (Durkheim), use of Darwinian/evolutionary arguments or schemes (Marx, Durkheim, Weber), and concrete empirical analyses of resource or 'environmental' issues (Marx, Weber).

Classical environmental sociology

Marx and Engels worked from a materialist ontology, which should be understood to mean not only a structural/non-idealist posture and an emphasis on the conditions of production and labor, but also an understanding that, in principle, the predominance of the sphere of production and social labor cannot be understood apart from nature. This is particularly clear if one reads the early 'philosophical' works by Marx (those published in 1844 or earlier), in which the notions of 'nature' and the material world are employed frequently, and in a non-deterministic or dialectical manner (Parsons, 1977). It is thus no accident that contemporary works in environmental sociology that are explicitly neo-Marxist (e.g., Dickens, 1992) tend to draw inspiration disproportionately from the works of the early Marx. But while the work of the young Marx was quite ecological in some respects, this does not imply that the later work of Marx (and Engels) was devoid of references to nature and the natural world.[2] Marx and Engels, for example, frequently referred to the penetration of capitalism as a cause of massive air pollution and other threats to the health and welfare of workers, and to the need for political economy to treat relations between society and nature (Dickens, 1992; Parsons, 1977). Parsons (1977) and Burkett (1996) have noted how Marx was fascinated by agronomy and the biology of agricultural production, and in *Capital*

Marx often used agronomic examples to demonstrate the role of nature in social production. Burkett (1996), in his elegant defense of Marx against claims that his methodology is unecological, makes particular note of the fact that Marx's analysis of modes of production includes not only class (or people-people) relations but also the relations of material appropriation (people-nature relations). Marx and Engels' schema positing the contradictory development of class societies and the revolutionary transformation from one mode of production to another contains an evolutionary component based on Darwin's work (Lopreato, 1984). But as I will note later, environmental social science theories have tended to incorporate one (and sometimes both) of two constitutive elements: a materialist outlook, and an emancipatory orientation. Thus, among the reasons for the particular relevance of the young Marx to environmental sociology is that Marx's work prior to 1945 or so tended to reflect both materialist and emancipatory postures.

Like Marx, Durkheim also set forth a modified evolutionary schema and relied heavily on metaphors from Darwinian evolution and organismic biology. While Durkheim questioned Spencer's argument that evolutionary change led to continuous progress, his theory was based on an evolutionary view of social change (Turner, 1994). The master direction of change was from primitive societies with a low division of labor to modern societies with a complex division of labor. Durkheim, however, differed from Spencer in emphasizing the disruptive qualities of change. The transition from primitive to modern societies was accompanied by anomie and a breakdown of social solidarity and regulation. While Durkheim anticipated that modernizing societies ultimately would exhibit new, more effective organic solidarity, he regarded the establishment of adequate integration and solidarity to be problematic.

Durkheim freely used biological concepts in presenting his theories of social evolution and solidarity, as is evident in the concept of organic solidarity. *The Rules of Sociological Method* (1895) referred to various types of societies along the continuum from traditional to modern as 'species' or 'societal species'. Moreover, his most famous work, *The Division of Labor in Society* (1893), set forth the major elements of a theoretical perspective that has come to be known as (classical) human ecology. The *Division*'s famous schema for explaining the transition from mechanical to organic solidarity was rich in imagery about population density, resource scarcity, and competition for survival that bears a strong resemblance to more modern notions of human ecology.

A final classical theorist widely considered to be among the most influential in Western sociology was Max Weber. While Marx and Durkheim largely assumed that there was some knowable *a priori* direction of social change, Weber firmly rejected the theoretical viewpoint that there was a unilinear course of societal development. Social change was determined by shifting constellations of subjective, structural, and technological forces that ultimately were rooted in human motivations and history. Moreover, Weber (1922) was an outspoken opponent of Social Darwinism, and he frequently stressed how social science differed from biological sciences and that the methods and concepts of the former must be different than those of the latter. Weber's work thus has been taken to be the first decisive break from nineteenth century evolutionism anchored in biological analogies (see Sanderson, 1990; Burns and Dietz, 1992; Dietz and Burns, 1992, for comprehensive overviews of 'social evolutionism'). Interestingly, Weber's works that most clearly reflect the break with biological analyses are those such as *The Agrarian Sociology of Ancient Civilizations* and his *General Economic History*, in which material on the impacts of social structures on natural resources or the impact of natural resources on social organization is most prominent.

The multiple respects in which Weber's work can be seen to accommodate an environmental or human-ecological dimension can be most dramatically illustrated by contrasting the neo-Weberian environmental sociologies of Patrick West (1984) and Raymond Murphy (1994). These two scholars draw on what might be regarded as entirely different 'corners' of Weber's work. According to West (1984), from Weber's historical sociology of religion and his empirical research on ancient society one can distill a human ecology that is rich and provocative for contemporary environmental sociology. Weber's historical and comparative method rested on environmental factors playing casually important roles at times in history. Weber treated environmental factors as interacting in complex casual models, and they 'frequently affect complex societies through favoring the "selective survival" of certain strata over others' (West, 1984: 232). It is arguably the case that Weber's (probably unintended) use of Darwinian imagery was truer to Darwin's theory than was that of Spencer and others. Thus, for West, the relevance of Weber's work to environmental sociology is not only that Weber analyzed concrete instances of struggles over resources (e.g., control of irrigation systems), but also that Weber's causal logic was essentially a Darwinian one that still has the potential to help build bridges between sociology and the ecological sciences.

While West's (1984) account of Weber's sociology of environment and natural resources is anchored in Weber's comparative macro-historical sociology, Murphy (1994) has developed a Weberian environmental sociology based largely on a completely different literature – Weber's concept of rationalization and his ideal-types of rationality and orientations to action drawn primarily from *Economy and Society* (1968). Murphy (1994) argues that rationalization and the expansion of formal rationality have involved tendencies to an ethic of mastery over nature (or the 'plasticity' of the relationship between humans and their natural environment), to the quest for technologies to realize this mastery, and to a lack of attention to human threats to the environment. Similar to Weber's notion of charismatic authority, Murphy suggests that the ecological irrationalities caused by rationalization will stimulate social movements that aim at 'de-rationalization' or 're-rationalization' of modern institutions.

Classical tradition in contemporary environmental sociology

Before proceeding to discuss some of the more significant recent trends in environmental sociology, it is useful to begin by noting some of the historical tendencies of environmental sociology from a classical sociological perspective. Two sets of observations deserve mention here. First, environmental sociology has tended to have affinities with two different – and, in some respects, contradictory – components of the classical tradition. On the one hand, environmental sociology has tended to be materialist – in its stress on the natural world and the material substratum of human life – and objectivist – in its stress on the negative consequences of science and technology and its endorsement of the project of the ecological and environmental sciences. On the other hand, environmental sociology has been sympathetic with the emancipatory or liberatory impulse as articulated by ecology movements of various stripes. The most creative and influential works in the field have aimed to incorporate both materialism and emancipation and to synthesize the two outlooks. Materialism, objectivism, and the emancipatory impulse are amply represented in the significant corners of the classical tradition. Emancipation, for example, is a key theme in the philosophical works of the 'young Marx' as well as in Weber's work on charismatic authority.

Several other aspects of the classical tradition have reinforced environmental sociology's affinities with it. Scholars such as Swingewood (1991) have stressed that the nineteenth century classical tradition was essentially pessimistic in that it stressed the negative social implications – anomie, alienation, the 'iron cage' and disenchantment – of bourgeois society. Giddens (1987) has likewise stressed the fact that the classical tradition was preoccupied with changes in the economy and in the structure of production as the prime movers in shaping social change in modern societies. It is significant that mainstream environmental sociology has tended to be a relatively pessimistic sociology (because of its being anchored in arguments about the intrinsic tendencies to environmental degradation) and has stressed the importance of the economy and the structure of production. It is thus no accident that as much as environmental sociologists have been critical of the shortcomings of classical theory, their work has tended to have very definite affinities with many of the concepts, methodological principles, and presuppositions of the classical tradition.

It is useful to conclude this discussion of classical sociology and the environment with a few observations, some of which depart in some respects from conventional wisdom on the topic. One must indeed recognize the radically sociological (and thus the 'exemptionalist') standpoint of Marx, Weber, Durkheim, Simmel, and other nineteenth and early twentieth century classical thinkers. As will be noted later, one of the characteristics of the sociological project as it emerged from the key figures of nineteenth century thought was that a sociological explanation is more elegant to the degree that its underlying concepts abstract beyond the natural world. But while classical theory did tend toward a radically sociological outlook, it by no means completely neglected the biophysical world. We need to recognize that the 'exemptionalism' of the classical tradition can be exaggerated, and thus the neglect of the biophysical environment in twentieth century sociology cannot be accounted for only by the 'exemptionalism' of the classical tradition. Indeed, as stressed several times earlier, a large share of the contemporary environmental sociological theory literature has been explicitly anchored in the works of one or more classical theorists. One of the ironies of modern environmental sociology is that while there is often ritualistic criticism of Marxism for its neglect of the environment, neo-Marxism is perhaps the most pervasive influence on environmental sociology across the world today. This is not surprising, however, when we consider Marxism's materialism, its critical posture toward bourgeois society, its focus on the economy and productive institutions, and its compatibility with social – emancipatory position – postures with which environmental sociology has long had strong affinities.

The lack of attention to the environment in sociological inquiry has arguably had as much or more to do with the historical conditions of the establishment of sociology, especially the imperative to distinguish sociology from other natural and social science disciplines (especially psychology, biology, and economics), and also the reaction against Social Darwinism and geographical-environmental determinism. In addition, the lack of attention to the environment was as much a reflection of Western culture – of its Enlightenment influences, its consumerism, its having been mesmerized by the 500-year boom, its colonial and postcolonial expansionism, and so on – as it was a prescription from the classical theorists. In fact, at mid-century classical theory was much more 'ecological', or much more likely to recognize the materiality (Buttel, 1996) of social life, than was mainstream sociology. The dominant forms of sociology at mid-century – Parsonian functionalism, post-industrial society theory, logic of industrialization theory, modernization theory, pluralism – were arguably far less cognizant of the natural world than were the perspectives of Marx, Durkheim, Weber, Simmel, Tönnies, and so on.

The classical tradition should by no means be seen as above criticism, however, both for sociology at large (see, for example, Giddens, 1987) as well as for its role in guiding environmental sociology. There are clearly some definite limitations of classical theory for understanding the environment. Interestingly, among the limitations of the classical tradition for understanding the environment are two shortcomings – its Eurocentricity and innocence about globalization – that it shares with much contemporary environmental sociology.

The theoretical development of North American environmental sociology

To the extent that the classical tradition can be regarded as exemptionalist or unecological, I consider the exemptionalism of the classical theorists to have been a necessary step in the development of a nondeterministic sociological tradition. This is to say, the classical tradition represented the liberation of social science from a series of reductionisms and chauvinisms – of biological analogies, utilitarianism, the German (neoclassical economics) analytical school, psychology, clericalism, nationalism, and so on. Overcoming these reductionisms and prejudices enabled the development of an overarching or encompassing social science discipline which would be able to accommodate a variety of explanatory schemes and methods, and explore a variety of empirical problems.

Likewise, I believe that we need to be respectfully critical of 1970s and 1980s North American environmental sociology as having been a necessary, but incomplete, step in the development of thought about societal-environmental relations. I believe that North American environmental sociology should be seen as a point of departure, rather than as an end-point for a period of 'normal science', primarily on account of the fact that this literature was developed as much in reaction to mainstream sociology and in metatheoretical terms as it was developed as a source of specific hypotheses to explain societal-environmental relations.

Though the underlying strategy for promulgating a new ecological paradigm in sociology was not often clearly articulated, the pioneers of North American environmental sociology can be seen to have reacted to, and to have striven to influence, mainstream sociology by employing three major arguments. First, the authors of the major works in environmental sociology tended to stress documenting the seriousness of the environmental crisis (e.g., Catton, 1980; Catton and Dunlap, 1978; Schnaiberg, 1980), albeit by relying primarily on secondary literature such as popularized treatments of ecology by persons such as Paul Ehrlich and Barry Commoner. This tendency remains very strong today (e.g., Benton, 1996a; Murphy, 1997). Second, these major works tended to theorize about how and why the regular or customary institutional dynamics of modern industrial societies – market processes, capitalist relations, industrial relations, urbanization, political democracy and corporatist structures, social norms and cultural values, science and technological innovation – have tended to involve *intrinsic or necessary trends to environmental degradation and crisis.*[3] Third, while the overall stress of these theories was to explain degradation, and thus was essentially pessimistic, at the same time these theories all held out hope that these dynamics can and would be negated or overridden through public value change and/or environmental movement mobilization. Thus, the public value change and the environmental movement were portrayed as being a rational and necessary response to the environmental crisis, and social policy change resulting from movement mobilization and pressure was seen to be the principal mechanism of environmental improvement (e.g., Milbrath, 1984; see also the discussion in Yearley, 1996).

Catton and Dunlap have articulated particularly clearly each of these three rationales for a specialized field of environmental sociology and for why these premises should be reflected in mainstream sociology. During the early years of North American environmental sociology, Catton (1976, 1980) documented in particularly comprehensive fashion the ominous implications for the human species of ongoing trends toward environmental destruction. In so doing, he showed that the core institutions of contemporary industrial societies were developed during a period of (apparent) abundance – the '500-year boom' dating from the heyday of mercantilism and the beginnings of capitalism – though Catton has hastened to stress that much of this abundance was ephemeral because it was based on past or future 'ghost acreage'. Catton suggested further that these institutions have persisted up to this day and have had remarkable momentum and inertia. Dunlap and Catton (1994) have also stressed the continuing force of the 'dominant Western world view' among publics in terms of how this world view props up the institutional practices portrayed in Catton's earlier work (e.g., Catton, 1976). Finally, Dunlap and Catton (1994) have developed very strong arguments about why environmentalism is a progressive force that should be supported by the social science community.

Allan Schnaiberg's environmental sociology (Schnaiberg, 1980; Schnaiberg and Gould, 1993), probably second only to Dunlap and Catton's in its influence throughout North America, has likewise been anchored in a conceptualization of the very powerful momentum behind environmental destruction. Schnaiberg has argued that the environmental crisis is due primarily to there being a very strong tendency to an environmentally destructive 'treadmill of production'. By treadmill of production, Schnaiberg means that the competitive character of capitalism and the imperative for states to underwrite private accumulation while dealing with the social dislocations of private accumulation combine to virtually compel private and public policies and practices that lead to exponential, capital-intensive, environment-degrading economic growth. Thus, the actions of private firms and state managers, as well as consumer-citizens and labor groups, all combine to reinforce the treadmill character of industrial accumulation. Schnaiberg has also portrayed the environmental movement as the principal countervailing social mechanism for societies to improve their environmental performance (Gould et al., 1996).[4]

There are certain shortcomings of the predominant conceptualizations of environment-society relations in the North American environmental sociology literature, however. One limitation of the traditional North American environmental sociology literature is that it has devoted far more attention to theorizing environmental degradation than to theorizing environmental improvement (see Buttel, 1996). As obvious as this limitation may seem, it needs to be recognized that shifting to a stress on environmental improvement is by no means unproblematic. Since environmental sociology's justification within the larger discipline is that it is better able than mainstream sociology to recognize why environmental problems are serious, getting worse, and could imperil human survival, the sub-discipline has historically had something of a self-interest in emphasizing degradation and crisis over improvement. Further, degradation is relatively easy to theorize because one can draw on historical as well as contemporary data in order to show how various institutional ensembles affect environmental degradation processes. The historical instances of environmental improvement (or of structures and processes of environmental improvement with contemporary relevance), however, are far fewer than those of degradation. Also, environmental improvement processes are very likely more subtle and more complex phenomena than are environmental degradation processes; while there are arguably some overarching institutional processes that lead to degradation, improvement is not likely to be so constituted. For these and other reasons, the traditional North American literature has not devoted a great deal of attention to improvement processes – and to the degree that improvement is discussed it is presumed that environmental movement mobilization resulting in state policy change is the master process.

A second shortcoming of the North American environmental sociology literature is that it has arguably overestimated the coherence of environmental movements, and exaggerated the degree to which environmental improvement will ultimately derive from environmental movement mobilization. For example, there has by no means been a strong association between environmental movement mobilization and environmental legislation in the US. Despite the growth of US environmental movements since the early 1970s, these groups have had very few major accomplishments since then. Virtually all of the landmark US environmental legislation – the National Environmental Policy Act, the Clean Water Act, the Toxic Substances Control Act, the Endangered Species Act, and so on – was passed in the early 1970s or earlier. Since the mid-1970s, the best this growing movement could do was to engage in rear-guard actions to preserve as much of this legislation as possible. Further, much of the environmental improvement that has actually occurred since the mid-1970s has been due to market-induced conservation, behavioral changes, technological advances, and responses to civil-legal incentives that discourage exposure to civil litigation on nuisance and liability grounds (Mol, 1995). Finally, there are now numerous critiques of modern environmentalism's track record in selecting issues and tactics that can lead to broadly shared environmental improvements (see, for example, Gottlieb, 1994; Sachs, 1993). Thus, North American environmental sociology has stood in need of revision. I will now turn to three relatively new environmental-sociological frameworks to which many sociologists have turned to complement the perspectives of the pioneering 'new human ecologists'.

The debate over social constructionism

Social constructionism – or, more broadly, the 'cultural turn' of sociology and environmental sociology – has been highly controversial, arguably more so than has been justified. While much has been made between the apparent debate between Dunlap/Catton (1994) and myself (Buttel, 1992; Taylor and Buttel, 1992) on this matter, I do not regard this debate as being particularly fundamental to the field. I suspect that the only truly significant difference between my perspective on the social construction of environmental knowledge and concepts and that of Dunlap is that I see environmental science as being less coherent, and as exhibiting a greater level of diversity and controversy, than he does. Further, it is my observation that there is a fairly strong consensus in the field that neither an unquestioning realist posture toward environmental knowledge nor a radical relativist posture is 'sustainable' within sociology, much less environmental sociology (see Freudenburg et al., 1995, for a related argument).

Social constructionism and related discourse-analytic scholarship (e.g., Hajer, 1995; Hannigan, 1995; Lash et al., 1996; Wynne, 1996; Yearley, 1996) have played a useful role in environmental sociology. But its utility has not been in making environmental sociology a branch of the sociology of scientific knowledge (SSK), i.e., in casting doubt on the degree to which global warming claims or global environmental knowledge are a faithful reflection of the realities of the natural world. Indeed, relativization of environmental science or global environmental change has in general not been, nor should it be, the major goal of social constructionist environmental-sociological studies. Rather, their purposes have been to make several important points. First, a one-sided realist account of environmental science and knowledge – which environmental sociology often comes perilously close to embracing – is just that – one-sided, and inconsistent with any reasonable sociological perspective which must take into account both 'structure' and 'agency', and the material and the symbolic. Even from a realist perspective, any concept such as global environmental change must be regarded as an abstraction which, if it proves to be a good one, will help to illuminate the processes of the natural world but in so doing will obscure others; and such an abstraction should be regarded as being important only on a provisional basis until its validity can be established through repeated replications and experiments. Second, environmental-scientific discourses matter. Interpretive-discursive sociologies of environmental knowledge claims, environmental risks, environmental ideologies, and environmental politics can contribute a great deal to environmental sociology by illuminating how and in what ways discourses are structured and renegotiated. These interpretive sociologies are not intrinsically inconsistent with structural or materialist views from more mainstream quarters of environmental sociology (as Hajer, 1995, masterfully demonstrates). Third, it is critical for environmental sociology to recognize the crucial role that the scientization of protest, social movements, and politics is playing (Buttel, 1992). In other words, while not detracting from the fact that scientific knowledge tends to more or less mirror the realities of the natural world, the crucial role of science often lies in how it is 'represented' and how it is employed within social movements, interest groups, regulatory agencies, epistemic communities, international organizations and 'regimes', and so on. Scientific knowledge thus often tends to be enmeshed with social symbols, political ideologies and discourses, social movement 'frames'. How this occurs makes an enormous difference in terms of environmental policy and politics.

In addition to social constructionist perspectives on the relations between environmental knowledge/science and environmental movement, some scholars (DuPuis and Vandergeest, 1995; Goldblatt, 1996; Hannigan, 1995) have strived to take the social constructionist agenda a step or two further, mainly by bringing the tools of cultural sociology into environmental sociology. In most such instances, the relativization of environmental knowledge is not a major or even a minor goal. Doing so, in fact, is a fairly mindless, trivial exercise. Further, several scholars from otherwise mainstream sociology have begun to enter the terrain formerly known as environmental sociology by applying cultural-sociological tools to phenomena such as environmentalism, ecology movements, green parties, and so on (MacNaughten and Urry, 1995). With environmental sociology becoming more important to the mainstream of the discipline because of their own initiatives rather than ours strikes me as being much more positive than negative.

Reflexive modernization, ecological modernization, and environmental sociology

Over the past half dozen or so years environmental sociology in the advanced countries has been very appreciably reshaped by the application of a second set of theories developed in Northern Europe, primarily for application to phenomena such as green parties, ecology movements, the relationships between science and environmental movements, industrial ecology, and so on. While these theories are in a sense quite diverse, they all tend to see environmental improvement as being as or more important to explain than environmental degradation. Each also embraces the notion that there are ongoing complex changes in the social values and behaviors deriving from environmental degradation and crisis. Each of these theories also tends to downplay the role of certain forms of environmentalism, especially radical, 'expressive' environmentalism, in environmental improvement. In my view, two of these emerging traditions in environmental sociology – risk society and ecological modernization theories – are playing particularly important roles in rectifying some of the shortcomings of North American environmental sociology. These two theories are both built around Ulrich Beck's (and, more recently, Anthony Giddens') concept of reflexive modernization (see Beck et al., 1994, and the commentary by Goldblatt, 1996).

Beck's notion of reflexive modernization is very closely linked to his theory of the transformation from 'industrial society' to 'risk society'. His concept of reflexive modernization is based on the notion that the further the process of 'simple modernization' (essentially capitalist rationalization and industrialization, undergirded by science and technological development) proceeds, 'the more the foundations of industrial society are dissolved, consumed, changed and threatened'

(Beck, 1994: 176). Modernization is seen to have led to a set of risks and hazards that not only threaten current generations, but might also prejudice the quality of life, and possibly the very survival, of future generations. For Beck, the 'ecological crisis', the risks of mega-accidents such as Bhopal and Exxon Valdez, the environmental risks of recombinant DNA and genetic engineering, the personal risks entailed by medical high technology, and the declining capacity of states to regulate directly the production practices that give rise to these risks are examples of the outcomes of simple modernization. The growing public recognition of these hazards and risks is one of the two principal precipitants of 'reflexive modernization', and ultimately of the risk society. The other major precipitant of reflexive modernization is the decline of parliamentary institutions, the individuation of politics, the decline of class, and the growing role of 'subpolitics'.

At one level, one can think of reflexive modernization as being both 'reflex' and 'reflection'. The 'reflex' or structural component of reflexive modernization essentially holds that the risks and hazards generated by modernization, or by industrial society, call the central institutions of society (science, the legal system, parliamentary democracy, the market economy) into question. The 'reflection' or more actor-centered component of reflexive modernization includes the 'freeing' of actors relative to social classes and other social-structural categories, the consequent construction of new identities and roles, and the formation of new social movements around these new identities and concerns about risks. In particular, the trend in the advanced societies has been for their citizens to scrutinize science, technology, and scientific institutions by making moral claims to rationality that are equal to those of modern science. 'Reflexive modernization', says Beck (1992: 14) 'means not less but more modernity, a modernity radicalized against the paths and categories of the classical industrial setting'.

The second major stream of work flowing from reflexive modernization is that of ecological modernization theory. The link between reflexive modernization and ecological modernization is less direct than it is to the theory of the 'risk society'. Some proponents of ecological modernization (Spaargaren and Mol, 1992) are actually quite critical of how Beck has applied reflexive modernization to risk society theory. But most sociological proponents of ecological modernization strongly concur with the two constituent notions of reflexive modernization: first, human and institutional choices are not simply the reflections of master-structural forces of capitalism, industrialization, and so on; and second, the solutions to environmental problems will lie in a progressive modernization of societies (rather than in the 'de-modernization' or 'counter-modernization' that is advocated within radical environmentalism).

As Mol (1995: Chapter 3) and Hajer (1995) have stressed, there are several different streams of thought within ecological modernization. Mol shows that Karl Polanyi's (1957) work can in some sense be seen as a precursor to ecological modernization. Most sociologists who work in the field, however, grant that the notion was most systematically developed by the German social scientist, Joseph Huber (see Mol's, 1995, summary of Huber's work). In contrast to the predominant thrust of mainstream US environmental sociology having been devoted to explaining environmental degradation, ecological modernization theory grew out of social research, environmental movement involvement, and ecological research on practical, non-utopian means of achieving environmental improvement. Among the principal claims of ecological modernization is that under conditions of state-and-society regulated market capitalism, which have become dominant since at least World War II in OECD countries in particular, contemporary capitalist enterprises have proved to be able to accustom themselves to ecological constraints to a certain extent up to now, without being deprived of favorable production conditions, new markets, or growing profits. In addition, environmental protection and reform has proved to be a profitable market for the expanding eco-industry. Consequently, there is reason to believe that – in economic terms – the incorporation of nature as the third force (in addition to capital and labor) of production in the capitalist economic process has become an increasingly feasible proposition (Mol, 1995: 41).

The heart of ecological modernization theory is a relatively optimistic view of the potentials for technological change to lead to solutions for environmental problems. In particular, it is argued that, at least in some societies, the socioeconomic influences on technological R&D and industrial technology choice will over time tend to lead to improvement in the efficiency of conversion of raw materials into finished products and to reduction in the quantity and toxicity of the waste stream from industry. Huber's work essentially incorporated an evolutionary neo-Schumpeterian view of long cycles of technological innovation. Other ecological modernizationists such as Mol (1995) and Spaargaren (1996) tend to stress a more contextual approach based on (1) the specificities of national and local regulatory and policy regimes, and (2) the social and physical nature of the sectors, products, and production processes involved.

Ecological modernization theory also holds a distinctive posture on the role of state environmental policy. Much mainstream environmental sociology has theorized that the traditional bureaucratic-capitalist state has performed poorly in terms of environmental protection. By contrast, many ecological modernizationists have argued that new trends in the structure of states (including the decline of 'command-and-control' environmental regulation and the concomitant 'modernization' of the state) are actually stimulating environmental 'self-regulation' within civil society via legal mechanisms (e.g., civil liability law), economic mechanisms (market-induced efficiency and environmental policy incentives), and citizen-movement pressures.

Those familiar with Northern Europe will recognize that risk society theory seems to be an apt portrayal of the rise of the German Green Party and other radical environmental movements in Germany. Ecological modernization theory also seems strongly consistent with many of the restructurings and technological trends in industry (and the consumption sphere) in Germany, The Netherlands, and elsewhere in the region.[5]

As risk society and ecological modernization theories have become more influential, they have been subjected to several criticisms.[6] Some of these criticisms have been directed at the larger reflexive modernization perspective. For example, these theories have been criticized from the perspective of developing countries because they tend to be Eurocentric and lack generalizability to most societies in the world. It has long been recognized that modernizationist theories of social change are not very useful in understanding developing-country dynamics. Many of the assumptions of modernization theories have limited applicability to low-income countries, and this is no less the case with reflexive modernization perspectives. Other theories have been specific to either risk society or ecological modernization theory. Some of the typical criticisms of risk society theory, for example, stress its quasi-constructivist conceptualization of technological and ecological risks, and note that in most societies the pervasiveness of public preoccupation with such risks is far less than portrayed in Beck's scholarship (see, for example, Murphy, 1994).

Perhaps the most problematic aspect of risk society theory is its being anchored in the notion of the 'equality of risk' – that regardless of one's class one cannot be exempt from or escape large-scale hazards and risks. The equality of risk is seen by Beck as contributing to the demise of social class and as facilitating new forms of subpolitics across traditional class lines. It is apparent, however, that to the degree that there is an equality of risk in the countries of Northern Europe this is because social inequality and residential segregation are less there than in the laissez-faire capitalist societies such as the US, and very substantially less than in most developing countries.[7] In most societies in the world today, in fact, there is relatively little 'equality of risk'. If anything, 'environmental inequality' (Szasz and Meuser, 1997) is the rule rather than the exception. And environmental inequality is not confined to the low-income countries of the South. For example, the US environmental justice movement, which arose from the civil rights movement to seek redress for the *inequalities* of exposure to environmental hazards along racial and class lines – has been a crucial innovation in American environmentalism. Environmental justice – the notion that all persons have a right to a clean environment as an entitlement of citizenship – may well be a strategy with considerable applicability to environment quality (and social justice) efforts in developing countries. At a minimum, environmental justice-oriented thinking in the international environmental movement could play a significant role in constraining institutions such as the World Bank to presume that it is a socially acceptable policy to encourage and facilitate movement of heavily polluting and toxic-producing production processes to the South. Likewise, Murphy (1994), recognizing the highly unequal distribution of the costs and benefits of environmental extraction and destruction across social groups in North America, has anchored his acclaimed environmental sociology treatise in the notion of 'environmental classes' – a concept that is premised on the inequality of risk, hazards, access to resources, consumption capacity, and other environmental processes.

Of the two reflexive-modernization-related theories, ecological modernization has proved to be the most influential in relatively mainstream environmental-sociological circles. Among the reasons for its influence is that it is (1) a particularly affirmative response to the imperative for environmental sociology to be socially useful and to be able to shed light on possibilities for environmental improvement, and (2) an apt portrayal of the culture, processes, and political structures of state-induced environmental reforms in much of the West (Hajer, 1995). Ecological modernization has thus risen to prominence in the field very rapidly. A good indicator of its influence is that Mol's (1997) paper in the recent Redclift and Woodgate's (1997) anthology on environmental sociology is one of the few chapters that is built around advocacy for a particular theoretical approach. Further, several of the chapters in the Redclift and Woodgate anthology devote considerable attention to ecological modernization.

The ecological modernization approach is much more likely to have lasting influence on environmental sociology than will risk society theory. But while ecological modernization has received growing attention in environmental sociology, it has also begun to receive criticisms. Here I will briefly discuss some of these criticisms and evaluate them along with ecological modernization's current and potential strengths.

The most significant criticisms of ecological modernization are three-fold. First, there is the matter of its having been based on the Northern European experience, and its concomitant lack of generalizeability. Thus, it has been argued that ecological modernization processes as depicted in the ecological modernization literature are premised on the existence of social and political institutions that can provide strong signals to industries and consumers that pro-environmental behavior is expected, and that the absence of environmentally sound behavior would result in costly public or private sanctions. This is not to say that in the OECD countries outside of Northern Europe, and especially in the newly industrializing countries of the developing world, there is an absence of forces pushing in this direction. Indeed, virtually all countries, including those in the South, have environmental laws and regulations on the books. There also exist customary social relations and practices (e.g., common property regimes) that serve to conserve resources (Lipschutz and Mayer, 1993). But across the world today the absence of state environmental-regulatory powers, including the inability to provide incentives to induce pro-environmental practices in civil society, is a widespread phenomenon (Lipschutz, 1996).

A second criticism of ecological modernization and related approaches (e.g., notions of industrial ecology and environmental Kuznets curves) is that the types of environmental reforms that are seen as progressive (e.g., pollution control and source reduction in the Dutch chemical industry (Mol, 1997)) tend to involve relatively little decrease in the human toll on environmental resources. As Bunker (1996) has stressed, despite global improvements in the efficiency of the use of materials over the past two decades, aggregate consumption of minerals and other raw materials has continued unabated.

A third and related criticism (e.g., by Blühdorn in this volume) is that ecological modernization as a theory and discourse can serve to legitimate a political culture of environmental policy-making that basically absolves industrial corporations and other agents of environmental destruction of their responsibilities, and serves to ignore how the 'modernization' of the state is contributing to growing inequalities of wealth and power within countries and across the world-system. Furthermore, Blühdorn has also been critical of how reflexive modernization theories in general tend to dismiss the significance and the potential long-term efficacy of 'radical' environmentalism.

These criticisms are serious, but ecological modernization's potential contributions are also very significant. The many advantages of reflexive-modernization-based approaches to environmental sociology suggest that it will be well worth the effort to address their shortcomings. Ecological modernization has the potential to temper the pessimism and the lack of attention to concrete processes of environmental improvement that has tended to be endemic to the materialist core of environmental sociology. Ecological modernization also avoids falling victim to a presumption of the 'equality of risk', and can be complemented by incorporating concepts such as Murphy's (1994) notion of 'environmental classes'. As I note below, ecological modernization potentially can add a comparative dimension, building on some of the insights of Hajer (1995). A more comparative ecological modernization perspective can help it to address its Eurocentricity and lack of generalizeability. Perhaps most importantly, ecological modernization theory suggests provocative hypotheses, the resolution of which will help to make environmental sociology a stronger empirical science.

While this is not the time or place for laying out an improved version of ecological modernization theory, I will sketch some notes about possible avenues for enhancing the contribution of ecological modernization and related approaches. First, instead of (at least implicitly) presuming that there is a common type of political culture and state structure that is most conducive to ecological modernization, ecological modernization scholars could begin with a comparative perspective. Thus, while ecological modernization is the prevailing discourse of environmental politics in much of the West, this is by no means the case in other societies and national states. This observation could lead to interest in comparative analysis of political cultures of environmental reform; further, to the extent to which there is diffusion of these political cultures through globalization processes – a project that Mol has already initiated in his contribution to this volume – ecological modernization could be explored in a broader world-systemic perspective.

It also strikes me that ecological modernization and its notions of political modernization and its appropriation of notions such as 'issue advocacy coalitions' are highly compatible with another new and rapidly growing area of sociological scholarship: the work of Peter Evans (1995, 1996) on 'embedded autonomy' and 'state-society synergy'. Evans' concepts are highly relevant to environmental sustainability issues, as is evidenced by the fact that his most recent research interest has been in urban sustainability and 'livable cities' in the South (Evans, 1998). Evans' concepts of embedded autonomy and state-society synergy can also provide much of the framework for deepening the comparative project of ecological modernization research. Evans' work represents an effective response to the need for a realistic model of states and power relations which is not innocent of coercion and domination, while at the same time it is able to identify salient respects in which states can have positive impacts on, and through, civil society processes. Evans' (1998) observations on the role of state-society synergy at the sub-national level also demonstrate how political cultures and configurations which have little in common with those depicted by Mol (1997) for the Netherlands can have positive impacts on environmental improvement in the South.

In addition to the need for a more comparative approach to ecological modernization research, it is also necessary for its practitioners to address the various literatures, such as those on postmodernization and development studies, that raise profound questions about modernization theories. Ecological modernization theories have yet to take account of some of the observations from the three or so decades of critique of modernization theory in development studies. Modernization theories of all kinds need to address the tendency to the teleological presumption that there are essentially common paths of development and change that all national societies must inevitably traverse.

I opened this paper by saying that ecological modernization is unlikely to become the dominant theory in environmental sociology. In part, this is because of some of its weaknesses, as just discussed. But beyond these weaknesses, it is arguably the case that ecological modernization has several characteristics – an optimistic view of the potential for environmental reform and 'ecological restructuring' and thus certain 'exemptionalist' overtones, skepticism about the role of radical environmentalism, and a lack of attention to stressing explanations for environmental degradation – that cause it to be viewed with some amount of suspicion in environmental sociological circles. These very characteristics of ecological modernization, however, cause it to be a provocative challenge to received wisdom in environmental sociology.

Conclusion: further theoretical exhortations

I have strived to advance a perspective on theory in environmental sociology that is appreciative of the continued relevance of the classical tradition but which is prepared to recognize the many shortcomings of classical social theory. The classical theorists' works can always be read in two different ways – as being hopelessly dated works that do not have much to say about an epoch that departs so much from that of the nineteenth century, and as a continuing source of insights and methodological approaches that are applicable to almost any empirical problem we might come up with. My guess is that environmental sociology will always essentially consist of a set of perspectives which have definite lineages to the classical theorists while also having an ecological-material ontology of some sort (e.g., a notion of 'ghost acreage', treadmill of production, ecological modernization) superimposed on them. Another aspect of classical theory that is very positive for environmental sociology is that which Wardell and Turner (1986) have referred to as the 'classical project'. The future of environmental sociology may well lie in whether it is able to restore the 'classical project' of uniting sociological abstraction with a meaningful role in addressing moral and political issues.

It is my view that environmental sociology broadly construed has continued to stray from the new human ecology or materialist core that predominated in the late 1970s and 1980s. A number of theoretical views that may be regarded in one respect or another as 'exemptionalist' – social constructivism, risk society, critical theory, and ecological modernization – have appropriated some of the turf formerly occupied by the new human ecology. On the whole, this has been a positive development. I feel that environmental sociology has been more than capable of assimilating these divergent views while avoiding environmental sociology becoming reduced to answering the question of how and why 'ecology' has come to be culturally significant. Environmental sociology is strong and well established enough that it can tolerate, and even thrive within, theoretical pluralism. The diversification of the field is opening up avenues of theoretical innovation and synthesis that were not present a decade ago (Buttel, 1996). Further, this more diversified theoretical base is increasing the opportunities for environmental-sociological theory to be integrated more closely into concrete empirical research. The overall trends thus strike me as being very positive.

The reflexive modernization-related frameworks portrayed above have been particularly creative responses to some of the limitations of the traditional North American environmental-sociological literature. They permit serious analysis of environmental improvement and take a less singular and utopian view of environmental movements than is typical of much of environmental sociology. They allow that citizen-actors are not just passive recipients of the overarching forces of modernity/modernization, and that they can themselves affect the 'modernization' process. Reflexive modernization theories recognize that it is possible that modernization can be 'turned back on to itself' in order to address the problems which it has itself created. Risk society theory, along with related interpretive, constructivist, and discourse-analytical theories (e.g., Hajer, 1995; Lash et al., 1996), has particular strengths in helping us to recognize that while there are social-structural forces that facilitate solutions to environmental problems, there also needs to be a sounder conceptualization of the multiple (and contradictory) processes by which these forces can become reflected in institutional practices and social behaviors (see especially Hajer, 1995). Most importantly, these theories – particularly ecological modernizationism – raise important questions and suggest areas of research that will be of relevance to environmental sociology as a whole.

Notes

1. It is striking to me that the first major work of contemporary (post-Earth Day) environmental sociology, Burch's *Daydreams and Nightmares* (1971), reflects the eclectic (or agnostically respectful) posture toward the classics which I see in ascendance in environmental sociology and which I believe is the most appropriate orientation toward the classics.
2. Also, two of the most influential recent neo-Marxist environmental sociology anthologies, by O'Connor (1994) and Benton (1996b), include contributions that derive largely from the work of the late Marx (particularly his *Grundrisse* and the three volumes of *Capital*).
3. The emphasis on the inherent or intrinsic tendency to environmental destruction or degradation should also be seen as one of the principal ways in which environmental sociologists distinguished their work from that of the neo-Durkheimian 'human ecologists'. Among human ecology's central concepts were those of adaptation and equilibrium, which implied a tendency for social institutions over time to come into symmetry or stability with respect to environmental resources. Mainstream North American environmental sociology practitioners thus tended to regard human ecology as having a benign or unrealistic view of the stresses humans were placing on the natural environment, and an overly optimistic assessment of the capacity of institutional processes for responding to environmental problems (Buttel and Humphrey, 1987).
4. Note that I am not disagreeing about the fundamental importance of 'environmental movements'. My concern is that one cannot theorize the role of environmental movements in dealing with ecological issues apart from the contradictory roles of states in this process. That is, while environmental conservation and protection are directly or indirectly state-regulatory practices and the role of states in a societal division of labor is to 'rationalize', states face conflicting imperatives (accumulation [or the responsibility for satisfactory aggregate economic performance] and legitimation) and formidable pressures from various groups in civil society (especially capital) that cause them to be internally conflicted over or very reluctant to invoke environmental protection policies and practices.
5. Note, though, that most ecological modernization research focuses on the production sphere, particularly heavy-industrial production. Spaargaren (1996: Chapter 6) has ar-

gued that its principles should also apply to consumption, and he has sketched out a schema for understanding the ecological modernization of household consumption.

6. Despite their common rooting in notions of reflexive modernization, there are very significant differences between risk society and ecological modernization theories. Risk society theory is far more 'apocalyptic' than ecological modernization theory. Ecological modernization theory tends to find states up to the task of making possible environmental improvement, whereas risk society theory is much more skeptical. While neither risk society nor ecological modernization theory has much use for the claims of radical environmental groups, risk society theory has a far more positive view of 'subpolitics' and new social movements than does ecological modernization theory. As Mol (1995: 48) puts the matter, 'subpolitics hardly emerge as an option in ecological modernization theory'.
7. The lack of residential and spatial segregation in Northern Europe does lead to a certain equality of risk vis-à-vis hazards such as industrial accidents and toxic wastes. This relative equality of risk as depicted by Beck is largely made possible by the fact that Northern European welfare-statist institutions place a floor under wages, and thus level class differences in income and living standards. The working and dominant classes of Northern Europe are far less residentially and spatially segregated than is the case in the developing world, and even in the US.

Bibliography

Beck, U. (1992), *Risk Society*. Beverly Hills, CA: Sage.

Beck, U. (1994), The reinvention of politics: towards a theory of reflexive modernization, in: U. Beck et al. (eds), *Reflexive Modernization*. Cambridge: Polity.

Beck, U. (1995), *Ecological Politics in an Age of Risk*. Cambridge: Polity.

Beck, U., A. Giddens and S. Lash (eds) (1994), *Reflexive Modernization*. Cambridge: Polity.

Benton, T. (1996a), Marxism and natural limits: an ecological critique and reconstruction, in: T. Benton (ed.), *The Greening of Marxism*. New York: Guilford, pp. 157-183.

Benton, T. (ed.) (1996b), *The Greening of Marxism*. New York: Guilford.

Bookchin, M. (1971), *Post-Scarcity Anarchism*. Berkeley: Ramparts.

Bunker, S.G. (1996), Raw material and the global economy: oversights and distortions in industrial ecology. *Society and Natural Resources*, 9: 419-430.

Burch, W.R., Jr. (1971), *Daydreams and Nightmares*. New York: Harper & Row.

Burch, W.R., Jr., N.H. Cheek and L. Taylor (eds) (1972), *Social Behavior, Natural Resources, and the Environment*. New York: Harper & Row.

Burch, W.R., Jr. and D.R. Field (1988), *Rural Sociology and the Environment*. Westport, CT: Greenwood Press.

Burkett, P. (1996), Some common misconceptions about nature and Marx's critique of political economy. *Capitalism-Nature-Socialism*, 7: 57-80.

Burns, T.R. and T. Dietz (1992), Cultural evolution: social rule systems, selection, and human agency. *International Sociology*, 7: 259-283.

Buttel, F.H. (1986), Sociology and the environment: the winding road toward human ecology. *International Social Science Journal*, 109: 337-356.

Buttel, F.H. (1992), Environmentalization: origins, processes, and implications for rural social change. *Rural Sociology*, 57: 1-27.

Buttel, F.H. (1996), Environmental and resource sociology: theoretical issues and opportunities for synthesis. *Rural Sociology*, 61: 56-76.

Buttel, F.H. and C.R. Humphrey (1987), *Sociological theory and the natural environment*. Paper presented at the annual meeting of the American Sociological Association, Chicago, IL, August.

Catton, W.R., Jr. (1976), Why the future isn't what it used to be (and how it could be made worse than it has to be). *Social Science Quarterly*, 57: 276-91.

Catton, W.R., Jr. (1980), *Overshoot*. Urbana: University of Illinois Press.

Catton, W.R., Jr. and R.E. Dunlap (1978), Environmental sociology: a new paradigm. *The American Sociologist*, 13: 41-49.

Cotgrove, S. (1982), *Catastrophe or Cornucopia*. Chichester: Wiley.

Cottrell, F. (1955), *Energy and Society*. New York: McGraw-Hill.

Dickens, P. (1992), *Society and Nature*. Philadelphia: Temple University Press.

Dietz, T. and T.R. Burns (1992), Human agency and the evolutionary dynamic. *Acta Sociologica*, 35: 187-200.

Dunlap, R.E. and W.R. Catton, Jr. (1979), Environmental sociology. *Annual Review of Sociology*, 5: 243-73.

Dunlap, R.E. and W.R. Catton, Jr. (1983), What environmental sociologists have in common (whether concerned with 'built' or 'natural' environments). *Sociological Inquiry*, 53: 113-35.

Dunlap, R.E. and W.R. Catton, Jr. (1994), Struggling with human exemptionalism: the rise, decline and revitalization of environmental sociology. *The American Sociologist*, 25: 5-30.

DuPuis, E.M. and P. Vandergeest (eds) (1995), *Creating the Countryside*. Philadelphia: Temple University Press.

Evans, P. (1995), *Embedded Autonomy*. Princeton: Princeton University Press.

Evans, P. (1996), Government action, social capital, and development: reviewing the evidence on synergy. *World Development,* 24: 1119-1132.

Evans, P. (1998), *Liveable cities?* Unpublished manuscript. University of California, Berkeley: Department of Sociology.

Firey, W. (1960), *Man, Mind, and Land*. New York: Free Press.

Freudenburg, W.R., S. Frickel and R. Gramling (1995), Beyond the nature/society divide: learning to think about a mountain. *Sociological Forum,* 10: 361-392.

Giddens, A. (1987), *Social Theory and Modern Sociology*. Stanford, CA: Stanford University Press.

Giddens, A. (1994), *Beyond Left and Right*. Stanford, CA: Stanford University Press.

Goldblatt, D. (1996), *Social Theory and the Environment*. Boulder, CO: Westview Press.

Gottlieb, R. (1994), *Forcing the Spring*. Washington, DC: Island Press.

Gould, K.A., A. Schnaiberg and A.S. Weinberg (1996), *Environmental Struggles*. New York: Cambridge University Press.

Hajer, M. (1995), *The Politics of Environmental Discourse*. New York: Oxford University Press.

Hannigan, J. (1995), *Environmental Sociology*. London: Routledge.

Hughes, J.A., P.J. Martin and W.W. Sharrock (1995), *Understanding Classical Sociology*. London: Sage.

Klausner, S.A. (1971), *On Man in His Environment*. San Francisco: Jossey-Bass.

Lash, S., B. Szerszynski and B. Wynne (eds) (1996), *Risk, Environment and Modernity*. London: Sage.

Lipschutz, R.D., with J. Mayer (1996), *Global Civil Society and Global Environmental Governance*. Albany: State University of New York Press.

Lipschutz, R.D. and J. Mayer (1993), Not seeing the forest for the trees: rights, rules, and the renegotiation of resource management regimes, in: R.D. Lipschutz and K. Conca (eds), *The State and Social Power in Global Environmental Politics*. New York: Columbia University Press.

Lopreato, J. (1984), *Human Nature and Biocultural Evolution*. Boston: Allen & Unwin.

MacNaughten, P. and J. Urry (1995), Towards a sociology of nature. *Sociology,* 29: 203-220.

Martell, L. (1994), *Ecology and Society*. Amherst: University of Massachusetts Press.

Milbrath, L. (1984), *Environmentalists: Vanguard for a New Society*. Albany: State University of New York Press.

Mol, Arthur P.J. (1995), *The Refinement of Production*. Utrecht: Van Arkel.

Mol, A.P.J. (1997), Ecological modernization: industrial transformations and environmental reform, in: M. Redclift and G. Woodgate (eds), *The International Handbook of Environmental Sociology*. London: Edward Elgar, pp. 138-149.

Morrison, K. (1995), *Marx, Durkheim, Weber*. London: Sage.

Murphy, R. (1994), *Rationality and Nature*. Boulder, CO: Westview Press.

Murphy, R. (1997), *Sociology and Nature*. Boulder, CO: Westview Press.

O'Connor, M. (ed.) (1994), *Is Capitalism Sustainable?*. New York: Guilford.

Parsons, H.L. (ed.) (1977), *Marx and Engels on Ecology*. Westport, CN.: Greenwood Press.

Polanyi, K. (1957), *The Great Transformation*. Boston: Beacon.

Redclift, M. and G. Woodgate (eds) (1997), *The International Handbook of Environmental Sociology*. London: Edward Elgar.

Sachs, W. (ed.) (1993), *Global Ecology*. London: Zed Books.

Sanderson, S.K. (1990), *Social Evolutionism*. Oxford: Basil Blackwell.

Schnaiberg, A. (1980), *The Environment*. New York: Oxford University Press.

Schnaiberg, A. and K.A. Gould (1993), *Environment and Society*. New York: St. Martin's Press.

Spaargaren, G. (1996), *The Ecological Modernization of Production and Consumption*. PhD thesis. Wageningen: Wageningen Agricultural University.

Spaargaren, G. and A.P.J. Mol (1992), Sociology, environment, and modernity: ecological modernization as a theory of social change. *Society and Natural Resources,* 5: 323-344.

Swingewood, A. (1991), *A Short History of Sociological Thought. Second Edition*. New York: St. Martin's Press.

Szasz, A. and M. Meuser (1997), Environmental inequalities: literature review and proposals for new directions in theory and research. Manuscript. University of California, Santa Cruz: Board of Studies in Sociology.

Taylor, P.J. and F.H. Buttel (1992), How do we know we have global environmental problems? Science and the globalization of environmental discourse. *Geoforum,* 23: 405-416.

Turner, J.H. (1994), The ecology of macrostructure, in: L. Freese (ed.), *Advances in Human Ecology, Vol. 3*. Greenwich, CT: JAI Press, pp. 113-138.

Wardell, M.L. and S.P. Turner (1986), Introduction: dissolution of the classical project, in: M.L. Wardell and S.P. Turner (eds), *Sociological Theory in Transition*. Boston: Allen Unwin, pp. 11-18.

Webb, W.P. (1952), *The Great Frontier*. Boston: Houghton Mifflin.

Weber, M. (1922/1968), *Economy and Society*. Berkeley: University of California Press.

West, P.C. (1984), Max Weber's human ecology of historical societies, in: V. Murvar (ed.), *Theory of Liberty, Legitimacy and Power*. Boston: Routledge & Kegan Paul, pp. 216-234.

Wilkinson, R.G. (1973), *Poverty and Progress*. New York: Praeger.

Wynne, B. (1996), May the sheep safely graze? A reflexive view of the expert-lay knowledge divide, in: S. Lash et al. (eds), *Risk, Environment, and Modernity*. London: Sage.

Yearley, S. (1996), *Sociology, Environmentalism, Globalization*. London: Sage.

3

Ecological Modernization Theory and the Changing Discourse on Environment and Modernity

Gert Spaargaren

Introduction

The central task for environmental sociologists is to relate the changing environmental profile of modernity to the changing character of modern societies themselves. In doing so environmental sociologists can profit from insights which result from the debate on 'modernization theory' within the social sciences. This holds true even when the confrontation with the environmental crisis brings out the need for a partial de- and reconstruction of modernization theory in its conventional form, be it Parsonian or not. In this contribution – and in constant debate with modernization theory – I try to show the relevance of various theoretical perspectives in environmental sociology in the context of the broader environmental discourse in modern societies from the seventies onward. It will be shown that changing theoretical perspectives go hand in hand with changing views on the relationship between environment and economic growth, on the role of science and technology and on the role of both governmental and non-governmental actors. We will distinguish between three different periods or phases in the environmental discourse, corresponding with three different perspectives on environment and modernity: the period of the 'limits to growth debate', when de-modernization theories dominated within environmental sociology; the period of the 'sustainable development debate', mirrored within environmental sociology by the dominant position of the ecological modernization theory; and the period of the 'global environmental change debate', bringing to the fore theoretical perspectives which can be referred to as reflexive modernity theory.

De-modernization theory and the 'limits to growth debate'

At least during the seventies and early eighties, almost every introduction to environmental social science not only contained a chapter on Malthus and the Club of Rome report but also a discussion on 'the Blueprint'. 'The Blueprint for Survival' (Goldsmith et al., 1972) was published as a special issue of the magazine *The Ecologist* and it contained not only a warning for the future of mankind but also a model of an alternative, green society. A society consisting of numerous small scale units, where people live their lives close to nature and to each other, where technology was of the proper scale e.g. 'adapted' to its social and natural context (Schumacher, 1973) and where all cells from the cell-tissue society were to decide as autonomous political units about their own future (Bookchin, 1980). This image of a future green society has been discussed by numerous authors with either a sympathetic or critical attitude towards the green case (see among others: Frankel, 1987; Dobson, 1990; Goodin, 1992; Martell, 1994).

The body of literature that has been developed around the (im)possibility of a small scale society of eco-communities, to our opinion belongs to the basic readings in environmental sociology. However, the literature is rather diverse with regard to its disciplinary origins and sometimes it has a strong political ring to it. Although Robyn Eckersley too presents her *Environmentalism and Political Theory* as a kind of political manifest in defense of ecocentrism, she at the same time made a successful attempt to embed the discussion on eco-communities and small-scale societies in some of the main streams of thought which together represent the sociological and political science tradition. Why is it, she asks herself, that a marriage between Marxism and green political theory must be seen as rather inconceivable? Is the failure of Frankfurter critical theory to incorporate green issues irreparable or not? Are eco-socialist thinkers prepared to think through a process of downscaling of society which seems to contradict the central role that socialist thinkers contribute to the nation state? (Eckersley, 1992). By raising these kind of questions Eckersley is reviewing the sociological legacy with regard to its (in)ability to theoretically deal with environmental issues.

Although our analysis is less outspoken with regard to its political background and also less ambitious and encompassing than Eckersley's book, we share with her an interest in exploring the theoretical streams of thought which underpinned the strategies and ideologies of the environmental movement from the early seventies onward. In Europe especially the works of Otto Ullrich (Germany), Ivan Illich (France), Fritz Schumacher (UK), Rudolf Bahro (Germany), André Gorz (France), Barry Commoner (US), Hans Achterhuis (the Netherlands) and other 'theorists of counter-productivity' have been very influential within the environmental movement. Although writing in the tradition of neo-Marxism, they were critical of Marxism because its critique of modernity was directed primarily and exclusively at the social relations of production. Marxism failed by not taking sufficiently into account the nature of the production forces, the character of the (growth) machinery itself. Counter-productivity theorists were critical about Marxism, but at the same time discarded industrial society theory – as developed by Daniel Bell and others – because of its naive belief in the essentially benign character of technology and its lack of class analyses. Arthur Mol summarized the position taken by the counter-productivity school of thought vis-à-vis Marxism and industrial society theory as depicted in Figure 1.

The essence of eco-socialist thinking as developed by these counter-productivity theorists was the concept of 'net-balancing' (*Total-bilanzierung*). For a correct measurement of the productivity of a technology or a certain sector of industry, one has to take on board all the 'real costs' that are involved, including the harm that is done to the environment. When modern industrial society is judged according to these 'realistic' criteria, its productivity will be much underneath the levels

Schools of thought	(Neo-)Marxist	(Post-)industrial society	Counter-productivity
Kind of theory	conflict theory	consensus theory	conflict theory
Institutional trait	capitalism	industrialism	triangle of capitalism, industrialism and surveillance
Prime cause of environmental crisis	relations of production	unadapted industrial development	forces ánd relations of production
Solutions	socialization of production	ecological adapted industry and post-materialism	decentralized organization and convivial technology

Figure 1 *General characteristics of three schools of thought in environmental sociology* (Mol, 1995: 16)

that are indicated by the accepted standards like GNP. Although in an early stage of societal development, the development of productive forces may enhance welfare both in the sense of material prosperity and socio-environmental well-being, at a certain stage of its development, industrial society reaches a 'socio-critical point'. When this critical point is passed over, the rewards of sustained growth in the material dimension are outweighed by the costs in the socio-environmental dimension, in which case the technology or industry is said to be 'counter-productive'. A technology or sector of industry runs the risk of becoming counter-productive especially in those situations where the original traits of industrialism (the scale of production; the machine-density of production; the corresponding division of labor) are pushed too far. In some cases this situation of over-development of the industrial traits mentioned, can be said to be a pre-condition for the existence of that industry or technological system. Those technological systems or sectors (with nuclear energy and the chemical industry being mentioned as the most typical examples) are referred to as slum-technologies (Sack-gasse/ dead end technologies) (Ullrich, 1979).

Readers who are familiar with the work of environmental economists will have noticed some striking similarities between the net-balance approach of the counter-productivity theorists on the one hand and the 'external costs' postulate within environmental economics on the other. Within the Netherlands, the work of Hueting has been path-breaking in its attempt to 'internalize' the costs of using environmental goods into neo-classical economic theory (Hueting, 1974). The major difference however between environmental economists and other environmental social scientists seems to be their basic attitude towards the modern industrial project. Where generally speaking economists tried to improve and correct the system of production and consumption, leaving its basic structure intact, most of the counter-productivity theorists were close to the opinion that the external (environmental) costs referred to structural design faults of the industrial system which could not be repaired overnight. Because of their insisting on the partial or total dismantling of the industrial system, we refer to the counter-productivity stream of thought as de-modernization theories.

Tellegen (1983), Cramer (1989) and Leroy (1983) analyzed the ideology and strategy of the 'new' environmental organizations and movements in the Netherlands and Belgium and they were joined by Lowe and Goyder (1983), Cramer in co-operation with Jamison (Cramer et al., 1988) and many others who described the situation in other (western) European countries. From their work a clear picture arises of the environmental movement during the period of the 'limits to growth debate'. They all sketched the basic ideology underlying the strategies of the environmental movements during this period as a 'de-modernization' perspective. This perspective was spelled out in some detail also within manifests published by the movements themselves, as we can see not only from the 'Blueprint' but also from the 'Program of the German Green Party', on which Goodin based most of his outline of a 'green political theory' (Goodin, 1992). Within the German movement of 'Bürger-initiativen' as well as in the Dutch movement for 'man and environmental friendly enterprises' (MEMO), the central focus was on experiments which could provide an alternative to modern industrial society. The environmental movements were characterized not only by their alternative-exemplary grass-roots initiatives, but also by their antagonistic relationship with the state and with industry. In that sense, the environmental movement was a child of her time, being one of the 'new social movements' challenging the existing order or the political paradigm of that time (Van der Loo et al., 1984; Offe, 1986; Van Noort, 1988).

Ecological modernization theory and the debate on sustainable development

When in 1983, Egbert Tellegen in his book on the (Dutch) environmental movement arrived at the conclusion that 'it is time for the environmental movement to positively reconsider its relation with the state', he hardly will have suspected that 'these times' were already under way and would be there to stay for at least the next fifteen years (Tellegen, 1983: 65). There were good reasons for the environmental movement to reconsider its relationship with the state, not only on theoretical but also on empirical grounds. During the seventies and the beginning of the eighties, the environment not only settled itself on the top of the political agenda, one could also witness the substantial growth of the body of environmental legislation. Environmental politics had been 'steadily growing into adulthood' (Biezeveld, 1985) not only from a quantitative but also from a qualitative point of view (Van Tatenhove, 1993).

At his retreat as the Dutch minister of the environment in 1986, Pieter Winsemius published a book summarizing the main concepts and strategies for environmental policy in the eighties. His book, *Gast in eigen Huis*, would turn out to become a very influential manuscript both inside and outside government circles. The reason for its popularity was that it in fact contained all of the main elements of the 'new politics of pollution' (Weale, 1992). And this new politics would successively gain support from major parts of the environmental movement and from business circles as well.

This process of a renewed policy outlook emerging with regard to environmental problems, was not restricted to the Netherlands only. On the international level the Brundtland report – of 1987 – signaled the definite breakthrough of the new policy approach. With the notion of sustainable development gaining ground, the concepts of economy and ecology were no longer regarded to be antithetical. According to Albert Weale, the broad and enthusiastic support that the Brundtland report received was a major sign of the fact that 'there was a new belief system emerging that might be named "ecological modernization"' (Weale, 1992: 31). Summarizing the description provided by Albert Weale, this new belief system can be said to include the following propositions:

- challenging the conventional idea of a zero-sum trade-off between economic prosperity and environmental concern (to be popularized later on in slogans like 'creating win-win-situations'; 'Doppelnutzung'; 'Pollution Prevention Pays' (PPP) etc.),
- redefining the relationship between the state, its citizens (including those organized in social movements) and private corporations and,
- a recognition of the fact that most of the pressing environmental problems exceed the level of the national state, making a supra- or transnational approach to the problem a fundamental necessity.

Besides Weale, both Maarten Hajer (1995) and Peter Wehling (1994) refer to the new environmental policy approach or discourse with the term 'ecological modernization'. Both Weale and Hajer based their conclusions regarding the new policy approach on their analyses of pollution and acid-rain politics in different countries in Europe (Germany, the Netherlands and the UK). As far as the periodization is concerned, Hajer notices that the ecological modernization debate started to emerge in Western countries and organizations around 1980. Around 1984 it was generally recognized as a promising policy alternative. With Brundtland it became 'dominant in political debates on ecological affairs' (Hajer, 1996: 249). Both authors also seem to be in agreement about the fact that Germany and the Netherlands at that time provided the most exemplary or prototypical models of the new approach.

The new 'belief system' can be said to contradict in some crucial respects the belief system which had been dominant within circles of the environmental movement and related sectors of the academic community during the seventies and early eighties. How did a de-modernist perspective give way to a paradigm that placed the further modernization of production and consumption at the heart of its intentions? How did it come about that the incidental and antagonistic relationships between environmental movements on the one hand and government and business circles on the other, were evolving into a critical dialogue with rather intensive frequency? In what way did the changes we signaled in the broader socio-political field influence the theoretical work within the socio-political sciences?

In fact it is one of our main arguments throughout the book that the development of the ecological modernization approach within environmental sociology has profited from and in its turn has contributed to the new policy approach that emerged. Two German authors can be regarded as the founding fathers of the ecological modernization approach, Joseph Huber and Martin Jänicke. Both authors developed the ecological modernization approach from a slightly different perspective. Jänicke laid great emphasis on the new role of the state that was emerging both within environmental policy and within politics more in general. Huber primarily elaborated the ecological modernization approach into a more encompassing theory of environment induced social change. We will provide a short introduction to both the variant of Jänicke and Huber and then go on to discuss our own emphasis with regard to the ecological modernization approach.

Martin Jänicke on ecological modernization as the 'modernization of politics'

In his earlier work, Martin Jänicke describes the environmental crisis as a crisis of the modern state (Staatsversagen). The modern state's inability to properly react to the environmental crisis originated from and further contributed to the so-called legitimation and steerings crisis of the national state. This crisis was not restricted to the domain of environmental politics alone but pertained also to domains that are historically associated with the rise of the welfare-state (Jänicke, 1986).

In contributing to the debate on the 'retreat of the state', Jänicke in his earlier works stresses the fact that the environmental crisis can and should provide a new rationale for state-intervention, although organized along different lines. Without state-intervention, the greening of production and consumption is an impossibility. The ecological modernization process must be actively supported by the state in the form of a green industrial policy (Jänicke, 1986, 1993, 1995; Wehling, 1994). Where, as we will discuss later, Joseph Huber in his elaboration of the ecological modernization approach mainly regards state-intervention as an obstacle to effective environmental reform, Jänicke – in his early works at least – firmly sticks to the need of an expanding state to support the ecological modernization process. Enlarging the steering capacity of the state is regarded as a necessity because the steering potential of markets and market actors is structurally weak. Although ecological modernization is targeted primarily at market actors and the industrial sector, its main bearer should still be the state.

In his more recent work there is a slightly different emphasis regarding the manner in which state intervention in the field of environmental politics is analyzed by Jänicke. Now the focus is no longer on the need of state intervention per se and on the enlargement of steering capacities as enforced by environmental problems. In his latest work the focus is mainly on the way in which politics are 'modernized' by translating the experiences acquired within the environmental field into other arenas of politics. Environmental politics bring about new forms, principles and instruments which imply the reshaping of the relationship between the state and civil society (actors).

The two most significant examples of the new political forms that have arisen are, firstly, the so called *target-group approach*, where on an ad-hoc basis civil actors and the state enter into several rounds of negotiations, trying to agree on norms and measures that are contextually relevant and that are accepted on a voluntary basis and, secondly, the approach that was labelled in the Netherlands as the *region-oriented, integrated approach*, where these characteristics are employed within the context of a specific spatial unit. Both in Germany and in the Netherlands these initiatives, as they originated especially in the environmental field, resulted in a vast body of (political science) literature describing a kind of paradigm shift in the making of politics. Classical, hierarchical and universalistic in the sense of de-spatialized politics gave way to a form of steering which is characterized by the principles of horizontal cooperation, consensual- and dialogical decision making, less formal institutionalization and a growing importance of actors at the de-central level. As Jänicke describes it:

> Es geht im Grunde um ein Politikmodell jenseits von liberalem "Laissez faire" und bürokratischem Staatsinterventionismus: um ein dezentraleres und konsensbetontes Politikmodell, das den Zentralstaat auf strategische Aufgabe konzentriert und Detailregelungen stärker auf dezentrale Akteure verlagert. Auf das Feld der Umweltpolitik übertragen, kommen dem Zentralstaat danach vor allem die Sicherstellung ökologischer Minima und "strategische" Gestaltungsfunktionen zu. Seine Aufgabe wäre es nicht zuletzt, die langfristischen Umweltprobleme zu definieren. Sache dezentraler Akteure wäre es, unter Nutzung ihrer spezifischen Innovationspotentiale über die nationalstaatlichen Grundbedingungen und Minima hinauszugehen. (Jänicke, 1993: 167)[1]

In the Netherlands these policy changes are discussed under the heading of the so called 'policy-networks-approach' (Glasbergen, 1989; Huppes, 1989; Nelissen, 1992; Godfroy and Nelissen, 1993). Policy networks are 'constructed' to deal with circumscribed problems at a de-central, regional level. Participating in the network are both public/state and private actors which are mutually interdependent in the sense that none of the actors can enforce a solution to the problem on its own. Networks are constituted in order to get an interaction-process going which may result in a shared 'definition of the problem' and a consensus on the most appropriate set of policy-instruments to be applied in their specific regional situation. The relative successful employment of the 'region-oriented, integrated approach' is documented for about ten specific problem-areas in the field of environmental, agricultural and transport politics (Glasbergen, 1989; Van Tatenhove, 1993; Van Goverde, 1995; Van Tatenhove and Leroy, 1995).

The central feature of the ecological modernization approach as a theory of political modernization is its focus on new forms of political intervention, e.g. the changing role of the state. Within Huber's variant of the ecological modernization approach to be discussed next, the role of the state is analyzed as just one element among a variety of initiatives and strategies developed in modern society to bring about environmental reform.

Joseph Huber on ecological modernization as a theory of social change

According to Joseph Huber, the eighties witnessed all kinds of initiatives and strategies for environmental reform which cannot be grasped within the theoretical framework of the 'modernization of politics'. Several forms of 'direct negotiations' between the environmental movement and circumscribed segments of industry, sometimes even separate firms, gained popularity. Many environmental arrangements resulted from intra-chain-(power)relations within sectors of industry, retailer pressure being a good case in point. Consumer pressure in some cases proved effective in enforcing producers to abandon products which were regarded as especially harmful to the environment etc. In short, the process of environmental social reform gained its own, specific momentum.

With the concept of an 'ecological switch-over' Huber (1982; 1985b) tried to grasp the essence of this process of environmental reform: a set of relatively rapid changes taking place from the eighties onward, together result in a long-term momentum of development. A long-term development which is only possible if certain key institutional adaptations and transformations are accomplished initially. This long term development is called the 'ecological modernization of production and consumption'. Notice the fact that Huber's theory implies a radical break with the de-modernization ideology in the sense that he calls for a further modernization of the existing institutions of industrial society.[2] His theory was developed to make a more detailed comprehension of this modernization process possible. I will first discuss the ecological modernization theory of Huber in relation to the postwar modernization theories within sociology. Then we go on to summarizing Huber's formulation of the two main propositions implied in the ecological modernization process conceived of by him as the twin processes of the 'ecologizing of economy' on the one hand and the 'economizing of ecology' on the other. We conclude this section by making a distinction between different meanings of the phrase 'ecological modernization' and by summarizing the way in which we ourselves would like to conceptually perceive of this (ecological) modernization process as a process of ongoing rationalization.

Ecological modernization theory and industrial society theory

In the environmental discourse of the seventies and the beginning of the eighties, the question of the relationship between modernity and the environment was interpreted primarily as a question concerning either the *industrial* or the *capitalistic* character of modern societies being the dominant or most relevant factor in bringing about environmental problems. Here we have (neo)Marxist theories pointing to the capitalistic character on the one hand and industrial society theorists holding responsible the industrial character of modern societies on the other. Having sketched the main theoretical streams of thought[3] and the positions that were taken in this debate, we arrive at a conclusion which is twofold. First, as Anthony Giddens has shown in his work on the character of modern societies, there are four institutional clusters which together characterize modernity. Both capitalism and industrialism are necessary for understanding modernity, the other institutional dimensions referring to surveillance and military power. Having said this, we secondly conclude that environmental problems are inherently connected with those institutions which give modern societies their industrial character: the industrial organization of production and consumption should be the core object for environmental sociologists. By stating that the environmental design fault of modernity refers to its industrial dimension, we in principle agree with Huber's basic idea that the dynamics of capitalism can also (be made to) work in the direction of sustainable production and consumption. What counts for the dynamics of capitalism can also be maintained with regard to the industrial dimension of modern production and consumption. There is no principle or theoretical argument making a 'modern' organization of production and consumption and its technology antithetical to sustainability.

When viewed against the background of the capitalism-industrialism debate as it was sketched above, Huber's theory must be labeled as a theory directed at the industrial dimension of modernity. His focus is on the industrial rather than on the capitalistic institutional dimension, as can be seen from the set of concepts Huber employs for analyzing historical change in modern societies. Here he distinguishes between the 'industrial' or 'techno'system on the one hand and the 'socio-sphere' and the 'bio-sphere' on the other (Huber, 1985b). When viewed from a long term perspective, modern industrial society has arrived at the situation in which the socio-sphere as well as the bio-sphere are 'colonized' by the industrial- or techno-sphere. In order to remedy this situation, the industrial system will have to (be made to) adapt to the demands stemming both from the socio-sphere and the bio-sphere.

Throughout the history of industrial society, three long-term phases or waves can be detected. After the *breakthrough* of the industrial system in Western Europe (dated from 1789 to 1848), its phase of *construction* lasted to 1980, to be followed by the phase of the ecological *reconstruction* of the industrial system from about the mid-eighties onward. Every new (sub)cycle and the social transformations that go along with them, is witnessed through and made possible by a new key technology to arise (steam-engine; railroad construction; electrification; mass motorization; super-industrialization). The key technologies have to be picked up by innovative entrepreneurs who, with the help of foresighted financiers, bring about a new wave of industrial innovation (Huber, 1982, 1985a).

When summarized at this general or abstract level, Huber's theory[4] can be said to display the same essential characteristics as other modernization theories analyzing the future of (post)industrial society. Its similarities with the models put forward by Daniel Bell, Alvin Toffler, Geoffrey Jones and others refer to the evolutionary view on social change, technology being its prime mover. As far as these similarities exist, Huber's theory of social change is vulnerable to the same kind of criticism that was formulated with regard to modernization theories that were developed within sociology after the Second World War, when structural functionalism became one of the dominant streams of thought. With the work of Talcott Parsons, modernization theory came to denote not only the rather 'neutral' concepts as rationalization, functional differentiation and the development of subsystems like economy, politics, law and religion. It also came to represent a model of social change in which social systems were driven from lower to higher stages of development by factors stemming (only) from within the social system itself. A process of evolutionary development which would result – in the 'Third World' as well as in the 'developed countries' – in a society that resembled to a considerable degree Parsons' homeland, the USA. So modernization theory in its guise of the seventies, represented both a set of 'neutral' in the sense of formal concepts as well as a set of substantive, normative concepts and procedures. Both aspects of modernization theory have been under attack within the social sciences ever since they were initially formulated. Giddens for example has criticized what he labeled as 'unfolding models of social change' not only for the 'teleological' view of system reproduction that results when human agency is left under-theorized. He also pointed to the rather inconsiderate way in which the formal concept of social system was equaled with that of the western nation-state, which can explain the 'euro-centrism' of many modernization theories.

When commenting on Huber's central thesis, we discuss both its lack of human agency and its supposed or declared euro-centrism. We think his use of the metaphor of the caterpillar developing into a ('green') butterfly can be seen as a perhaps unintended but apt illustration of the determinism included in his theory of social change. Caterpillars just have to sit, lay down or hang and dormantly wait till the hormonal guided process of the metamorphosis befall on them. As will be argued in more detail below, we think the determinism and passivity implied in this metaphor does not suit the ecological modernization process. The process of 'adaptation' the industrial system has to go through, cannot be grasped in an adequate way without taking into account the fact that this adaptation or switch-over is brought about by knowledgeable and capable agents. And also the second main point of critique, regarding the euro-centrism contained in much of the 'classical' modernization theories, should be taken seriously when elaborating Huber's variant of ecological modernization. In his initial formulation, the geographical scope of the theory was (deliberately) restricted to Western, industrialized societies. Later on, Huber himself has discussed the possibility of extending the theory to former Eastern European countries (Huber, 1993). On the issue of the 'western' character of the theory and its applicability to non-western contexts, I refer to the elaborate treatment that is provided by Arthur Mol (Mol, 1995: 54-57).

This being said, the theoretical approach of Huber can be regarded as a fertile starting point for analyzing environment induced social change mainly for two reasons. First, when comparing Huber's theory with other theories of industrial society, he more than other authors analyzes major changes in the organization of production and consumption in a direct connection with environmental problems. Second, his elaboration of the theory does not evolve into a cultural critique of modernity or into a post-industrial or even postmodernist perspective, but instead puts in the center of attention the institutions which are most important in bringing about the switch-over into more sustainable production and consumption cycles: economy and technology.

Economy, technology and ecology

On a more concrete level, the ecological modernization of production and consumption can be analyzed by looking in some detail at two kinds of mechanisms or processes which according to Huber are at the heart of the modernization process: the 'ecologizing of economy' is joined by the twin process of the 'economizing of ecology'. The phrase 'ecologizing the economy' he uses to refer to the process of 'internalizing' external costs or, in a more encompassing formulation, to the anchoring of environmental concerns in the organization of production and consumption. The back side of this process will be the economizing of ecology. This expression refers to the fact that, according to Huber, the ecologizing of the economy is only possible when ecology 'loses its innocence'. Ecology has to get rid of the innocence it displayed in the form of a 'romantic and holistic' critique of modernity. In order to make an impact in the 'rational' world of business and industry, it has to develop into a full blown, 'hard' science of the sustenance base.

This 'scientification of ecology' is in fact what is meant by the phrase 'economizing of ecology'. Both processes are summarized in the quote 'Wenn die Ökologie eine Zukunft hat, dann nur in industrieller Form, und die Industrie kann nur eine Zukunft haben, wenn sie ökologisch wird' (Huber, 1982: 12).[5]

In Huber's thesis of an environment induced rationalization of production and consumption, using science and technology as its main carriers, several types of criticism have been raised by different authors (Martell, 1994; Wehling, 1994; Hajer, 1995). The two most frequent objections are the 'green capitalism' argument and the 'technological fix' argument. We will shortly address both arguments.

In the debate on 'green capitalism', we can recognize a return of the capitalism-industrialism debate as discussed before but now on a more concrete level. The question is raised – for example by Luke Martell (1994) in his *Ecology and Society* – whether environmental problems can be solved within capitalism. Is it theoretically possible and/or empirically feasible that – under capitalist and market conditions – business firms and consumers are pursuing environmental goals and objectives because they think it's in their 'interest' to do so? Martell claims that capitalism and sustainable production and consumption are impossible to reconcile and that 'collectivist intervention rather than economic liberalism is necessary for securing sustainability' (Martell, 1994: 63). Although his presentation of the debate on green capitalism is stimulating, we cannot go along with Martell in his main conclusions. We think that most of the arguments Martell uses to criticize the behavior of capitalists – their shortsightedness and the fact that they do not take into account the general interests or the interests of future generations – are not confined to or specific for the behavior of 'capitalists' or market actors alone but can also be said to apply to, for example, state actors. Furthermore, in arguing for the impossibilities of green capitalism, Martell is referring primarily to the empirical situation during the 'construction-phase' of modern societies. When discussing the objectives and instruments of the 'new environmental politics', Martell demonstrates a rather unreflected belief in the superiority of centralized state intervention over market regulation. He thereby seems to overlook the fact that the ecological modernization approach for a considerable part was developed as a reaction to the 'old', interventionists and ineffective politics of the seventies.[6]

The second main argument against a process of environment induced modernization of production and consumption, refers to the role attributed to science and technology. Within Huber's theory, there is indeed a lot of confidence with respect to modern science and technology as propellers of environment induced change. He refers to modern, advanced technologies as being the key technologies in the process of change. He does not seem to be worried about modern science and technology being sometimes complex, high-scale and even 'hard' technologies. When talking about the switch-over process in terms of a new round of 'superindustrialization', Huber does not pay any serious attention to the debate on 'soft-technology' as put forward in the USA by Lovins (1977), in the UK by Schumacher (1973) or in Germany by Von Gleich (1991), to mention just a few of the propagators of the soft-technology approach. Should one always prefer – as it is claimed by the soft-technology perspective – technologies which are more 'natural' or 'soft' in the sense that they make use of natural processes instead of digging deep into nature, synthesizing or (genetically) recombining it? For Huber the fact that science and technology, during the phase of construction of the industrial system, have been the source of many of the contemporary environmental problems, does not necessarily mean that there is an inherent connection involved here. We should always be aware of the side-effects of science and technology, regardless of whether they are 'soft', 'conventional' or even super-industrial in character. Science and technology should, in their development as well as their use, be judged on their side-effects and unintended consequences both for the natural and the social system.

In the debate on soft or alternative technology one can witness some interesting developments from the seventies onward. The initial close connection between the 'soft' character of technologies on the one hand and their small-scale application in the context of a decentralized organization of production on the other, seems to be broken up now that for example windmills and solar energy are incorporated into the environmental policies of electricity companies and now that methods of biological and integrated pest-management are propagated and diffused within the mainstream agricultural sector. When soft-path technologies are propagated 'outside' the market, their chances for survival seem to be less in comparison to the situations in which they are incorporated in the strategies of major industrial actors, as Arthur Mol has shown in his study on the chemical industry (Mol, 1995). These examples illustrate the fact that the former strict opposition between soft-path and hard-path technologies (as characteristic for the de-modernization ideologies discussed above) seems to be overtaken by developments in present-day environmental politics. Instead of the dichotomy hard versus soft technology we elsewhere proposed a different approach to the relation between technology and environmental change which can be said to be in line with Huber's theory but illustrating at the same time that the ecological modernization perspective cannot be reduced to a so called 'technological fix' approach (Mol and Spaargaren, 1991). There we argued that the debates on the social construction of technology and on the shift from add-on or end-of-pipe technology to preventive technologies (Pinch et al., 1987; Cramer and Schot, 1990) are more relevant than the debate on the hard or soft character of environmental technologies.

Ecological modernization and rationalization

We have discussed so far the emergence and main content of the ecological modernization theory as it was developed in close connection to developments within the field of environmental politics in the eighties. Thereby we did not pay much attention to the need to make strict distinctions between ecological modernization as a program, a perspective, a belief system, a program of political reform and a social theory. We think it is useful to maintain the following main distinction. A difference should be made between ecological modernization as a political program on the one hand and ecological modernization as a theory of social change on the other. Behind the distinction of socio-political program on the one side and theory of social change on the other there is the division generally made within sociology between so called 'substantive' and 'formal' approaches. Substantive approaches refer to historical empirical developments using theoretical concepts as vehicles for description of social reality. Within the 'formal' approach main emphasis is laid on the further development of theories or conceptual models as an end in itself, resulting in a conceptual model which can be said to be relevant for doing empirical research across specific socio-political contexts. Although of course we are dealing with an analytical distinction implying that one should not strive to uphold a dogmatic division or separation between empirical and theoretical work, we do think the distinction between substantive and formal work is relevant for environmental sociology.

Ecological modernization as a socio-political program refers to the historic-empirical developments in the field of environmental policies and politics during the seventies and the eighties in some western European countries. It describes the different ways in which in practice policy makers, managers, financiers, environmental activists and households were (not) dealing with environmental issues and dilemmas. The main conclusion here can and must be that environmental issues moved from the periphery to the center of concern for a great number of different social groups and organizations. This can be interpreted as a process of gradual institutionalization of environmental concerns both within the media and its public, within different levels of the governmental administration and within business circles.

Ecological modernization as a theory of social change reflects on this process of institutionalization of environmental concerns in terms of the need to conceptually refine the existing models that are used within social science to analyze processes of modernization or rationalization. Here we start from the premises of Huber's theory of social change to develop it into the thesis that the process of ecological modernization can be best conceived of in terms of an independent sphere of ecological rationality to arise. When Huber refers to the ecological switch-over in the context of his substantive analysis of the western, industrial mode of production, this process can be interpreted theoretically in terms of ecological concerns developing into an autonomous, independent factor which has to be taken into account and to be dealt with in the restructuring of production and consumption. Instead of the dichotomous model – counterpositioning society and nature – which is most commonly used within the social environmental sciences, we propose a model in which four dimensions or spheres are analytically distinguished (see Figure 2).

What this picture tries to illustrate is the fact that ecological rationality can and should no longer be conceived of only or exclusively in relation to the economic rationality as it prevails in the present-day organization of production and consumption. In want of a more appropriate term, Arthur Mol expresses the growing autonomy of the ecological sphere by using the phrase 'the emancipation of ecology' to express the fact that ecology is no longer 'contained' or 'enclosed' by the economic sphere (Mol, 1995: 30). The institutions involved in the industrial mode of production are analyzed and judged using criteria which have their own right of existence in the sense that they cannot entirely be reduced to or deduced from economic criteria. On the need of the closing of substance chains in production and consumption we cannot argue or decide by referring to economic criteria.

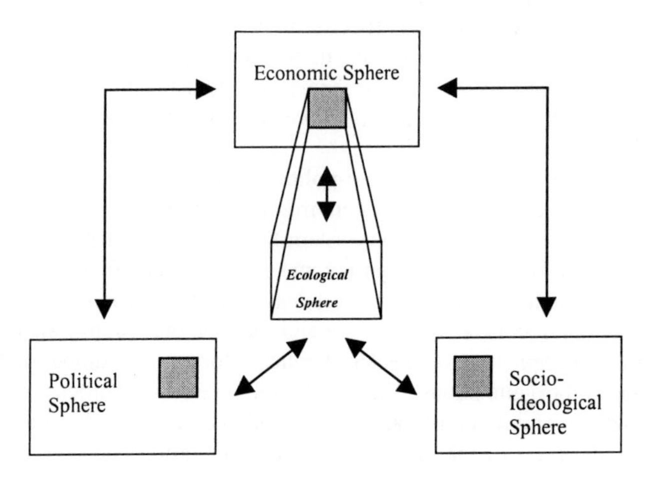

Figure 2 *Growing 'independence of the ecological sphere'*

It has to be decided upon using criteria which stem from ecology. The implications of this line of reasoning should be explored somewhat further with regard to both its practical and theoretical consequences.

On the practical level, its implications can be illustrated by providing some examples. Instead of working toward a 'greening of the conventional GNP' we are in need of a 'green Gross Nature/environmental Product' as an independent indicator of the environmental performance of a national economy. Instead of integrating 'green points' or other environmental product-labels with existing labels on, for example, health information, we should work on separate environmental performance indicators on all relevant products or product cycles. Like all forms of eco-book-keeping, employed by individual companies or farms or households, these indicators have to be developed as part of an independent information structure for the environmental aspects of production and consumption cycles. Programs of political parties should be screened and evaluated during election time not only on their economic but also on their ecological consequences. An independent environmental planning office will play an important role, as well as an environmental minister who is safeguarding the environmental figures with the same rigor that is accepted from his or her colleagues in the department of finances, etc.

But there are also consequences on the theoretical level. In terms of the literature on modernization theories, we are suggesting indeed that the environmental crisis becomes the vehicle for a further rationalization process, where new subsystems arise to 'deal' with (ecological) issues because they cannot be properly dealt with within the existing institutional make up of modern societies. By doing so, we find ourselves working in the tradition of modernization theory in sociology which started with Weber and Parsons and which is given contemporary expression by

Habermas, Luhmann and others. As for example Mommaas (1993, part I) and Touraine (1995) have emphasized in their analyses of modernity, the debate on rationalization, from its inception on, contained an awareness of the conflicting relationship between formal, technical rationalization and the more encompassing, substantial rationalization. The critical tension between both forms of rationalization stems from the tendency of technical or functional rationality to intrude into domains that used or ought to be reigned primarily by subjective, substantive rationality. Both the members of the Frankfurter Schule, most notably Jürgen Habermas, as well as for example Alain Touraine devoted a great part of their work to analyzing the conflicting relationship between objective (science, technology) and subjective (culture, subjectivity) reason or rationality.

In analyzing the status or primacy of the ecological sphere and ecological rationality in relation to the other spheres/dimensions and rationalities, two main conclusions can be drawn. Firstly, by locating its origin in the economic sphere, we try to indicate that the ecological rationality not only developed mainly in relation to economic rationality but also that there is a kind of 'Wahlverwantschaft' between the two rationalities. This can be illustrated by the prominence of concepts like 'environmental productivity' and 'environmental efficiency' as used by Huber to illustrate the analogy with labor productivity and capital productivity. This intertwinement of economic and ecological rationalities should not prevent us from noting that concepts like 'the polluter pays principle' or the 'precautionary principle' stem from the interrelation with the political and socio-cultural sphere respectively. When it comes to defining sustainability, an interesting debate has arisen in the Netherlands on the limits of technical rationality vis-à-vis substantive, socio-political rationality (WRR, 1994; Van Hengel and Gremmen, 1995; Van den Belt, 1996). We regard as one of the most important conclusions of this debate the claim defended by Van Hengel and Gremmen that although there are political choices involved in defining and defending criteria for sustainable development, sustainability is not just a matter of political choice. In other words, there can be said to exist a set of ecological criteria which has gained relative autonomy also with respect to the political sphere.

Secondly, by putting the ecological sphere analytically on a par with the economic sphere, we do not follow the suggestion made by John Dryzek as well as some 'deep ecology' theorists (like for example Eckersley) to perceive of ecological rationality as a rationality which should be *above* or *prior* to all other rationalities. In doing so, Dryzek for example reaches the conclusion that none of the major social choice mechanisms of modern society (the market being one of those mechanisms or subsystems) are able to deal with ecological issues and for that reason should be revised or partly dismantled. Like Eckersley in her critique of mainstream anthropocentric social theory, he ends up with a plea for a radical decentralized society as the only possible way to properly deal with environmental issues (Dryzek, 1987; Eckersley, 1992). Although we regard Dryzek's book as a very stimulating contribution to environmental sociology, we are not convinced by his argument that 'modern' social choice mechanisms like markets, administrative systems and the system of modern law can not (in principle and in practice) take on board ecological issues. As Mol has argued, the ecological switch-over should be interpreted in terms of ecological rationality catching up with the long standing dominance of the economic rationality without concluding from this the need for an abolition or abandoning of the economic rationality. He points to the fact that Huber (1991b) for example criticizes ecological sound production processes which do not meet criteria of economic rationality (Mol, 1995: 33). So far, we have discussed the ecological modernization approach as a reaction to both 'old' politics and 'old' theories. As a political program, it meant a break with the environmental policies as they were developed from the seventies onward. Politics in which there was a strong reliance on state interventionism and on ad-hoc, pragmatic, end-of-pipe solutions to environmental problems which were conceived of to an important degree as problems to be dealt with at national levels of political administration. The environmental movement played an important role in putting environmental issues on the political agenda but refrained from taking an active part in the elaboration and implementation of environmental policies. When ecological modernization became the dominant paradigm during the eighties, this implied a lot more than supplanting the antithetical view on environment and economic growth by the concept of sustainable development. It also implied a redefining of the role of the state vis-à-vis civil society, with both market actors and environmental movements redefining their former roles.

As a theory of social change, ecological modernization meant a break with demodernizing perspectives which had dominated the environmental discourse until then. As against both counter-productivity theory and radical ecological thinking, the ecological modernization theory starts from the proposition that the environmental crisis can and should be overcome by a further modernization of the existing institutions of modern society. As a formal theory, it is an attempt 'to define nature as a new and essential subsystem' and to develop a specific set of social, economic and scientific concepts that make environmental issues calculable and – by doing so – facilitate the 'integration of ecological rationality as a key variable in social decision making' (Hajer, 1996: 252).

Within ecological modernization theory at least in its original formulations, there is a strong emphasis on the role of institutional actors within the sphere of production in bringing about sustainable development. In a sense, the theory in this respect mirrors the new environmental policies of the eighties, in which institutional actors or the so called 'target groups' formed the most central objects of concern. Although formally included in the list of target groups, consumers and also small company holders were not among the central concerns of environmental policy-makers at that time. These groups were thought to be 'difficult to reach' because of their diffuse, heterogeneous character. While environmental policies were developing rapidly with regards to both their instruments and ideologies, there were some blind spots with regard to non-institutional actors. These blind spots, we will argue, were shared by the ecological modernization approach in its initial form, as we will argue in the next section.

Ecological modernization theory and the role of citizen-consumers

Because we regard it as among our main contributions to the further development of the ecological modernization approach, we will discuss the conceptualization of the role of citizen-consumers in the process of ecological modernization from two perspectives. Firstly, the 'productivist' orientation of ecological modernization theory should be corrected by analyzing the role of citizen-*consumers* in the context of production-consumption cycles. In doing so, we think that several relevant themes can and should be derived from the debate on the sociology of consumption. Secondly, it is important to take a closer look at what is meant by the role of *citizens*, individuals or human agents in relation to institutional developments. Although the theme of the relationship between 'actor and structure' refers both to the production sphere and the consumption sphere, we will center our attention on the sphere of consumption to illustrate the interplay between action and structure.

The dynamics of consumption

The ecological modernization approach is focused primarily on the reconstruction of those institutions of modern society which are involved in production and consumption. Being directed at this industrial dimension of modernity, its core concepts and main assumptions are all related to the sphere of production. Within the environmental science in general, core concepts like environmental management systems, (product)life cycle analyses and integrated chain management all derive their meaning from the production sphere. Although it is formally acknowledged that consumers are an integral part of the production-consumption chains, their behavior – when taken into account at all – is analyzed in a very instrumental way, as units at the end of the chain, in most cases just processing products into waste.

It could be expected from sociologists to make a vital contribution to the environmental sciences by pointing out and analyzing in detail the specific dynamism of consumer behavior and the sphere of consumption when compared to the sphere of production. Unfortunately, sociologists for a long time saw it as their principal concern to develop extensive critiques of the consumer society, hardly paying any attention to the actual process of consumption itself. Consumers were depicted as passive agents who are – willfully or reluctantly – seduced by the advertisements of big companies to keep the endless treadmill of consumer society going.

In trying to supplant this superficial way of analyzing consumer behavior by an approach in which the immense variety of manners in which people relate themselves to products, two themes are at the center of our concern: the relation between production and consumption and the 'meaning' of goods and services for people.

Within neo-Marxists' critiques of consumer culture, consumption is analyzed as a derivative of production. Instead of analyzing consumer practices as 'determined' by the production sphere, we follow Fine and Leopold when asserting that the concept of 'system of provision' combines the recognition of the specific dynamics of the consumption sphere with the need to include in our analysis of consumer behavior the differences that spring from the production sphere (Fine and Leopold, 1993). Systems of provision thereby resemble the concept of chain or vertical sector as employed by economists. With the introduction of this concept Fine and Leopold want to correct or complement the horizontal approach they see as the dominant approach within the sociology of consumption. A horizontal approach in which the similarities in our dealing with different groups of commodities are stressed. While of course people may use cars, shoes, clothes and houses all as signs of good taste, our consumption theories should not ignore the major differences in the way commodities and services are handled in for instance the traffic system, the fashion industry or the housing industry. I think the most fruitful approach to consumption will be a combination of both vertical and horizontal theories, designed to analyze characteristic features both within and between different groups of products and services respectively. Though the process of the use or consumption of water, electricity and energy by households share some basic similarities, it can also be shown that major differences exist between the corresponding segments of the public utility sector.

The second main theme that we think to be important for environmental sociologists dealing with consumption, refers to the meaning people attach to goods or commodities and services. Of course we can agree that people are not purchasing goods and services exclusively for their intrinsic qualities or for their so called use-functions. On the other hand we should not be tempted to reduce the complex process of consumption to a distinction-game only. A game in which the symbolic meaning or the identity value of products and services is thought to be much more important than the objective qualities of the products themselves. A bridge needs to be built between the technical environmental sciences focusing on the intrinsic qualities of products on the one hand and the general sociologists looking only or primarily to the identity-value of products and services on the other. The work of Douglas and Isherwood can be used as an important pillar in bridging this gap. In their book *The World of Goods* (1979) they introduce the notion of 'keeping to the level', referring to the fact that people are synchronizing their consumption behavior with or tuning it to the activities of those they consider as relevant others. Standards of comfort and hygiene are not decided upon by individuals in isolation and neither are the ways in which these standards are to be met by (group and class specific) social arrangements and institutions. While agreeing with Bourdieu and others that there is always a social group dimension to domestic consumption, Douglas and Isherwood also point to the fact that people use goods to 'rationalize' their households from a specific perspective. They 'modernize' their household using both technological and organizational devices in order to realize for example a higher 'personal availability'. In accordance with this line of reasoning, we argue that this rationalization process – both in theory and in practice – can and should be directed also at the environmental consequences of group or class specific levels and forms of domestic consumption. I will discuss this environment induced modernization of domestic consumption in somewhat more detail, paying special attention to the way ecological modernization theory can be combined with an actor-centered approach.

An actor-centered approach to the ecological modernization of domestic consumption

When trying to formulate an actor-centered approach to the modernization of domestic consumption, we should be careful not to reproduce the dualism that exists also within the social environmental sciences between micro analyses of environmental (un)friendly behaviors on the one hand and the macro analyses of institutional developments on the other. Such a dualism does indeed exist in the attitude-behavior approach that for so long dominated environmental sociology. The attitude-behavior paradigm is characterized by both theoretical and empirical imperfections. Among the theoretical problems are the neglect of the influence of 'structure' on action and the emphasis on action as a matter of constant and conscious choices. When applied in empirical research, these faults are shared by a third problem namely the fact that most of the attitude-behavior research is conducted with respect to isolated strings of behavior like choosing between paper or plastics, bicycle or car etc.

In an earlier contribution I have reviewed some of the major streams of thought within environmental sociology which could serve as an alternative for the primarily socio-psychological attitude-behavior model, concluding that the structuration theory of Anthony Giddens offers the best perspectives for solving these problems at least at the conceptual level (Spaargaren, 1994). One of Giddens' major 'innovations' within sociology is the concept of the 'duality of structure'. Within the reproduction of social practices, human beings as knowledgeable and capable agents make use of sets of rules and resources which are constituent for their behaviors. By drawing upon these rules and resources, they are at the same time reproducing these rules and resources. These rules and resources are to be conceived of as structures having a virtual existence: they are only real or visible during the moments of their instantiation, within the process of structuration. To illustrate the working of this 'mechanism', Giddens himself often refers to the example of speaking a language: we (have to) make use of certain rules of grammar, which are reproduced during the speech act.

Per Otnes has applied Giddens' notion of the duality of structure to the 'structuration of domestic daily life' (Otnes, 1988). He points to the way in which people, during their daily domestic routines, are 'making use' of technical and organizational devices (tap-water, telephone, heating, electricity etc) which at the same time are reproduced. The relation between households and the numerous so called Socio-Material-Collective-Systems (SMCS) can best be conceived of – according to Otnes – in terms of a process of 'serving and being served'. When using water or electricity we connect ourselves to both other users and to technical expert-systems. The principle of organization of these expert-systems (the 'structures of our daily life') seem to be far beyond the reach of every individual or separate household. Together with the fact that most of our actions are routines – we hardly pay attention to the functioning or the (environmental) performance of the water system when using the tap – it might look as if social actors are 'determined' by social structure. However, instead of depicting human agents as the passive recipients and users of external devices, we can analyze the process of 'serving' by deploying Giddens' notion of practical consciousness. This concept refers to the fact that we can be steering our behavior in a skilful manner without being aware of it on a discursive level. We just do it on automatic pilot. However, once this automatic pilot is turned off or put out of use by internal or external factors (for example brown water coming from the tap, a very high bill from the water company or a removal to another house) we are able to examine and assess these routines or behavior patterns from a specific perspective. While this 'de-routinization' will be followed sooner or later by a 're-routinization', there results a new way of organizing the process of serving and being served.

The ecological modernization of domestic consumption can be conceived of as a series of de- and re-routinizations with regard to a great number of domestic routines or domestic practices, ranging from child rearing to gardening or doing the laundry. When elaborating upon the role of actors in the process of de- and re-routinization the concept of lifestyle of individuals seems relevant. Again following Anthony Giddens, I use the concept of lifestyle to refer to the systemness or coherence between the social practices which together make up our lifestyle on the one hand and to the 'narrative of the self' agents attach to these practices on the other. Instead of analyzing isolated strings of action or separate individual choices, the concept of sustainable lifestyle takes into account the environment induced modernization of different segments or sectors of the lifestyle of an individual as well as the way this modernization process is connected to the identity of the actor.

We tried to show that, with some of the core concepts of the structuration theory, ample room can be given to the notion of human agency also with regard to ecological modernization theory, without lapsing into the dualisms like voluntarism-determinism or actor-structure. The ecological modernization of domestic practices are not either actor driven or system-imposed. They are both at the same time. The 'theoretical' solution of the actor-structure dilemma by Giddens is often misunderstood and criticized for not taking into serious account the structural constraints to social action. People cannot decide – so it can be argued – to run away from the tap-water system overnight or to simply quit the services of the powerful electricity companies from one day into another. These empirical facts are then supposed to illustrate the fact that the institutions involved in the water and electricity sector leave actors very little room for maneuver. Giddens has met these kinds of criticism time and again by pointing to the crucial distinction to be made between formal, conceptual exercises on the one hand and historical empirical matters on the other. Structures are always both enabling and constraining and the room for maneuver which is granted to citizen-consumers by for example the public utility sector in a certain country within a certain period of time, has to be decided upon by historical empirical research.

Two examples of historical research are very illustrative in this respect. First, in her impressive study *More Work for Mothers* Ruth Schwartz Cowan (1983) provides a detailed historical analysis of the changing room for maneuver for women in the USA during the process of industrialization of the households from about the second half of the 19th century up to the Second World War. She illustrates the fact that within distinct technical systems like the food system, the clothing system or the transport system, different roads were taken with regard to the ways in which they were to be connected to the individual households. Second, in the Netherlands Johan Schot and others investigated not only the history of (among others, 'domestic') technology of the 19th century but they also try to develop their 'history of choices' well into the 20th century (Schot, 1992; Schot et al., 1995).

What can be learned from these examples of historical research is that the principle which is documented for the period which Huber describes as the construction phase of the industrialization process, can also be said to be relevant for the phase of ecological reconstruction of the industrial system from about the mid-eighties of the 20th century onward: (ecological) modernization will profoundly affect and alter the organization of domestic consumption as well as the interrelations between households and socio-material-collective-systems.

So far, we have been more or less defending the ecological modernization approach with regard to its basic premises, trying to correct and complement the approach with regard to some of its basic weaknesses and imperfections. We discussed the temporal and the socio-political context in which the approach emerged and sketched the way in which de-modernization perspectives of the seventies and early eighties were successfully questioned and superseded by the ecological modernization approach. As was stated before, the modernization perspective in its turn would become challenged by new insights that were to emerge within the environmental discourse. With global change and reflexive modernity entering the stage, some specific elements were added to the ongoing debate on environment and modernity. Once again the role of (environmental) science and technology became one of the central focuses of concern, albeit from a different perspective, as we will discuss in the next section.

Global change and reflexive modernity: the end of simple modernization (theories)

With the advent of the nineties, some major changes can be witnessed in the environmental discourse. For Goodin the shift to thinking about 'new' environmental issues like global warming, destruction of tropical rain forests and the growing hole in the ozone layer may even be seen in terms of a shift from the 'first' to the 'second' environmental crisis (Goodin, 1992: 5). The changes did not only relate to new problems moving to the center of attention. We can also point to the new roles that were assigned to both national governments, scientists and civil society actors in the context of the global change discourse. Fred Buttel has pointed to the fact that in a relatively very short time a rather firm coalition developed between environmental scientists, environmental movements and major parts of government administrations (Buttel and Taylor, 1992). All of the participants seemed willing to cooperate on these new problems and they all shared some basic assumptions with regard to the urgency of the problems and the need for a truly international or even global approach. And, last but not least, these new problems were requiring new ways of thinking and theorizing about the relationship between global environmental problems and global modernities.

We can take the writings of Ulrich Beck and Anthony Giddens as a starting point for discussing global modernity and its relation to environmental problems and policies. In some respects, the arguments presented by Ulrich Beck in the book *Risikogesellschaft* – a book that turned out to become one of the most influential books within sociology in recent times – seemed to contradict some of the major assumptions of the ecological modernization approach. The sub-title of the risk-society book runs like this: 'Auf dem Weg in eine *andere Moderne*' (emphases added). And while both Huber and Beck use the language of a shift into an other, new phase of modernity, both authors seem to have radically different views on the character of the new modernity to come. Against the optimistic, confidential view of Huber that we are on our way to a 'better' in the sense of a more sustainable modernity, Beck argues that, with the Chernobyl disaster, we, at a blow, became collectively aware of the fact that we are living in a risk-society, a society where fear and anxiety are reigning both in politics and in our daily lives and where environmental problems are inherently out of control. Without enforcing a kind of flabby synthesis between both perspectives, we try to incorporate some of the important arguments of Ulrich Beck into ecological modernization as a theory of social change.

A distinction can be made between Beck's substantial theory – he mostly refers to it with the term risk-society – on the one hand and his theoretical insights on the changing character of modernity, which later on developed into a theory of reflexive modernity (Beck, 1986; Lash et al., 1996), on the other. Within both the substantial and the formal parts of his work, environmental issues are used as illustrations and sometimes as proofs of Beck's line of reasoning or argumentation. With regard to the risk-society theme, the illustrations from the field of environmental policies and politics are – according to me – most of the time ill chosen and sometimes hard to sustain. The main objections that result in a critical view of Beck as a commentator of environmental politics are threefold. First, the impressive body of environmental norms and regulations did *not* come down as a house of cards by the shake that Chernobyl caused but was instead further developed and strengthened during the eighties. Second, by the time Beck was writing the book, business circles already had discovered and picked up the fact that environmental problems can work back on you as a boomerang. Third, environmental scientists, especially from the natural sciences, in general were not engaged in white-washing practices of environmental deterioration but instead they were among the first to build up a network for counter expertise. In short, I argue that the empirical developments within the field of environmental policies during the eighties seem to fit better into the description provided by ecological modernization as a political program than into the frame of reference of the risk-society. Because all (normative) descriptions which sociologists or journalists provide of social reality will always, in a certain sense and to a certain degree of course, fuel back into society to become itself part of the ongoing process of the (re)production of that society, (environmental) sociologists have a responsibility with regard to the kind of futures that they depict e.g. the kind of windows they open up with regard to the future development of society.[7]

While being rather critical of Beck's notion of the risk-society and Giddens' version of it in the form of the juggernaut society, I consider their contributions on the theme of reflexive modernity very stimulating and also relevant for the ongoing debate within environmental sociology on environment and modernity. The concept of reflexivity refers to the 'self-confrontation' of modern society due to its self-endangerment. Especially the emergence of global environmental risks or so-called High Consequence Risks (HCRs) has triggered the transition within modernity from its 'simple modernity phase' to its 'reflexive modernity phase'. In a similar way, Giddens refers to the latest phase of modernity in terms of 'high', 'late' or 'radicalized' modernity, thereby keeping a distance from the debate on postmodernity, in which the radical break with the institutions of modern industrial society is widely discussed and sometimes 'celebrated'. Both Beck and Giddens are working from a 'modernist' perspective, as Bauman concludes in his discussion on the risk-society (Bauman, 1993: 186-222) because their strategies for coping with the risks of the risk-society would imply 'more, not less modernity'.

While the quarrels about being in or belonging to the modernist or postmodernist camp perhaps can throw some light on an author's position in the academic field, it tells us little about the main argument itself. Beck's concepts of (the end of) simple modernity and its corresponding 'halfway modern' institutions is a much better way of approaching the central issues than quarrelling about postmodernity. With respect to two basic institutions, science and technology on the one hand and the political system on the other, both authors convincingly describe and document the kind of changes that have taken place and thereby they argue that these changes must be interpreted as making a qualitative difference when compared with earlier phases of modernity.

In the context of a post-traditional society,[8] science and technology cannot be treated in the same way as before because they are 'disenchanted'. The role of science and technology in society, the relationship between lay-actors (non-scientists) versus experts (scientists) and the relationship between science and for example environmental policy making has undergone fundamental changes now that scientific uncertainties are no longer an internal matter, to be recognized and dealt with by the scientific community itself. Technology and science as half-modern institutions – keeping uncertainties for themselves and displaying authority in the relationship with the outside world – are 'dethroned' by the ongoing process of modernization. In Giddens' words: there has come an end to providential reason and everybody knows it.

When talking about the pivotal role of science and technology in bringing about the ecological modernization of production and consumption, neither Huber nor Jänicke paid much attention to the changing role of these institutions. Perhaps more than in any other field of policy, (solutions to) environmental problems are intertwined with science and technology and the changes discussed in the reflexive modernity debate may have profound consequences for the environmental field. The debate on 'social constructionism' that has developed recently within environmental sociology has as its main parameters the 'authority of science' and 'the uncertainties surrounding especially global environmental change (GEC) issues'. When Buttel concludes from his analysis of the relationship of science and global change that more attention should be paid within environmental sociology to the role of science, he resembles closely the issues put forward by Beck and Giddens. Though perhaps arriving at the same conclusion, I think the reflexive modernity perspective provides a more fruitful ground for discussion than the (epistemological) debate on 'realism' or 'naturalism' versus 'constructivism' (Buttel and Taylor, 1992; Dunlap and Catton, 1994).

Post-traditional politics are the second main theme in the debate on reflexive modernity. Here again environmental sociologists who are working from a simple modernization perspective, will have some difficult questions to answer. Within environmental politics and environmental sciences as well, the main unit of analysis has been the nation-state. Environmental politics has been connected so far primarily with national arrangements, brought about by political parties, labor unions, representatives of industry and social movements working on a national level. The politics of 'unambiguous modernity' (Beck et al., 1994: 17) was a politics of national actors, following well circumscribed rules of the game and supported by such well founded and accepted dichotomies as 'left and right' and 'public and private'. This unambiguous politics has come under attack by two sets of processes: individualization and globalization. Both Beck and Giddens analyze in some detail the dialectic relation between the global and the personal that has emerged under conditions of reflexive modernity. Old, generative or emancipatory politics give way to new sub-politics of lifestyles and life-agendas as far as the individual level is concerned. At the same time however lifestyles are no longer unambitious in the sense that people 'know' that settled careers and fixed patterns of domestic relations are something of the past. On the global level, nation-state policies and policy regimes which used to be settled at the national level, give way to transnational arrangements and international policy regimes (Liefferink, 1995). When trying to think through the possible consequences of the changing nature of politics for the ecological modernization approach, two themes for reflection and research in the future seem essential. First, connections should be made between the debate on reflexive modernization on the one hand and the program of 'political modernization' as discussed above on the other. In what way and to what extent are the new environmental politics as discussed by Weale, Jänicke, Van Tatenhove and others to be viewed as illustrative or representative for the institutional and non-institutional political forms Beck is looking for? Second, more attention should be given to the question about what is left of the old, institutional politics of simple modernity. Which role do nation states play in transnational (environmental) politics, and what are the major forms of non-institutional or sub-politics that are emerging on different levels? We consider the arguments provided by the so called 'Lisbon group' for rethinking the role of both institutional and non-institutional actors in the context of a globalizing world, to be promising starting points for a debate on these issues (Petrella et al., 1995).

Epilogue

The debate on environment and modernity as sketched above may further our understanding of the socio-political and temporal context of ecological modernization theory as a new perspective that originated within environmental sociology during the last two decades. The ecological modernization approach was developed mainly in reaction to de-modernization theories and 'old' environmental politics which dominated the seventies. Although ecological modernization theory in its original Huberian form is in principle susceptible to the criticisms which have been raised in the past with regard to modernization theories, its recent formulations seem to deal with most of these deficiencies. The main achievement of the ecological modernization perspective as we have it, is that the theory provides new concepts to think through the relationship between economy and ecology, between society and its sustenance base. The concept of ecological rationality has been put forward above as such an innovation, referring to ecological criteria, procedures and norms which are gaining relative autonomy vis-à-vis economical, socio-cultural and political rationalities. In advancing the route into an ecological more rational modernity, science and technology can indeed be said to fulfill an important and in some respects essential role. Recognizing this fact of science and technology being important vehicles in the ecological modernization process, does not imply however that one would automatically or inevitably lapse into a technological fix approach or become 'hobbled by an unflappable sense of technological optimism' (Hannigan, 1995: 184). Instead of the dichotomy optimism-pessimism, we think the counterposing of the ecological modernization perspective with the risk society or reflexive modernization perspective, to be more challenging for the debate. The challenges posed by the reflexive modernization debate help to take the ecological modernization theory beyond the traditional modernization perspectives. The contribution which the ecological modernization perspective in its turn can make to the broader sociological debate on reflexive modernity, consist of taking into account notions of reflexivity that specifically refer to the (environmental) sub-politics that come along with and help to shape the socio-technical changes in production and consumption cycles that are required to carry us over into a more sustainable modernity.

Notes

1. 'Basically what is at stake is a political model which surpasses the old division between a liberal "Laissez faire" model on the one hand and a model of bureaucratic state-interventionism on the other: a decentral and consensual political model in which the central state concentrates itself on strategic tasks, leaving more detailed regulations to actors at decentral levels. Within the field of environmental politics the role of the central state in this model pertains to the securing of basic ecological values and the performing of 'strategic' functions. Among the state's first responsibilities is the definition of long term environmental problems. Decentral actors are assigned the task of using their specific innovative potentials for surpassing the basic conditions and minimum values set at the national level'.
2. Arthur Mol has discussed in some detail Huber's autobiography, which in this case can be said to be especially relevant because Huber himself made a kind of 'switch-over' from an active supporter – both in theory and in practice – of 'Selbsthilfe Netzwerke' in Berlin during the seventies into a staunch supporter of a modernization perspective during the eighties (Mol, 1995: 35-36).
3. For a very stimulating and well-organized overview of these debates, see also the book by Krishan Kumar (Kumar, 1995).
4. When discussing mainly Huber's initial formulations of the ecological modernization theory, one should not overlook the fact that the initial approach has been enlarged and refined in Huber's subsequent works, while other authors also contributed significantly to the elaboration of the ecological modernization approach as a theory of social change (Huber, 1989a, 1991a, 1991b, 1993; Jänicke, 1988, 1993, 1995; Mol, 1995; Zimmermann et al., 1990; Simonis, 1988; Hajer, 1995; Weale, 1992; Wehling, 1994).
5. 'If there is to be a future for ecology, it will be in its industrial form, and industry will only have a future when becoming ecologically sound'.
6. For example when discussing the so called economic or market regulation instruments (like pollution taxes, energy levies or charges on resource depletion) Martell argues that, in order to be really effective, they should be raised to levels that are so 'punitive' or coercive that – in their consequences – they would match state coercion. And from this he concludes: 'we may just as well then go for non-market restrictions on environmental harm such as state regulation' (Martell, 1994: 71). Huber would turn this argument just in the other direction: when equally effective, we should indeed rely on the 'voluntary' actions of capitalists and consumers rather than 'pressing them into line by coercive state legislation from above' (Martell, 1994: 67).
7. We did dwell a bit on our critical appraisal of certain aspects of the work of Ulrich Beck because some commentators found our contribution of 1993 (Mol and Spaargaren,

1993), too critical on Beck and somehow misinterpreting his basic intentions. We have been critical especially on those parts of his work that are dealing with (environmental) science and technology in a way which resembles the de-modernization perspective. We have been supportive however in his overall effort to rethink modernity and the changing role of science and technology. That Beck does not embrace de-modernist perspectives – be it in its classical or postmodern form – becomes clear in the comments of Zygmunt Bauman on parts of his work. When for example Bauman discusses Beck's notion of the risk society, he does so under the heading of 'the risk society being technologies last stance' and he in fact blames Beck for not going far enough in his critique of the Enlightenment project (see Bauman, 1993 and the interview with Bauman included in Munters et al., 1998).

8. With the concept of post-traditional society, Giddens refers to the latest phase of (globalized) modernity. The concept does not mean however that traditions no longer play a role in social life. Post-traditional refers to the fact that, after the phase of simple modernization, both grand traditions (religion, science, the nation-state) as well as 'down-to-earth' traditions (family, gender) now have to be defended and explained (Giddens, 1994: 5).

Bibliography

Andersen, J.G. (1990), 'Environmentalism', 'New Politics', and Industrialism: Some Theoretical Perspectives. *Scandinavian Political Studies no. 2*, pp. 1-20.

Bauman, Z. (1993), *Postmodern Ethics*. Oxford: Blackwell.

Beck, U. (1986), *Risikogesellschaft. Auf dem Weg in eine andere Moderne*. Frankfurt: Suhrkamp.

Beck, U. (1987), The Anthropological Shock: Chernobyl and the Contours of the Risk Society. *Berkeley Journal of Sociology*, 32.

Beck, U. (1988), *Gegengifte. Die organisierte Unverantwortlichkeit*. Frankfurt: Suhrkamp.

Beck, U. (1992), From Industrial Society to the Risk Society: Questions of Survival, Social Structure and Ecological Enlightenment. *Theory, Culture & Society*, 9, pp. 97-123.

Beck, U., A. Giddens, S. Lash (1994), *Reflexive Modernization; Politics, Tradition and Aesthetics in the Modern Social Order*. Cambridge: Polity Press.

Belt, H. van den (1996), De grammatica van duurzaamheid. *Kennis en Methode*, no. 2, pp. 187-202.

Biezeveld, G.A. (1985), Naar eenheid in verscheidenheid. *Milieu en Recht*, no. 8, pp. 256-263.

Bookchin, M. (1980), *Toward an Ecological Society*. Quebec: Black Rose Books.

Buttel, F.H. (1986), Sociology and the Environment: the Winding Road Toward Human Ecology. *International Social Science Journal*, vol. 38, no. 3, pp. 337-356.

Buttel, F.H., A.P. Hawkins and A.G. Power (1990), From Limits to Growth to Global Change. Constraints and Contradictions in the Evolution of Environmental Science and Ideology. *Global Environmental Change* (December), pp. 57-66.

Buttel, F.H. and P.J. Taylor (1992), Environmental Sociology and Global Environmental Change: A Critical Assessment. *Society and Natural Resources*, no. 5, pp. 211-230.

Buttel, F.H. (1996), Environmental and Resource Sociology: Theoretical Issues and Opportunities for Synthesis. *Rural Sociology*, vol. 61, no. 1, pp. 56-76.

Callon, M., J. Law and A. Rip (eds) (1986), *Mapping the Dynamic of Science and Technology*. London: Macmillan.

Catton, W.R. and R.E. Dunlap (1978), Environmental Sociology: A New Paradigm. *The American Sociologist*, XIII, pp. 41-49.

Catton, W.R. and R.E. Dunlap (1980), A New Sociological Paradigm for Post-Exuberant Sociology. *American Behavioral Scientist*, XXIV (1), pp. 14-47.

CLTM (1990), *Het Milieu, denkbeelden voor de 21ste eeuw*. Zeist: Kerckebosch.

Cowan, R. Schwartz (1983), *More Work for Mothers*. New York: Basic Books.

Cramer, J. (1989), *De groene golf; geschiedenis en toekomst van de milieubeweging*. Utrecht: Jan van Arkel.

Cramer, J., R. Eyerman and A. Jamison (1988), Intellectuelen en Milieubeweging. *Kennis en Methode*, 12, no. 4, pp. 334-351.

Cramer, J. and J. Schot (1990), *Problemen rond innovatie en diffusie van milieutechnologie. Een onderzoeksprogrammeringsstudie verricht vanuit een technologiedynamica perspectief*. Rijswijk: RMNO.

Dickens, P. (1991), *Society and Nature. Towards a Green Social Theory*. New York: Harvester Wheatsheaf.

Dietz, T. and T.R. Burns (1992), Human agency in evolutionary theory, in: B. Witrock (ed.), *Agency in Social Theory*. London: Sage.

Dietz, T. and R.S. Frey (1992), Risk, technology, and society, in: R.E. Dunlan and W. Michelson (eds), *Handbook of Environmental Sociology*. Westport, CT: Greenwood Press.

Dobson, A. (1990), *Green Political Thought*. Unwin Hyman: London.

Douglas, M. and B. Isherwood (1979), *The World of Goods: Towards an Anthropology of Consumption*. London: Allen Lane.

Dryzek, J.S. (1987), *Rational Ecology. Environment and Political Economy*. Oxford/New York: Basil Blackwell.

Dunlap, R.E. (1980), Paradigmatic Change in Social Science. *American Behavioral Scientist*, XXIV (1), pp. 5-14.

Dunlap, R.E. and W.R. Catton (1979), Environmental Sociology. *Annual Review of Sociology*, V, pp. 243-273.

Dunlap, R.E. and W.R. Catton, Jr. (1992/93), Toward an Ecological Sociology: the Development, Current Status, and Probable Future of Environmental Sociology. *Annales de l'institut international de sociologie*, vol. 3, pp. 263-284.

Dunlap, R.E. and W.R. Catton, Jr. (1994), Struggling With Human Exemptionalism: The Rise, Decline and Revitalization of Environmental Sociology. *The American Sociologist,* vol. 25, pp. 5-29.

Dunlap, R.E. and W. Michelson (eds) (1992), *Handbook of Environmental Sociology*. Westport, CT: Greenwood Press.

Dunlap, R.E., G.H. Gallup and A.M. Gallup (1993), Of Global Concern: Results of the Health and Planet Survey. *Environment,* vol. 35, no. 9, pp. 7-39.

Eckersley, R. (1992), *Environmentalism and Political Theory: Toward an Ecocentric Approach*. London: UCL Press.

Featherstone, M. (1991), *Consumer Culture and Postmodernism*. London: Sage.

Feenberg, A. (1979), Beyond the Politics of Survival. *Theory and Society*, 7, pp. 319-360.

Fine, B. and E. Leopold (1993), *The World of Consumption*. London: Routledge.

Frankel, B. (1987), *The Post-Industrial Utopians*. Cambridge: Polity Press.

Giddens, A. (1977), *Studies in Social and Political Theory*. London: Hutchinson.

Giddens, A. (1984), *The Constitution of Society*. Cambridge: Polity Press.

Giddens, A. (1990), *The Consequences of Modernity*. Cambridge: Polity Press.

Giddens, A. (1994), *Beyond Left and Right: The Future of Radical Politics*. Cambridge: Polity Press.

Glasbergen, P. (1989), *Beleidsnetwerken rond milieuproblemen* (inaugurele rede). Den Haag: VUGA.

Glasbergen, P. and P.P.J. Driessen (1993), *Innovatie in het gebiedsgericht beleid. Analyse en beoordeling van het ROM-gebiedenbeleid*, Den Haag.

Gleich, A. von (1991), Über den Umgang mit Natur. Sanfte Chemie als wissenschaftliches, chemiepolitisches und regionalwirtschaftliches Konzept. *Wechselwirkung* 48, pp. 4-12.

Godfroy, A.J.A. and N.J.M. Nelissen (eds) (1993), *Verschuivingen in de besturing van de samenleving*. Bussum: Dick Goutinho.

Goldsmith, E. et al. (1972), A Blueprint for Survival. *The Ecologist,* vol. 2, no. 1, pp. 1-44.

Goodin, R.E. (1992), *Green Political Theory*. Cambridge: Polity Press

Goverde, H.J.M. (1995), *Macht om het maaiveld: een 'koude oorlog' politicologisch beschouwd*. Wageningen: LUW.

Hajer, M.A. (1995), *The Politics of Environmental Discourse: Ecological Modernization and the Regulation of Acid Rain*. Oxford: Oxford University Press.

Hajer, M.A. (1996), Ecological Modernisation as Cultural Politics, in: S. Lash, B. Szerszynski and B. Wynne (eds), *Risk, Environment and Modernity: Towards a New Ecology*. London: Sage, pp. 246-268.

Hannigan, J.A. (1995), *Environmental Sociology: A Social Constructionist Perspective*. London: Routledge.

Harmsen, G. (1974), *Natuur, Geschiedenis, filosofie*. Sun-schrift 89, Nijmegen: Sun.

Hengel, E. van and B. Gremmen (1995), Milieugebruiksruimte: tussen natuurwet en conventie. *Kennis en Methode* jrg. 19, no. 3, pp. 277-303.

Hofstee, E.W. (1972), *Milieubederf en milieubeheer als maatschappelijke verschijnselen*. Amsterdam: Noord-Hollandse Uitgevers Maatschappij.

Huber, J. (1982), *Die verlorene Unschuld der Ökologie. Neue Technologien und superindustrielle Entwicklung*. Frankfurt/Main: Fisher.

Huber, J. (1985a), Ecologische modernisering, in: E. van den Abbeele (ed.), *Ontmanteling van de groei*. Nijmegen: Markant.

Huber, J. (1985b), *Die Regenbogengesellschaft. Ökologie und Sozialpolitik*. Frankfurt/Main: Fisher.

Huber, J. (1989a), *Technikbilder. Weltanschaulich Weichenstellungen der Technik- und Umweltpolitik*. Opladen: Westdeutcher Verlag.

Huber, J. (1989b), Social Movements. *Technological Forecasting and Social Change*, 35, pp. 365-374.

Huber, J. (1991a), *Umwelt Unternehmen. Weichenstellungen für eine ökologische Marktwirtschaft*. Frankfurt/Main: Fisher.

Huber, J. (1991b), Ecologische modernisering, weg van schaarste, soberheid en bureaucratie?, in: A.P.J. Mol, G. Spaargaren and A. Klapwijk (eds), *Technologie en Milieubeheer: tussen sanering en ecologische modernisering*. Den Haag: SDU.

Huber, J. (1993), Ökologische Modernisierung. Bedingungen des Umwelthandelns in den neuen und alten Bundesländern. *Kölner Zeitschrift für Soziologie und Sozialpsychologie*, 45, no. 2, pp. 288-304.

Hueting, R. (1974), *Nieuwe schaarste en economische groei*. Amsterdam: Agon Elsevier.

Humphrey, C.R. and F.R. Buttel (1982), *Environment, Energy & Society*. Belmont, CA: Wadsworth.

Huppes, G. (1989), De beperkte betekenis van de netwerkbenadering voor het milieubeleid. *Tijdschrift voor Milieukunde*, no. 5, pp. 153-158.

Jänicke, M. (1986), *Staatsversagen. Die Ohnmacht der Politik in der Industriegesellschaft*. München: Piper.

Jänicke, M. (1988), Ökologische Modernisierung. Optionen und Restriktionen präventiver Umweltpolitik, in: U.E. Simonis (ed.), *Präventive Umweltpolitik*. Frankfurt/Main: Campus.

Jänicke, M. (1993), Über ökologische und politische Modernisierungen. *ZfU*, no. 22, pp. 159-175.

Jänicke, M. (1995), *Green Industrial Policy and the Future of 'Dirty Industries'*. (Unpublished paper).

Korthals, M. (1994), *Duurzaamheid en democratie: sociaal-filosofische beschouwingen over milieubeleid, wetenschap en technologie*. Amsterdam: Boom.

Kumar, K. (1995), *From Post-Industrial to Post-Modern Society*. Oxford: Basil Blackwell.

Lash, S., B. Szerszynski and B. Wynne (1996), *Risk, Environment and Modernity*. London: Sage Publications.

Leroy, P. (1983), *Herrie om de Heimat; milieuproblemen, ruimtelijke organisatie en milieubeleid*. Antwerpen: Universiteit van Antwerpen.

Leroy, P. (1995), *Milieukunde als roeping en beroep*. Nijmegen: KUN.

Liefferink, J.D. (1995), *Environmental Policy on the Way to Brussels*. Wageningen: Landbouw Universiteit Wageningen (dissertation).

List, P.C. (ed.) (1993), *Radical Environmentalism: Philosophy and Tactics*. Belmont: Wadsworth.

Loo, H. van der, E. Snel et al. (1984), *Een wenkend perspectief? Nieuwe sociale bewegingen en culturele veranderingen*. Amersfoort: De Horsting.
Lovins, A.B. (1977), *Soft Energy Paths: Towards a Durable Peace*. New York: Harper & Row.
Lowe, Ph. and J. Goyder (1983), *Environmental Groups in Politics*. London: Allen & Unwin.
Martell, L. (1994), *Ecology and Society: An Introduction*. Cambridge: Polity Press.
Mol, A.P.J., G. Spaargaren and A. Klapwijk (eds) (1991), *Technologie en Milieubeheer. Tussen sanering en ecologische modernisering*. Den Haag: SDU.
Mol, A.P.J. and G. Spaargaren (1991), Introductie: technologie, milieubeheer en maatschappelijke verandering, in: A.P.J. Mol, G. Spaargaren and A. Klapwijk (eds) (1991), *Technologie en Milieubeheer. Tussen sanering en ecologische modernisering*. Den Haag: SDU, pp. 9-21.
Mol, A.P.J. and G. Spaargaren (1993), Environment, Modernity and the Risk-Society. The Apocalyptic Horizon of Environmental Reform. *International Sociology* 8, 4, pp. 431-459
Mol, A.P.J. (1995), *The Refinement of Production: Ecological Modernization Theory and the Chemical Industry*. Utrecht: Van Arkel.
Mommaas, J.Th. (1993), *Moderniteit, Vrijetijd en de Stad; sporen van maatschappelijke transformatie en continuïteit*. Utrecht: Jan van Arkel.
Munters, Q.J. (1998), *Zygmunt Bauman: Leven met veranderlijkheid, verscheidenheid en onzekerheid*. Amsterdam: Boom.
Musil, J. (1990), *Possibilities in Formulating New Approaches to Social Ecology*. Paper presented at the World Congress of Sociology, Madrid.
Nelissen, N.J.M. (1979), Aanzetten tot een sociologische theorie over het milieuvraagstuk, in: P. Ester (ed.), *Sociale aspekten van het milieuvraagstuk*. Assen: Van Gorcum, pp. 5-20.
Nelissen, N.J.M. (1992), *Besturen binnen verschuivende grenzen* (oratie). Zeist: Kerckebosch.
Nelissen, N.J.M. (1994), *De terugtredende overheid*. Pre-advies Vereniging Milieurecht.
Noort, W. van (1988), *Bevlogen bewegingen; Een vergelijking van de anti-kernenergie, -kraak- en milieubeweging*. Nijmegen: Sua.
Offe, C. (1986), Nieuwe Sociale Bewegingen als meta-politieke uitdaging, in: L.J.G. van der Maesen et al., *Tegenspraken, dilemma's en impasses van de verzorgingsstaat*. Amsterdam: Somso staatsdebat, pp. 27-92.
Opschoor, J.B. and S.W.F. van der Ploeg (1990), Duurzaamheid en kwaliteit: hoofddoelstellingen van milieubeleid, in: CLTM, *Het Milieu; denkbeelden voor de 21ste eeuw*. Zeist: Kerckebosch, pp. 81-128.
Otnes, P. (1988), Housing Consumption: Collective systems service, in: P. Otnes (ed.), *The Sociology of Consumption*. Atlantic Highlands, New Jersey: Humanities Press Int., pp. 119-138.
Otnes, P. (1988), The Sociology of Consumption: 'Liberate our Daily Lives', in: P. Otnes (ed.), *The Sociology of Consumption*. Atlantic Highlands, New Jersey: Humanities Press Int., pp. 157-177.
Otnes, P. (ed.) (1988), *The Sociology of Consumption*. Atlantic Highlands, New Jersey: Humanities Press Int.
Paehlke, R.C. (1989), *Environmentalism and the Future of Progressive Politics*. New Haven: Yale University Press.
Pepper, D. (1996), *Modern Environmentalism*. London: Routledge.
Petrella, R. et al. (1995), *Grenzen aan de concurrentie*. Brussel: VUBPRESS.
Pinch, T., W. Bijker and T.P. Hughes (eds) (1987), *The Social Construction of Technology*. London: MIT.

Redclift, M. and T. Benton (eds) (1994), *Social Theory and the Global Environment.* London: Routledge.
Redclift, M. and G. Woodgate (eds) (1995), *The Sociology of the Environment.* 3 Volumes. Aldershot: Edward Elgar.
Saunders, P. (1988), The Sociology of Consumption: A New Research Agenda, in: P. Otnes (ed.), *The Sociology of Consumption.* Atlantic Highlands, New Jersey: Humanities Press Int., pp. 139-156.
Schnaiberg, A. (1980), *The Environment: From Surplus to Scarcity.* Oxford: Oxford University Press.
Schot, J. (1992), Constructive Technology Assessment and Technology Dynamics: The Case of Clean Technologies. *Science, Technology and Human Values*, 17, 1, pp. 36-56.
Schot, J., A. Slob and R. Hoogma (1995), *De implementatie van duurzame technologie als een strategisch niche management probleem.* Den Haag: DTO.
Schumacher, E.F. (1973), *Small is Beautiful.* London: Blond and Briggs.
Simonis, U.E. (1988) (Hg.), *Preventive Umweltpolitik.* Frankfurt am Main: Campus Verlag.
Spaargaren, G. (1985), *Ekologie en maatschappij; de ekologische benadering in de sociologie.* Wageningen: vakgroep Sociologie.
Spaargaren, G. (1994), Sustainable lifestyles. Dutch Social Science and Policy Perspectives on Promoting 'Sustainable' Behaviour. *Tijdschrift voor Sociologie* 15, 2, pp. 29-66.
Tatenhove, J. van (1993), *Milieubeleid onder dak? Beleidsvoeringsprocessen in het Nederlandse milieubeleid in de periode 1970-1990; nader uitgewerkt voor de Gelderse Vallei.* Wageningen: WAU (PhD Thesis).
Tatenhove, J. van and P. Leroy (1995), Beleidsnetwerken: een kritische analyse. *Beleidswetenschap* nr. 2, jg. 9, pp. 128-145.
Tellegen, E. (1983), *Milieubeweging.* Utrecht: Het Spectrum.
Tellegen, E. (1984), Milieu en Staat; Essay over een haat-liefde verhouding, in: W. Achterberg and W. Zweers (eds), *Milieucrisis en Filosofie.* Amsterdam: Ecologische Uitgeverij, pp. 33-49.
Tellegen, E. and M. Wolsink (1992), *Milieu en Samenleving.* Leiden: Stenfert Kroese Uitgevers.
Touraine, A. (1995), *Critique of Modernity.* Oxford: Basil Blackwell.
Ullrich, O. (1979), *Weltniveau: In der Sackgasse der Industriegesellschaft.* Berlin: Rotbuch Verlag.
Warde, A. (1990), Introduction to the Sociology of Consumption. *Sociology*, 24, (1), pp. 1-4.
Weale, A. (1992), *The New Politics of Pollution.* Manchester: Manchester University Press.
Wehling, P. (1994), *Die Moderne als Sozialmythos; Zur Kritik sozialwissenschaftlicher Modernisierungstheorien.* Frankfurt: Campus Verlag.
Winsemius, P. (1986), *Gast in eigen huis: Beschouwingen over milieumanagement.* Alphen aan de Rijn: Samsom.
World Commission on Environment and Development (1987), *Our Common Future.* Oxford University Press: Oxford.
WRR-rapport nr. 44 (1994), *Duurzame risico's: Een blijvend gegeven.* Den Haag: SDU.
Yearley, S. (1991), *The Green Case; A Sociology of Environmental Issues, Arguments and Politics.* London: Harper Collins Academic.
Zimmermann, K., V. Hartje and A. Ryll (1990), *Ökologische Modernisierung der Produktion. Strukturen und Trends.* Berlin: Sigma.

4

Modern Theories of Society and the Environment: the Risk Society[1]

Eugene A. Rosa

Introduction

Momentous transitions like the twentieth century's *fin de siècle*, but also the end of a millennium, the *fin de millénaire*, have always heightened the demand on social thinkers to reflect on the past age and to characterize the one that will soon replace it. What is the spirit of our age? Recent sociological theorizing, reflecting on this question, is looking toward the end of the century and toward the end of the millennium, in seeking to characterize and name both momentous events. The theoretical goal is to provide the conceptual frame and an imprimatur that captures the spirit, outlines the character and gives a name that sticks to this age of transition.

A *sine qua non* of theories characterizing our age will, doubtless, be the environment. Few themes can compete with the environment – the growing recognition of human dependency and human impacts on the biosphere – as the key feature of our age and of its spirit. Few themes have captured such widespread, worldwide attention, and have endured. Indeed, Robert Nisbet – whose penchant was always to develop very large sociological ideas – made the following observation: 'When the history of the twentieth century is finally written, the single most important social movement of the period will be judged to be environmentalism' (1982: 101).

Risk is the principal analytic tool for assessing human impacts on the environment.[2] It comprises both an analytic orientation and a suite of methodologies for formally anticipating the untoward outcomes to the environment of technological and other human choices. It provides a disciplined basis for choosing among alternative courses of action – where the decision clearly implicates the environment. It is no coincidence that its emergence over the past three decades occurred *pari passu* with the rise in environmental consciousness, with the spate of laws and regulations passed by all levels of government, and with the sustained effort to manage the environment better. Risk and environmental assessment are inseparable partners.

Risk, too, is the keystone of recent theorizing about large-scale social forms and broad social change. Anthony Giddens and Ulrich Beck, two European social theorists, have developed theoretical frames for characterizing the current age – the one that is evolving beyond (high) modernity – and place risk at the core of this characterization. According to them our transitional age is witnessing social change that is huge, that is fundamental, and that is far-reaching – resulting in the evolution of worldwide institutions never seen before. Because of the central role that risk plays in their theories of change, and because risk is typically connected to environmental concerns, the theories of Giddens and Beck – perhaps unwittingly – should be considered theories of society and environment. The spirit of our age for them is the universal concern over hazards in the contemporary world and over the vulnerability of the environment. And owing to the great scale at which these theories are conceptualized, they, in effect, represent some of the most comprehensive theories of society and environment.

Our goal is to situate these theories within three different contexts at the same time. The first context is the centuries old intellectual tradition of naming historical eras along hermeneutic lines. The second context, about a century old, is that of classical sociological theory. The third context, only decades old and still on the road to maturity, is the modern field of analytic risk. The bringing together of the three contexts in one place, to our knowledge, has not been done before. The linking of the latter two contexts represents the joining of two disjoined bodies of thought, both putatively committed to the same topic. Combining all three contexts here should produce several useful results: a meta-theoretical framework for organizing the rapidly growing risk literature (including other social science approaches); an understanding of the presuppositions underlying the classic sociological traditions guiding theory about risk and environment; and a disciplined basis for comparing competing approaches and theories seeking to understand risk.

The idea of spirit

Historical context

The idea that an historical period could be identified by a 'spirit of the age' first received its most articulate expression in the works of Hegel. Indeed, it was Hegel who coined the term *Zeitgeist* in his native German language to capture this grand idea that each historical epoch has its own thematic tone.[3] Writing in the early nineteenth century Hegel believed, as did a number of his intellectual predecessors, that an era was coming to an end – that the civilization known to him was coming to a close. Through the lens of momentous historical change he, and others before and following, asked: what theme captures the essence of the social world? How shall the moment in history be marked? What to label this moment?

Later in that century, John Stuart Mill attempted to portray the spirit of the age in his native England (Wolfe, 1996). Mill believed, like the founder of sociology, Comte, that history was governed by laws that were reflected in clearly defined epochs that changed in an orderly succession: an age of Faith leads to an age of Reason which, in turn, leads to a scientific or positivistic age.

What is remarkable about juxtaposing these two thinkers is that they arrived at their common quest from vastly different intellectual traditions and intellectual dispositions. We could hardly find two thinkers from more contrasting traditions and styles. Hegel wrote in the German historical, romantic, tradition while Mill wrote from the English utilitarian, empiricist tradition. Each of these traditions survives in modified form to this day. Each tradition survives in contemporary theories about the environment, especially theories that recognize environmental threats as risks. The dominant perspective in contemporary risk studies, the rational actor paradigm (RAP), derives from a long intellectual tradition that runs through Mill. Recent sociological theory on the environment where risk is the axial principle, especially the work of Anthony Giddens (1990) and Ulrich Beck (1992 [1986], 1995 [1988], 1995 [1991]), derives from a similarly long tradition that runs through Hegel and other macro-theorists.

These two traditions provide the larger historical and intellectual context which underpin and shape both sociological theorizing and the field of analytic risk. This leads us to ask: 'What are the presuppositions and meta-theoretical elements of each tradition? And how do they speak to the 'spirit' of our age – our *Zeitgeist*?' We take up the second question first.

The spirit of our age

What is the spirit of our age? Automobile and plane crashes, toxic waste spills, nuclear accidents, food contamination, global warming, the spread of AIDS, and the persistence of nuclear arsenals – these are recurring events attracting headlines around the world. People nearly everywhere are increasingly worried about the hazards of the contemporary world. The headline events and the public worries they engender are conceptualized as risk. By their frequency and pervasiveness risks – environmental risks – are, doubtless, a central feature of the contemporary world.

Social fabric is a root metaphor for collective life. Coming to dominate the contemporary world are a variety of forces putting 'the social fabric at risk' (Short, 1984). These forces shape the growing variety of dangers that underpin risk itself. Though a feature of human existence from the beginning of time – literally – risk grows in importance as both an analytic focus and as an embedded feature of public conscience. Risks abound and people know it. People remotely distanced from one another often have little more in common than shared risks. People nearly everywhere are at risk of being the victim of nuclear holocaust, or at risk of being radiated by nuclear accidents or are at risk to the consequences of global warming. To paraphrase Kant's two central questions: 'how did things get this way?' and 'How can we understand what needs to be done about them?' Kant's questions beg of sociological analysis because worries about risks are not just individual problems, but problems of a growing collective consciousness.

The social fabric is a webbing of interdependencies. These interdependencies embed expectations, obligations, actions, and interactions – the mutualities and bonds of social life. The Hobbesian question undergirding our mutual dependencies is: what does the individual obtain in return for abrogating her extreme individuality, her total freedom of personal action? The answer seems obvious enough: it satisfies fundamental human existential needs as well as the requisite condition for social life. (Needless to say, this answer has engaged centuries-old debates on the types and range of need satisfied, and the just basis for the resultant social contract.) What are those existential needs? One such need, in Giddens' terms, is ontological security (Giddens, 1984). 'The phrase refers to the confidence that most human beings have in the continuity of their self-identity and in the constancy of the surrounding social and material environments of action' (Giddens, 1990: 92). A degree of regularity or orderliness is pivotal to the viability of a reflexive self-identity and to an anchoring of the social fabric. The reappearance of the sun each day, reflecting a probabilistic certainty, is a form of ontological security for humans everywhere. For pre-scientific societies an eclipse of the sun poses a serious threat to that security, often producing excited, worrisome, or appeasement (to gods, for example) reactions.

Regularity and orderliness, however, share time with an unavoidable feature of the physical and social world – uncertainty. While uncertainty cannot be avoided, it must be managed to prevent psychic overload and to regulate danger. Uncertainty is ever lurking as a threat to ontological security. Risk is a conceptualization of uncertainty that demarcates between situations or actions with outcomes involving human stakes and those that do not (Rosa, 1998). Risky outcomes, such as automobile accidents, occur with some probability – not certainty. The occurrence of outcomes produces consequences, such as the number of deaths from such accidents. Consequences with catastrophic potential are especially disquieting to ontological security.

The modern world, emerging from Europe in the seventeenth century and spreading nearly worldwide, refashioned the social fabric. The patterns of relationships and constitutive interdependencies became, due to industrialization, more distanced, more impersonal, and, in a manner of speaking, more tenuous. The defining features of industrialization – mass production via specialization of task and division of labor – refashioned the social world in its image. As a result, trust assumed a central role in coordinating social relationships. It had always been so, even for all pre-modern societies. But trust loomed larger in the modern world, much larger. For unlike the past, where the ontological security of face-to-face familiarity, interaction, and mutual expectation resulted in trust, the modern world required trust in order to sustain patterned social relationships. With modernity the seams of social dependencies were increasingly vested in distant, rather than face-to-face, relationships, resulting in a greater reliance on trust. The contemporary world has stretched the social fabric of modernity – and made it more vulnerable. Dominant patterns of the fabric's larger framing are the worldwide spread of industrial production, called by some theorists (Ross and Trachte, 1990) 'a new world manufacturing order', and the internationalization of the division of labor, called 'the new international division of labor' (NIDL) (Froebel et al., 1980). These production patterns are accompanied by the availability everywhere of a wide assortment of consumer goods; the market baskets of people widely distanced around the globe are becoming increasingly similar.

The international spread of production and consumption patterns is accompanied by an unshakable traveling companion: risk. One unavoidable feature of globalized interdependencies is the sharing of common risks among distant and divergent peoples. Another unavoidable feature is still a further increase in the importance of trust to the social order. Risk reminds us of vulnerability. Catastrophic risk is an even stronger reminder. Since risk is a central feature of the postmodern world, and since the imposition and management of many risks are the responsibility of others, trust is essential to our ontological security. To feel secure – that things are safe – requires our trust in distant and unknown others: not only in distant individuals, organizations, and institutions, but also in the impersonal rules, standards, and regulations they create (Porter, 1995). When risks abound, as they do in the modern world, and when public trust is low, as recent evidence suggests, the social fabric is itself at risk.

Postmodernity as the risk society

Abundant are the signs, as outlined above, of a transition from the world of the present era to some future one. This transition has engaged widespread sociological attention. Two leading European theorists, Anthony Giddens (1984, 1990, 1991) in Britain and Ulrich Beck (1992 [1986]; 1995 [1988]; 1995 [1991]) in Germany, have developed a theoretical frame for understanding the transition from the current world of modernity to a future social world – to a world of what I label postmodernity. While each of their theories was developed independently and while each theory is grounded in a fundamentally different theoretical orientation, both converge recently on a common underpinning: risk. The transition to the postmodern world is a transition in our collective consciousness about risk. The postmodern world is a risk world, a world that has made our vulnerabilities transparent.

The adoption of 'risk' as a central concept in sociological theory marks a significant refocusing of social thought. At the foundation of Western thought since the Enlightenment – whether from Comte, Spencer, Marx, Parsons, or Habermas – has been an expectation of a continued improvement in the social world, if not an uninterrupted march of progress. The emergence of a 'Risk Society', in Beck's terms, abruptly challenges that assumption. '"Risk society" means an epoch in which the dark sides of progress increasingly come to dominate social debate' (Beck, 1995 [1991]: 2). The focus of attention has shifted from the 'goods' of modernization, including not only economic spoils, but also social and political ones as well, to its many – often unintended – 'bads'.

Risk as an analytic lens

The pervasive public consciousness about risk and the theorizing it has attracted are recent developments in the quest to understand risk more generally. An entire body of thought and research preceded these developments. The field of modern risk analysis is now three decades and thousands of articles old (Renn, 1996). Its early and chief aim was to develop analytic techniques for estimating and managing risks, especially environmental, technological, and health risks. It was joined, a decade after its beginning, by social scientific efforts – principally from geography and cognitive psychology – to understand the 'human' side of risks. Subsequently the field attracted anthropology and sociology – with sociological macro-theorizing as sociology's most recent extension.

An analytic perspective on risk, thus, antedated social science interest in the topic. The analytic perspective is dominated by the Rational Actor Paradigm, or RAP. As a founding perspective in the field, RAP enjoys the advantages that typically accrue to pioneer developers; they establish first claim in the staking out of new terrain. Thus, the emergence of the social science of risk and recent sociological macro-theorizing about risk must, inevitably, confront the already laid claims of RAP. RAP is a tradition traceable directly in the line of thinking that includes John Stuart Mill in its stream. In contrast, major sociological theorizing about risk by Beck and Giddens is traceable to contrasting lines of thinking that has roots to Hegel and other Continental macro-social theorists. A comparison of these several lines of thought addresses the instrumental task of this paper – to provide a critical background of each of these traditions and to contrast their logic. The background of RAP, and its mapping into risk, is discussed first, followed by competing traditions.

Rational Actor Paradigm (RAP)

Briefly we can outline RAP as consisting of an orienting perspective that emphasizes the uniquely human qualities of what we typically associate with the term 'rational', namely, conscious, purposive human choice and action. Humans are *assumed* to be purposive (meaning they act with some purpose in mind and often explicit goals) and are *assumed* capable of determining alternative courses of action and outcome (meaning they can distinguish between strategies and can project the consequences) and are *assumed* capable of determining which alternative will produce the best overall outcome (happiness, utility, satisfaction, etc.)

The Rational Actor, or RAP perspective – having evolved since the Italian Renaissance to utilitarianism (in the work of Bentham, Mill and others) and then to modern versions of utility theory – was the singular perspective adopted by not only risk researchers, but also by regulatory and policy-making bodies in the defining stages of this field of study. Indeed, the classic article launching the modern field of risk analysis by Chauncey Starr (1969) borrowed its foundation and structure from rational actor theory in economics. Starr's argument that we could understand the risky choices made by society by looking at the actual risks it tolerated was a straightforward translation from RAP-based economics of the idea of 'revealed preferences'. Determining consumer preferences, a cornerstone of economic theory, is derived from what consumers actually purchase – presumably actual purchases reveal what their true preferences are. Arguing analogously for risks, Starr claimed that we could understand human preferences for risks by observing the risks they were willing to tolerate – their revealed risk taking.

The RAP perspective, quite expectedly, has been attractive to risk economists as well as risk scientists (e.g. nuclear physicists), risk engineers, and modern utility theorists (such as in the decision sciences) – and to those policy makers and regulators using RAP tools such as risk/benefit analyses. RAP was first to stake out the intellectual terrain of risk research and policy. Thus, RAP is the reigning conceptual monarch in the risk field. But, as the examples below will show, the monarch rules an ever-shakier kingdom: growing theoretical and policy limitations have exposed weaknesses in the monarch, and competing perspectives have arisen as pretenders to the throne.

RAP methods

Here are examples of three widely practiced methods for developing quantitative risk assessments, all derived from RAP: the actuarial approach, probabilistic risk assessment (PRA), and toxicological and epidemiological approaches. All of these RAP-driven approaches have the objective of providing precise quantitative measures of risk probabilities.

Actuarial approaches to risk estimation seek to provide expected values based upon the relative frequency of events, such as automobile fatalities. This approach embeds two assumptions consistent with RAP: first, that risks can be assigned a value of harmfulness independent of social, economic, political, or cultural context; and second, that a rational actor should seek to minimize exposure to risk just as rational economic actors seek to maximize personal utility. Probabilistic Risk Assessment (PRA), the second approach is a highly reductionistic, Cartesian method of componential analyses. It is consistent with RAP in that it contains both a rational orientation and an emphasis on micro, individualistic elements. PRA seeks first, via fault-trees or event-trees, to assess the probability of component failures in complex systems and then to aggregate the separate probabilities into a risk estimate for the entire system.

Thirdly, the toxicological approach seeks to establish dose-response curves for toxic and other harmful substances, while the epidemiological approach seeks to establish odd ratios based upon data from real populations. Embedded in both approaches and consistent with RAP, is that health is an uncontested desirable end state and the rational thing to do is to maintain or improve one's health.[4]

Shortcomings of RAP

Despite its pervasive presence in the risk field, and despite its continued importance in shaping debates about risk, RAP is not without serious shortcomings. These shortcomings, becoming ever more apparent in theoretical and policy debates, have been the source of growing criticisms of the RAP perspective, on the one hand, and the source of the urge to identify alternative perspectives for better understanding risk, on the other hand. While the list of criticisms of RAP has grown too large to be fully explicated here, we can identify a central shortcoming that frames nearly that entire list: total neglect of social context. RAP is an individualist-oriented account of the social world, abstracting actors from social contexts and treating them as cold, calculating, self-interested maximizers uninfluenced by family, friends, or other outside forces. The operational theory derived from RAP is the expected utility model which provides a normative basis (maximizing or optimizing) for making choices between alternatives under risk. The model is built upon a set of axioms that specifies the alternative a rational individual will choose in order to maximize expected utility.

Repeatedly, it has been found that many individuals violate the axioms underlying utility theory. One of the first – and now classic – studies to expose the flaws in the axiomatic structure of RAP, regarded the behavioral response to natural hazards (Kunreuther et al., 1978). Public responses to a serious earthquake in Alaska in 1964 and those following tropical storm Agnes hitting the northeastern United States in 1972 are described by the principal investigator, economist Howard Kunreuther:

> ...in September 1964 I undertook a study (of)...the economic problems facing Alaska following the Good Friday earthquake of March 1994.... (and was) surprised to find that behavior did not conform to the patterns prescribed by economic theory. Food prices and rents went down in the short run even though there were shortages in perishable goods and a limited supply of housing. Most residents and businesses did not have earthquake insurance.... It was not unusual to hear many victims remark that financially they were better off after the earthquake than prior to the event due to liberal disaster relief... in the summer of 1972... Tropical Storm Agnes caused approximately $2 billion in damage to the northeastern section of the United States. Few homeowners and businesses had flood insurance even though it was highly subsidized by the federal government... .(1992: 301)

These responses, subsequently corroborated by an entire research program in psychometrics (see, for example, Slovic, 1987), illustrated the fact that individuals, contrary to the assumptions of utility theory, incorporate important psychological, cultural, and social factors into their decisions. Values, emotions, social networks, civic duty – these and other non-rational forces formed the principal bases for the variety of actions taken. And the specific decision to purchase disaster insurance was best predicted, not by an expected-value calculation presumed in utility theory, but by past experiences with disasters or by the actions of community members, especially friends and neighbors. In short, social relations and social forces shaped decision making, not the abstractions of economic rationality.

The failure of RAP – via its operational theory, expected utility – to explain the response of victims to different natural disasters produced not only a challenging set of case studies, but also uncovered a more general shortcoming of the application of RAP to risk. Natural disasters are typically low probability events. Yet, once they occur they often produce high-level consequences: physical damage, injury and death, and geographical and social dislocations. By the expected-value utility calculus of RAP, individuals should be indifferent between low-probability/high-consequence events, such as major disasters, and high-probability/low-consequence events, such as an injury due to a fall, when the expected value between the two is identical.

But, this is seldom the case. A whole host of external forces – the degree of voluntariness of the risk, the level of knowledge about it, how much it is dreaded, the breadth of its consequences, and trust in the agencies responsible for its management – all conspire to shape perceptions and responses to risk (Slovic, 1987; Kleinhesselink and Rosa, 1991, 1994). As a consequence the calculated indifference predicted by the normative utility model is seldom confirmed in practice. These failures of RAP to predict risk behavior, as well as other problems that are becoming apparent, has drawn attention to its limitations in understanding risk.

Alternative social science of risk

In part due to the shortcomings of RAP to explain risk behavior, and in part due to the inherent biases in orientation across academic disciplines, a variety of competing perspectives to the social science of risk have emerged. The principal ones challenging RAP are culture theory, Weberian organizational theory, and social constructivism – and most recently the already noted work of Giddens and Beck. It is the aim of this article to provide a critical, comparative evaluation of each perspective. But before doing that, we must first develop a framework for disciplining the comparisons. In developing such a framework for comparison in the next section, we combine two meta-theoretical considerations: ontological foundations and the orientation provided by theoretical method.

Towards a framework for comparing risk theories

The history of Western thought since Classical Greece has been punctuated with a relentless recurrence of ontological dualisms. Idealism versus materialism, nominalism and relativism versus realism, romanticism versus skepticism, German romanticism versus Utilitarianism, Continental idealism versus British empiricism – these are some of the prominent dualisms, oppositional binaries, shaping Western thought as dialogue (Tarnas, 1990). These unrelenting dualisms have undergone substantive modification and refinement, show the ascendancy of one side of the binary to the subordination of the other, followed by a reversal in fortune, and have experienced long periods of apparent peaceful co-existence, followed by heated clashes. Whether seen as dialogue or dialectic, whether seen as conflict or as complementarity, dualism in Western thought has passed the test of long historical time, surviving as an underlying influence on modern thought and, accordingly, on the social sciences. As the various social scientific perspectives turn their analytical lenses to the topic of risk they, perhaps unwittingly and unconsciously, import this dualism into this theoretical domain. So, social scientific theories about risk generally fall into one or the other of these domains.

All social science approaches to risk are meta-theoretically laden. They are laden with ontological and epistemological presuppositions, unchallenged first principles, and orienting perspectives driving various theoretical methods. One useful way for characterizing meta-theory in the social science of risk is to categorize it along two independent, axial continuums. The first continuum, part of the imported dualism of Western thought described above, concerns a presupposition about the state of the world (ontology).[5] The second, imported from Enlightenment thinking, concerns the theoretical method adopted (the action component of epistemology) presumed to be most effective in providing knowledge about that world. The net result is a cross-classification grid – Figure 1 – where the horizontal dimension expresses the ontological nature of the physical world and the vertical dimension expresses the orientation to the study of the social world.

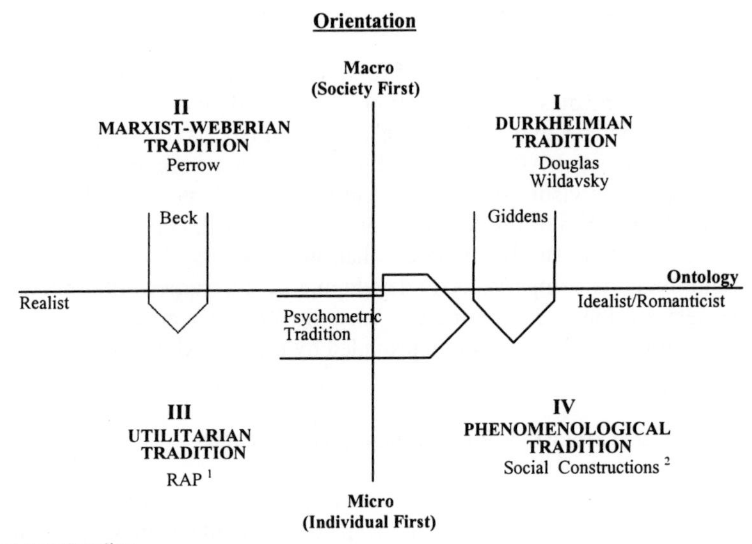

Figure 1 *Two Dimensional Dualism in the Study of Risk*

Meta-theoretical orientation: idealism-realism

Due to an extraordinary historical richness, there is no shortage of terms to characterize the dualism mapped by the first continuum. For convenience we refer to it as an idealism-realism dichotomy, which is mapped as the horizontal dimension of Figure 1. In self-consciously crude terms we can outline the meanings of the polar terms – idealism[6] and realism – that define this continuum.

For the idealist, it is the human mind that is the basic stuff of the world. The mind is not simply an extension of the body, a physical entity, but exists independent of the body: it is a transcendental reality. There is no world 'out there' independent of our apprehension of it.[7] The only real things are the perceptions and sensations felt by the senses, with ideas as the ultimate reality. Thus, the essence of the world is neither in physical surroundings (which can only be perceived through the senses anyway), nor in physiognomy, but in consciousness. Emphasized in the humanistic and German romanticist line of this thinking is the subjective side of social scientific analyses. Society is an ideational organism or construct – once created its essence is independent of individual human beings. Society and history represent the whole, and these are greater than the sum of the parts.

For the realist, the idealist has the world backwards. No mind, no ideas, no consciousness can exist independent of real, material conditions of existence. A world 'out there' exists independent of our perception and comprehension of it. Even if it is agreed that it is impossible to prove that an external world exists, it is equally impossible to prove that it doesn't. Indeed, given the difficulty of proving a negative hypothesis, the resulting and unavoidable logical asymmetry in how a hypothesis is stated leads us to favor the positive hypothesis, that an independent world does exist. Furthermore, our senses, our mind, our consciousness cannot be generated entirely internally, but must be stimulated by some external sources. Those external sources are found in material conditions and in history, that is, the march of real events, not simply our perceptions of those events. Society, by and large, is the sum of its parts and of its history. This, then, is the dimension of meta-theory that provides the ontological grounding of theory.

Theoretical method

Disciplined inquiry is driven by theoretical method, the other dimension of meta-theory. The idea of 'theoretical method' used here reflects a concern with the development of theory that is principled, systematic, coherent, and fruitful (see, for example, Freese, 1980). Theoretical method is aptly thought of as the orienting or action part of epistemology. It is derived from epistemology and should be logically consistent with the epistemology from which it emanates. It can be distinguished from epistemology by its emphasis on procedures for generating knowledge claims whereas epistemology emphasizes the basis, the logic, and the soundness of knowledge claims. Theoretical method is a framework for guiding inquiry that, appropriately applied, results in theoretical propositions corroborated by empirical regularities: theory is only useful to the extent it can explain features of the social world around us.

The second continuum of our grid, forming the vertical dimension in Figure 1, is the underpinning of theoretical method. It provides the basic sociological orientation toward theory that originally crystallized with the birth of the social sciences. Emerging from the Renaissance and streaming into the Enlightenment was an unparalleled striving to understand human existence better. For among the many accomplishments of the Enlightenment, the period from the mid-seventeenth to roughly the middle to late eighteenth century, one of its most significant – and most enduring – was that for the first time in history thinkers tried to provide a general or systematic explanation of the social world – of, in short, society. While an interest in the social world can be traced to the very beginnings of the written word, those writings were typically descriptive and ideographic – unmindful of systematic structure and processes. With the Enlightenment there was an urge to understand the essence of society, on the one hand, and the general processes of history, on the other hand. These were the most basic beginnings of the human sciences.[8]

The fundamental, persistent ideas of Enlightenment thought both shaped the foundations of the human sciences and prompted an unavoidable meta-theoretical dualism, one of foundational importance in conducting inquiry about human action. We can refer to this dualism, after Tarnas (1990: 366), as 'two temperaments' or 'general approaches' to human existence, often labeled *macro-micro*. Translated into the orienting perspectives of the emergent human sciences, these temperaments represented a clash of first principles: was the individual prior to society? Or was society prior to the individual? We can think of the first of these as an 'individualistic' perspective and the second as a 'sociolistic' perspective. A corollary of the respective perspectives was the appropriate theoretical method for proceeding with inquiry: a micro, bottom-up approach for the individualistic and a macro, top-down approach for the sociolistic – or methodological individualism, in the first instance, and methodological holism in the second.

Within certain contemporary theory circles – structuration theory being most prominent – it has become fashion to make a similar distinction by using the terms *structure and agency*. There is little quarrel with the observation that this distinction, its theoretical vocabulary, and its resultant continuum provide a useful analytic structure for the purchase of a variety of theoretical issues. The rub arises with the widespread propensity to view the 'macro-micro' continuum as identical, and therefore interchangeable, with the 'structure-agency' one. This inattentive merging of the two concepts results in a serious fallacy, for there is a subtle but very significant difference between the two.

In the simplest terms macro-micro and structure-agency map fundamentally different dimensions of multi-dimensional phenomena – philosophically and sociologically. Macro-micro, on the one hand, maps an epistemological orientation, while structure-agency, on the other hand, maps an ontological one. These (perhaps) orthogonal dimensions serve very different theoretical purposes.

For the macro-micro position, the question – macro versus micro – is one of orienting perspective and units of analysis: 'Do we start big? Or, do we start small?' We can choose to examine the social world from a macro orientation, one whose focus is on 'big structures, large processes, and huge comparisons'. These latter three entities, as effective illustrations of the word 'macro', are key elements in defining the term 'macro'. They are also consistent with Collins' (1988) three, irreducible, defining features of the definition of 'macro' as (1) the extent of *space* involved in the processes; (2) the extent of *time* the process takes; and (3) the *number* of people or situations involved. Macro orientations seek, in varying combination, to cover large territories, temporally durable social phenomena , and multiples of people. At the other side, we can choose to examine the social world by focusing on the constitutive elements of that world: humans – their perceptions, attitudes, dispositions, actions with others, and constructions of meaning. Typically this means a focus on the recursive interplay between the individual, as a social actor, and the social setting – the proximate environment, including other social actors. Since individuals are the raw materials of all social organization, they are also the main basis for large-scale social forms and institutions. And the decisions and actions of large-scale organizations and institutions, the focus of the macro approach, are ultimately traceable to interactions in micro contexts.

Viewed in this way, micro-macro is an analytic strategy.[9] It comprises presuppositions about the best strategies for getting the 'bloomin, buzzin' social world to yield to logical coherence, to organization, and to correspondence with empirical evidence. It is *a priori* distinct from ontological presuppositions. One need to make no claim about the ontological reality of macro phenomena in order to profitably pursue social phenomena in macro terms – the taking of large gulps in space, time, or multiples of people. So, much work takes place at the level of the nation-state because of the potential for understanding important social processes – not because it is driven by the presupposition that the nation state enjoys an unquestioned ontological reality.

The essential presupposition of the macro approach is the belief that interesting puzzles can best be approached with a wide-angle lens. The exact obverse provides the basic orientation for micro-directed theorists – the puzzles of social life can best be observed through telescopic lenses that provide close-ups of social life. It is real, on-the-ground people that gives social life its dynamic qualities.

The structure-agency position is based in ontology. For this reason, it is the focus of a debate, as old as the discipline itself, over the essence of social reality. It begins with Durkheim's classic work, *The Rules of the Sociological Method* (1982 [1895]), where collective phenomena is to be understood as a reality *sui generis*, independent of individuals but nevertheless real. This is essentially an assertion about the ontological state of the social world: about what, in fact, exists 'out there' as social realities. Virtually all of the classic sociological theorists joined Durkheim in adopting this viewpoint: Comte, Marx and Engels, Tonnies, Pareto, Weber in later writings, and then Parsons and other systems theorists.

This ontological presupposition was eventually challenged on many fronts. Beginning with the work of American Charles Horton Cooley, as well as George Herbert Mead and other pragmatists, the idea that society existed at all was categorically rejected. Indeed, its most radical manifestations claimed not only that society did not exist, but also that society only exists in our minds. Another challenge came from the phenomenological tradition in philosophy, whose lineage of thinkers – Hegel[10], Dilthey Husserl, Heidegger – led to a sociological translation of this school of thought by Alfred Schutz, 'who connected Weber's ideal types to Husserl's phenomenological philosophy' (Collins, 1988: 377).

Among the more piquant questions raised in the debate are: are social structures real, or merely the reified products of our theoretical constructions? If social structures are, indeed, real, are there causal relationships at that, the structural level? What role can volunteerism – agency, in short – play in macro sociological systems which abstract human actors away, leaving them as over socialized automatons? What is the essential ontological reality of human existence?

Four major traditions in risk theory

Thus, two meta-theoretical dimensions – a presupposition about ontology and an operational epistemology – form a useful framework for classifying theory: for classifying classical theory in sociology and for classifying the array of social scientific approaches that have evolved from the classics to understand risk. This is the framework of Figure 1. The framework is not to be viewed as a rigid structure of thought, but as a heuristic aid for organizing our understanding of the topic. The grid produced by cross classifying the two dimensions aligns itself nicely with four major traditions that have shaped social scientific investigation in general.[11] The configuration of quadrants is, thus, a disciplined way for comparing major bodies of thought. We briefly describe each quadrant, its major tradition, and the translation of that tradition to the field of risk.

The Durkheimian tradition (quadrant I)

The writings of Émile Durkheim (1965 [1915]; 1982 [1895]; 1984 [1893]), in the late nineteenth and early twentieth century, were an instrumental part of the very beginnings of sociology as a separate discipline and remain in the privileged position, at the core of the discipline. Durkheim, utilizing the then rapidly accumulating ethnographic evidence from non-Western cultures, was one of the first thinkers to codify precisely central questions emerging from the Enlightenment: how is society possible? What is the glue that holds society together? An implicit answer to these essential questions was already available in the long line of Enlightenment thinking on the social contract. From Hobbes via Locke, Hume and Rousseau to Kant: society was believed to exist because rational humans had entered into an original social contract. Durkheim objected to this immaculate view of rationality. Among other things, the social contract account of society was embarrassed by ethnographic evidence showing that pre-literate societies achieved social cohesion and social solidarity despite all the requisites implied by rational contract being absent: strategically acting individuals, markets, and legal codes.

Having eliminated rational, contractual relations as the basis of society, again the central question emerged: how is society possible? The answer for Durkheim was in emotive forces, such as trust, social bonding, moral obligation, and moral solidarity. These forces, the elements of cultural and institutional continuity, held the structure of society together. What is more, that structure was presumed to be a reality *sui generis* (uniquely of its own type), irreducible to individual thoughts and actions.[12] 'The first and most fundamental rule is: *Consider social facts as things*'(1982 [1895]: 14). We could only understand society, Durkheim argued, by looking at its macro-patterns, not by trying to aggregate the actions and behaviors of its individual members.

Key elements of the Durkheimian legacy are a commitment to an idealism/romanticism tradition that emphasizes symbols and cultural practices, on the one hand, and to a macro-sociological approach that sees society primary and individual behaviors as derivative. This is the legacy imported into risk analysis by those, following Durkheim, who have developed a culture theory of risk. The exemplary of this is the work of anthropologist Mary Douglas and political scientist Aaron Wildavsky in their 1982 book, *Risk and Culture*. Perceptions and concerns about risk, in this view, emanate from the structure of culture and one's position in that structure. Our thoughts about risk are not entirely free, but infused with the emotive power of the groups to which we belong.

Risks, then, are entirely relative. There is little room here for a match between claims about risks and actual risks in the world. Important in this perspective is the fact that, indeed, the shared meanings of culture shape our shared perceptions and that macro forces influence individual thoughts and behaviors. Yet, it can be objected that this hyper-relativism is both logically flawed, and ultimately a type of social reductionism: reality and our cultural belief systems are one and the same. Furthermore, even if one accepts this perspective's basic premise that structure shapes perceptions so that people conceptualize and think about risk differently, it doesn't necessarily follow that they do so in a totally relativistic way.

The work of the British social theorist Anthony Giddens on risk (Giddens, 1990), has deep roots in the Durkheimian tradition. It begins as a macro-sociological orientation of culture, but attempts to penetrate the subjective world of the individual actor with agency and self-identity. This balancing act can be understood by recognizing that Giddens (1977) has pointed to the existence of two separate, but related wings of phenomenology: the hermeneutic and the existential. Although both arise from the same epistemological foundations, and although both emphasize culture, they proceed along different theoretical methods. The hermeneutic is more macro in orientation, focusing – much like Durkheim – on the collective aspect of culture. The existential is more micro in orientation, focusing – much like Berger and Luckmann (1966) – on culture as it is internalized into the subjective consciousness of individual social actors. Giddens has been concerned with risk as a central feature of 'globalization'. By globalization he means 'the intensification of worldwide social relations which link distant localities in such a way that local happenings are shaped by events occurring many miles away and vice-versa' (1990: 64). A central consequence of globalization, as we noted above, is the creation of a postmodern world of growing interdependence: a global division of labor, a worldwide diffusion of consumption practices, and the spread of risks associated with production and consumption practices to every corner of the globe. Thus, the postmodern world will increasingly share a common culture, that culture will be increasingly textured with risks, and accompanying these risks will be globally shared anxieties. In short, the postmodern world will be a risk world (the hermeneutic) and risk will be increasingly internalized by its inhabitants (the existential). This combination of the sociology of risk and human responses to risk constitute, for Giddens, a double hermeutic: 'The development of sociological knowledge is parasitical upon lay agents' concepts; on the other hand, notions coined in the meta-languages of the social sciences routinely reenter the universe of actions they were initially formulated to describe or account for' (1990: 15).

Giddens' has outlined the structure for describing the fundamental and far-reaching social change leading to what he calls postmodernity. Taken this far, that structure is nearly mute about the explicit role played by the environment in this transformation. It is also mute, thus far, about the role the environment plays in the emergent worldwide culture of risk. The environment becomes a constitutive feature of the theoretical structure by virtue of its dominance over the types of risks that are becoming universally recognized and worrisome. So, for Giddens' the exemplars of the worldwide risk culture are: the continuing threat of nuclear war, radiation from nuclear power plant accidents or from nuclear waste, chemical pollution everywhere, threats to the oceans' phytoplankton (responsible for renewing atmospheric oxygen), global warming, the hole in the ozone layer, the destruction of large areas of rain forest, and the exhaustion of millions of acres of topsoil (1990: 127-128).[13] The core feature of all of these examples is risk to the environment: risk due to untoward alteration of the environment, such as from nuclear annihilation, or from resource exhaustion, and risk due to environmentally damaging side effects of production, such as chemical or nuclear pollution of the land, sea, and air, as well as other risks to sustainability. Thus, risk consciousness is tantamount to environmental consciousness. There is a growing ecological consciousness worldwide that rests upon a deep uncertainty about the environment that sustains us. There is the shared conscious awareness that humankind could be destroying the very ecological base upon which its survival depends.

As for theoretical method, is Giddens macro or micro? Neither, according to Giddens. He, in essence, proclaims a curse on both houses with his structuration theory. For too long theorists had been preoccupied with the primacy of one approach, macro or micro. Structuration theory attempts to go beyond this counterproductive dichotomy. The theory rests upon the notion that both the purposive actor or agency and a structured pattern of constraints via cultural rules and social recourses mutually constitute each other to form social life. This approach is one of duality (a better term might be mutuality), not dualism. Here is what Giddens says about the duality of structuration theory: 'The constitution of agents and structures are not two independently given sets of phenomena, a dualism, but represent a duality. According to the notion of the duality of structure, the structural properties of social systems are both medium and outcome of the practices they recursively organize' (Giddens, 1984: 25).[14]

While the social world may exist as an ontological duality, this duality is not mirrored in epistemology – nor in the theoretical methods that activate epistemology. The human mind is limited in its ability to think in terms of duality; it is difficult, if not impossible, to maintain a balanced consciousness of the macro and micro simultaneously. So, in practice, this means that even with a theoretical context of duality, like structuration theory, the theorist lays a foundation with either an orientation dominated by the macro or one dominated by the micro. For those theorists who presuppose the existence of social forms independent of individual actors – such as society, social structure, or culture as shared meaning, or culture as shared rules (like Giddens) – a pre-framed orientation awaits them. The presupposition means that these phenomena pre-exist for individuals. As such, it biases the theoretical orientation toward the macro. This appears to be the case with Giddens. In general, he tends to begin with the macro, before moving toward the micro. The effect is to have the macro dominate the micro. Collins, for example, observes: 'My conclusion is that Giddens has not attempted micro-translation seriously enough. His autonomous macro level includes not only time and space, but also a realm which transcends both of these' (Collins, 1988: 399).

What is true of Giddens generally is especially true for his work on risk. He begins the foundational development of a theory of postmodernity, his first work on the topic of risk, with these telling words: 'In what follows I shall develop an institutional analysis of modernity with cultural and epistemological overtones' (1990: 1). Totally missing from this sentence is the social actor – just as that actor is virtually missing from the rest of this book. So, in its own theoretic language structuration theory transcends the macro-micro dichotomy, by treating them as a duality and mutuality of social life. On this basis Figure 1 should show a two-headed arrow that extends equally between Quadrants I and IV. Yet, in practice, as argued above, there is a clear emphasis on the macro orientation – although it aims to incorporate the micro. It is this execution of theory that is the basis for positioning the work of Giddens in Figure 1. It is depicted with a single-headed arrow that extends from Quadrant I to Quadrant IV.

The Marxist/Weberian tradition (quadrant II)

The placement of Marx and Weber, and modern theorists such as Perrow (1984) and Beck (1992) who work within that tradition, on the realism side of the horizontal dimension reflects their emphasis on the reality of social history, social organizations, and social relations. Furthermore, the theoretical method of this social organizational perspective is, just as with the Durkheimian/cultural approach, macroscopic, focusing on structural relationships and their institutional underpinnings. Their placement above the rational tradition of the social contract, utilitarianism, and RAP is fitting since both traditions emphasize the reality of social actors or institutions and share a focus on types of self-interest and the importance of material economy.

Both Marx and Weber observed a fundamental and persistent rationality in society. For Marx capitalism was a system built upon a set of rational relationships inherent in the nature of industrial production, all quantifiable through the labor theory of value. But while this rational system produced rational output and wealth for the capitalist class, it produced inherent exploitation and often misery for the working class. For Weber history produced increased rationality, especially in the development of increasingly abstract means-ends calculations. The organizations and institutions of society, and society itself, therefore, became more rational: via accounting systems, via bureaucracies as rationalized organizations, and via rational-legal authority that legitimated the modern state (which Weber equated with society) itself (Roth and Schluchter, 1979; Schluchter, 1981).

The most explicit and developed macro-sociological theory of risk is that of Ulrich Beck (1992 [1986]; 1995 [1991]; 1995 [1988]). Beck is, on the one hand, a thoroughgoing realist and, on the other, a theorist whose thinking is steeped in the rich Germanic tradition in which he was trained (the Frankfurt School) and to which he contributes. While his orientation clearly reflects the influence of this tradition, he moves that tradition in new directions. Beck's theoretical frame consists of two interrelated theses: the preoccupation with risk in the contemporary world and with what he calls 'reflexive modernization'. The first thesis rests upon a fundamental distinction between industrial societies and contemporary societies: the former is concerned with the distribution of 'goods', especially economic goods, while the latter is concerned with the distribution of 'bads' (dangers and hazards) and is, therefore, aptly titled 'the risk society'. Moreover, the shift from a preoccupation with goods to a preoccupation with bads reflects a fundamental restructuring of social organization, from one of class positions to one of risk positions. The shift also produced a new culture of shared meaning where 'in class positions being determines consciousness, while in risk positions, conversely *consciousness (knowledge) determines being*' (emphasis in the original) (1992 [1986]: 53). In essence, just as Marx had turned Hegel on his head so Beck does the same to Marx.[15] This shift from concerns with production to concerns with the consequences of production represents a transformation of social order. Accompanying the shift is a decline in the importance of structures, like class, while at the same time individualizing social agents (actors) who, now forced to confront the many risks, are now conscious with risk decisions. This confrontation forces social actors to reflect on the social institutions responsible for those decisions. This confrontation draws into close scrutiny the role of institutional trust. And because there has been a globalization of risks, as we have already noted, the world itself has become a risk society.

Beck's second thesis is based on the central role of science in issues of risk, a problem noted by a number of other scholars (e.g. Dietz et al., 1989; Rosa and Clark, 1998). Science, on the one hand, is partly responsible for the growth of risks and hazards while, on the other hand, it is the principal institution entrusted with knowledge claims about risk. For example, the dawn of the postmodern era is often traced to the development and dropping of atomic bombs in 1945. This remarkable achievement of science planted the seeds of a wide variety of contemporary risks and hazards. It is this Janus-like role of science where the rub lies for Beck. Because risks are difficult to define, ambiguous and subject to competing interpretations and conflicting claims, 'in definitions of risks *the sciences' monopoly on rationality is broken*' (emphasis in the original) (1992 [1986]: 29). When science is no longer privileged, when institutions are no longer trusted, how are risk societies to make knowledge claims about the increased risks that define those societies. For Beck the answer is 'reflexive modernity'. By this he means 'the possibility of creative (self-) destruction for an entire epoch: that of industrial society' (Beck et al., 1994: 2)... (and)... 'This concept does not imply (as the adjective 'reflexive' might suggest) *reflection*, but (first) *self-confrontation*' (1994: 5). The purpose of this confrontation is neither crisis nor revolution, but the continual examination and self-reflection on meaning, on forms of social organization, on cultural norms and conventions, and on the nature of rationality itself. One key element of this reflexive modernization is the role of citizens *vis-à-vis* scientists and public officials. 'Its goal would be to break the dictatorship of laboratory science... by giving the public a say in science and publicly raising questions' (Beck, 1995 [1991]: 16). In the risk society knowledge claims about hazards and dangers need to be negotiated between scientists, political stakeholders and laypersons – in effect, a negotiation between different epistemologies. With this proscriptive element in his theory Beck joins a growing number of voices calling for the extension of public involvement and democratic processes in assessing and managing risks (Renn et al., 1995; Stern and Fineberg, 1996; Rosa and Clark, 1998).

What of the environment for Beck? Similar to the approach of Giddens, described above, the environment becomes a constitutive feature of Beck's theoretical structure because of the dominance of environmental concerns in the types of risks that constitute the Risk Society. It is environmental risks, principally, that are becoming universally recognized and worrisome. Beck alerts us to this agenda with two of his books following on *Risk Society*, both of which have 'ecology' as part of the title: *Ecological Enlightenment* (1995 [1991]) and *Ecological Politics in an Age of Risk* (1995 [1988]). That the Risk Society is dominated by environmental risks is made clear in the first sentence of *Ecological Enlightenment,* where Beck writes: 'The success of the ecological movement force us all to keep repeating ourselves. It has become a given in contemporary consciousness that species are dying out, that oceans are being contaminated, that climatic catastrophe is looming—these basic assumptions are shared by the Communists and the conservatives, by the chemical industry and fundamentalist Greens' (1995 [1991]).

Beck has created a multi-faceted theoretical frame whose foundation is classical sociological theory in the German tradition. Social structures and processes are real, are greatly influenced by history, and can best be understood with a macro orientation. But Beck goes beyond this tradition with the incorporation of the concept 'reflexive modernity' which gives social actors an active role virtually missing from the tradition upon which he builds. Individuals are no longer unreflective, passive observers of their contemporary world. Instead, armed with a consciousness about pervasive risks, they are active appraisers of the world of modernity, especially of its untoward consequences. Thus, while steeped in the macro tradition of German sociology, Beck's theory extends to the micro by incorporating social actors now weighing the costs of modernity: its bads, its risks. Accordingly, Beck's work is depicted with an arrow that extends from Quadrant II to Quadrant III.

Utilitarianism and RAP (quadrant III)

In the above discussion we outlined the fundamental features of the occupant of Quadrant III, the Rational Actor Paradigm, or RAP. The basic idea of RAP – whose beginnings are traceable to the Italian Renaissance, whose codification occurred during the Enlightenment, whose formalization was launched in classical economic theory by the nineteenth century utilitarians, and whose role is assured in modern utility theory – places the individual actor at center stage in a world of real choices. In RAP that actor is an isolated individual driven by self-interest to maximize satisfaction, utility, social position, or some other desideratum in life. The rational actor chooses a course of action based upon its expected utility: all alternative outcomes from the action are arrayed, the probability of each outcome is estimated, and the highest value from multiplying the probability times utility (that is, the action producing the highest expected utility) gives the rational alternative to be chosen.

RAP enjoys its longest tradition in economics due to that field's emphasis on the material world: alternatives and utilities are framed in terms of monetary costs and gains. Its adoption as the dominant perspective in risk research is a straightforward extension of material considerations, but not only of monetary ones but also of other things valued by humans: health, lives, safety, etc.

The theoretical method of RAP is tenaciously that of methodological individualism. Individual human action is the elementary unit of social life. But here the term 'individual' is to be broadly construed: "The term 'individual' (is) used in an extended sense that also includes corporate decision makers, like firms or governments" (Elster, 1989: 13). Thus, RAP theorists see in their structure of thought a tremendous power, the power to unify the social sciences. RAP risk researchers, likewise, see the RAP calculus as a way of addressing the entire spectrum of human concerns over risk.

The phenomenological tradition (quadrant IV)

The phenomenological tradition is relatively recent to modern philosophy, and even more recent to sociology. The appearance of the book, The *Social Construction of Reality* , by Peter L. Berger and Thomas Luckmann in 1966 was a landmark that pointed in two directions. Pointing backward, it drew into the sociological tradition the philosophy of the German phenomenologist Edmund Husserl, a tradition heretofore ignored in sociology. The principal quest of Husserl was to arrive at absolute certainty, to get at the true essences of the world, a presupposition that is quintessentially in the idealist tradition with a lineage back to Classic Greece. In effect, Berger and Luckmann took the fundamentals of phenomenology, Husserl's initial formulation and the refinements of his student Alfred Schutz, incorporated existential philosophy, a direct descendant of phenomenology, and wove them together with elements of classic sociology.

With this act, and expressed in the title of their book, the second direction to which this landmark pointed was toward the view that all reality was socially constructed and, accordingly, that society itself was a social construction.[16] A cornerstone of this perspective is a humanistic emphasis on the fluidity and unpredictability of social life. Society is constituted out of the actions of its individual members who are always negotiating meaning – thus structure is never ultimately achieved, it is always in the process of becoming. Society is not an abstraction that exists as a social fact, as the Durkheimians would have it, nor is it a tangible structure, as Marxists and Weberians would have it, nor is it the collection of rational actions by individuals, as the utilitarian tradition would have it. Society, in all these approaches, is a reification. Viewing society as a real entity, or as having independent ontological status, is fundamentally wrong, according to social constructivists, because society only exists inter-subjectively in our minds.

There is, for the most devoted social constructivists, no separation between reality and our perceptions of reality: our negotiated knowledge of the world is the functional equivalent of that world itself. A social constructivist variant of the phenomenological tradition speaks with an audible voice in the social science of risk. At its core, it translates the perception/world functional equivalency into the presupposition that risks do not exist in a reality separate from our perceptions of risk (Stallings, 1995). Our perceptions of risk, our choices of which risks to be concerned about, are equivalent to risk itself. One definition of the social construction of risk is provided by one of its leading proponents: 'A key shift required by the perspective described here is that definitions of risk, knowledge, and responses to information and uncertainty are based ultimately on the attempted maintenance of familiar social identities.... Physical risks thus have to be recognized as embedded within and shaped by social relations and the continual tacit negotiation of our social identities' (Wynne, 1992: 295).

The theoretical method of all variants of the phenomenological tradition, like the utilitarian one, is clearly directed at the individual actor in micro-settings. While permitting the existence of institutionalized rules, including such cultural foundations as legal codes, social norms, linguistic conventions, and customs, the focus is always on the local interplay between interaction and the generation of meaning. Definitions of reality emerge from this interplay. Indeed, the individual actor (even the self) does not exist as independent reality, but as the socially constructed outcome of meaning between actors negotiated within the confines of the institutionalized rules. A key shortcoming with this hyper-micro focus is that it allows little room for any reality external to the phenomenological setting itself. As such, it typically ignores the possibility that powerful exogenous forces may have a role in defining reality, or that untoward events can exist and impact human interaction in a very defining way.

Hybrid approach: the psychometric tradition

In addition to the perspectives that are well aligned with each of the grid's four quadrants, we can note a very influential (hybrid) approach that cuts across two or more perspectives. One of the most sustained social scientific efforts to understand risk, over two decades old, is a tradition developed by cognitive psychologists with a bent toward the quantitative representation of human thought processes – a psychometric approach to cognition. Not only is the psychometric tradition one of the most sustained social science efforts attempting to understand risk – especially perceptions of risks – but it is also one of the most sustained critiques of RAP. That it should be one of RAP's most consistent critics carries with it an historical irony, for the psychometric tradition emerged within a RAP framework where the efforts of cognitive and decision psychologists were toward the experimental test and elaboration of the axioms underpinning expected utility theory, RAP's operational arm.

Classical expected utility theory combines a normative model of decision making (what people ought to choose to maximize their utility) with a behavioral model (people are rational and actually make such choices). What turned the psychometric tradition from friend to foe of RAP was the appearance – from systematic, experimental tests of the axioms of expected utility theory – of a number of troubling choice anomalies. People's behavior frequently violated the most basic and most critical axioms of utility theory. Using a variety of mental shortcuts, 'heuristics' in the language of the tradition, people made decisions found to be acceptable and satisfying to them, but which were in clear violation of the formal expectations of utility theory. The cumulation of these anomalies led two of the leading researchers in the field to conclude that: '... the logic of choice does not provide an adequate foundation for a descriptive theory of decision making. We argue that the deviations of actual behavior from the normative model are too widespread to be ignored, too systematic to be dismissed as random error, and too fundamental to be accommodated by relaxing the normative system.... We conclude from these findings that the normative and descriptive analyses cannot be reconciled' (Tversky and Kahneman, 1987: 68).

Directing the same psychometric microscope to the study of risk produced a similar incongruity between the behavioral expectations of RAP and what people actually did. We already pointed out above that this finding originally appeared in the context of responses to natural hazards. Based upon those early hints of chinks in the armor of RAP, psychometricians Paul Slovic, Baruch Fischhoff, and Sarah Lichtenstein began a systematic program of investigation (see, for example, Lichtenstein et al., 1978) to discover how people actually thought about the risks to which they were routinely exposed. To what extent did perceptions and behaviors deviate from the predictions of utility theory that people would choose rationally based upon a calculus of the probability of a risk times its consequences? The cumulative results of that effort (see, for example, Slovic, 1987) demonstrated systematic deviations from theoretical prediction. People do not ordinarily think of risks in terms of probability times consequence. Instead, they incorporate a variety of qualitative features into their fears and worries about risks: whether the risk is voluntary or imposed, how well known it is, who is in charge of managing the risk and can they be trusted, whether the risk produces few casualties frequently versus many casualties all at once (catastrophe), and especially the level of dread that comes to mind when thinking about the risk.

In Figure 1 the psychometric tradition overlaps three quadrants. Its roots are, as we pointed out above, clearly in the utilitarian cum RAP tradition. While it began there it eventually, based upon the empirical results it produced, found itself in the Durkheimian tradition – although psychometricians are generally not cognizant of this alignment. The variety of qualitative aspects that people factor into their risk perceptions—dread, catastrophe, trust, disruption, etc. – are the very elements of moral solidarity that, in the Durkheimian view, holds society together. Finally, the psychometric tradition slips over into phenomenology by virtue of the fact that its leading practitioner takes this recent position on risk: 'One of the most important assumptions in our approach is that risk is inherently subjective. Risk does not exist "out there", independent of our minds and cultures, waiting to be measured. Human beings have invented the concept of "risk"... there is no such thing as "real risk" or "objective risk"' (Slovic, 1992: 119). One would be hard pressed to find a more succinct statement summarizing phenomenological nominalism.

Critical reflection and prospects for sociological theory

Theoretical reconciliation

With the intellectual history and theoretical contexts we have outlined above as a backdrop, we are now in a position to ask: is it possible to reconcile the various theoretical traditions that are pursuing the topic of risk? Our analysis distinguished the various traditions along two meta-theoretical axes: ontological presupposition and methodological orientation. Meta-theoretical presuppositions and orienting perspectives form the bedrock of theory, shaping its logic and scope of investigation. Theoretical elaboration on that bedrock, if the logic is to remain coherent, must be faithful to its meta-theoretical roots. Our basis of comparing the competing theoretical traditions is, therefore, at the abstract level of meta-theory. As such, our basis for comparing theoretical traditions is through a comparison of the building blocks which form the foundation of those theories.

At this meta-theoretical level of comparison, it seems highly unlikely that the competing theoretical approaches to risk can be reconciled. The differences in presupposition and orientation are too contradictory, too inconsistent, and too antithetical – at a very fundamental level – to yield to such a reconciliation. Indeed, the different theoretical approaches to risk reflect, in large part, the differences that separate intellectual fields and sub-fields and which separate sociology's specialties. It would be presumptuous, indeed, to expect these differences to dissolve over risk – or any other topic important to sociological theorizing.

Accomplishing such a grand task would first require the establishment of a set of meta-rules that lay the groundwork for reconciliation. The rules would need to explicate a logical structure for subsuming the conflicting presuppositions and orientations, and would need to delineate the scope of conditions for each of the subsumed theoretical orientations. To date such a task has not been completed. However, Jaeger, Renn, Rosa and Webler (forthcoming) have undertaken the preliminary spadework essential to developing the proposed meta-rules for reconciliation. In particular, they provide a critical evaluation of all the leading theoretical approaches to risk – RAP, culture theory, social constructivism, postmodernism, psychometric, macro-sociological (including critical) theory – in order to lay bare the strengths and weaknesses of each. In so doing, they provide the raw material for meta-rules since, perforce, these rules would need to build on the strengths of the various theoretical traditions while addressing their weaknesses within a coherent structure.

Theoretical pluralism

Is a unified theory of risk desirable? Perhaps a quest for a unifying theoretical structure is misguided. Perhaps it is a modern version of seeking a single grand theory, the holy grail of sociology two generations ago. Eventually sociology recognized this holy grail to be too elusive, even illusive, to pursue any longer and abandoned the task. This reorientation was due, in no small degree, to the arguments of the highly influential Robert K. Merton who argued against grand theorizing and for theories of the 'middle range'. Merton was deeply influenced by the science that was then, though perhaps less so now, the chief model of sociology: physics. He observed that physics had no single theory; yet its status as a science was never in question. On this point he quotes Richard Feynman, Nobel winning theoretical physicist: 'today our theories of physics, the laws of physics, are a multitude of different parts and pieces that do not fit together very well' (Merton, 1967: 48).

If we substitute the word 'risk' for the word 'physics' we could probably find no more succinct statement describing our knowledge of risk. Thus, despite the lack of an all-encompassing theoretical system, physics has arguably experienced considerable progress in its knowledge. If this is anything of a model for risk, we can expect to expand our understanding, even as we devote efforts toward more encompassing theoretical structures. This expectation is reinforced with a recognition that the topic of risk is so pervasive, so complex, so multifaceted, and so important that we can only hope to improve our understanding of it – short of a unified theoretical structure – by attacking it with multiple perspectives.

Risk and environment

That risk has become a central concept in the contemporary world is now well established. That risk studies are institutionalized is also well established, as evidenced by the large and growing intellectual infrastructure worldwide devoted to understanding and measuring risk. That risks are a means to define the end of the century and the end of the millennium has recently attracted sociological theorizing – and speculation on broad social change. This recent engagement of sociological theory was preceded by a variety of more specific risk theories and perspectives. Taken together the various theories and perspectives now available offer a multi-faceted vantage point for better understanding risk.

Risk, as pointed out above, is a valuable analytic lens for evaluating past and future impacts of humans on the environment. And recent macro-sociological theorizing, such as the work of Giddens and Beck, looks to the environment for paradigmatic examples of risk – big, postmodern risks especially. Since that theorizing is grounded in the classic traditions of the field it has had the effect of incorporating those traditions into concern for risk and the environment. The result is that sociological theorizing about the environment has been deepened. But, it has not been deepened nearly enough.

While the environment has entered the theoretical schemes of classically grounded sociological theorizing, it has been conceptualized as a key marker of social change or as a key theme, along with risk, of a new collective consciousness being spread worldwide. For Giddens, this phenomenon is intelligible through the lens of his structuration theory with its duality. He writes: 'According to the notion of the duality of structure, the structural properties of social systems are both medium and outcome of the practices they recursively organize' (1984: 25). What is needed now is a greater recognition of the environmental duality of human existence. Also needed is more effort toward understanding how social systems are both the medium and outcome of the production and consumption practices that transform the environment. Ultimately, sociological theorizing needs to understand how social forms transform the environment and how the transformed environment shapes social forms.

Notes

1. Various versions of this paper have benefited from the contributions of Robert Brulle, Fred Buttel, Tom Burns, Thomas Dietz, Riley Dunlap, Silio Functowicz, Louis Guay, Carlo Jaeger, Valerie Jenness, Amy Mazur, Aaron McCright, Ellen Omohundro, Otwin Renn, Jim Short, Paul Stern, Tom Webler, Steven Yearley, and Steve Zavestoski.
2. Another important approach is cost-benefit analysis, which most writers consider a separate tool. Indeed, many who consider it a separate tool would argue that it is certainly as important, if not more important than risk analysis. But, conceptually both tools have a similar foundation and can be subsumed under the same rubric risk.
3. Since the time of Hegel there has been little interruption in a steady stream of neologisms for describing society's dominant theme: Tawney's *Acquisitive Society*, Popper's *Open Society*, Gailbrath's *Affluent Society,* Etzioni's *Active Society*, Bell's *Post-Industrial Society*, to name some of the more well-known, and to which can be added such contemporary renderings as *Information Society*, *Litigious Society, Carbon Society,* and *Knowledge Society.*
4. We note an important connection between these methods and risk policy, although it is beyond our compass here to develop the issue fully. Virtually all historical risk communication programs have been based upon RAP. The underlying logic is that rational actors desire precise estimates of risks and once given those estimates they will take actions that are rational. Embarrassing to this singular RAP-driven approach is the reality that people often prefer social, political, and cultural information that contextualizes risks, not simply abstract estimates of supposed mathematical precision (National Research Council, 1989).

5. For the development of meta-principles of risk analysis based upon defining risk as a state of the world represented by ontological realism and epistemological hierarchicalism (or OREH), see Rosa (1998).
6. One could, without comprising the argument, substitute 'nominalism' for 'idealism'. Although somewhat at odds with a purist's distinction, we use 'romanticism' also as a synonym for idealism, to more accurately reflect some of the modern, refined versions of idealism, especially versions – such as German romanticism – that have impacted the sociological enterprise.
7. Owing to this presupposition a corollary of idealism is often the relativizing of knowledge claims.
8. This was a monumental legacy of the Enlightenment: with the writings of Thomas Hobbes and John Locke, the core sociological question – 'How is society possible?' – was raised explicitly. And with the writings of Giambattista Vico, history was now written as the interplay and unfolding of causal forces – not just the great feats of powerful actors.
9. This is consistent with the position taken by Alexander and Giesen in the introduction to about the only volume with a singular focus on this subject: 'We will argue that the micro-macro dichotomy should be viewed as an analytic distinction and that all attempts to link it to concrete dichotomies – such as "individual versus society" or "action versus order" – are fundamentally misplaced' (1987: 1).
10. In 1807 Hegel published *Phenomenology of Spirit.*
11. Left out of account in this discussion, but addressed in Jaeger, Renn, Rosa and Webler (forthcoming), is the contemporary body of thought that goes by the name of postmodernism. It would properly be placed in Quadrant I, reflecting its roots in classic German Romanticism and Continental idealism. As observed by Collins (1994: vii), 'In the 1980s the idealist and relativist side of this philosophical tradition /from the battle between Continental idealism and British empiricism/ has been enunciated in the highly polemical version under the label of postmodernism. This movement is especially hostile towards the positivistic approach...'.
12. Durkheim was keenly aware of and devoted to the task of distinguishing the sociological from the psychological level of analysis. He was also guided in this task toward a deep criticism of Herbert Spencer – one of the leading sociologists of the day with a wide following in Great Britain and the United States – because of Spencer's utilitarian thinking with its acceptance of the *laissez-faire* doctrine that modern society emerged from the competition of individuals in the market; that is, from economic exchange.
13. While Giddens integrates the environment into his theoretical argument, his position on sustainability represents a clear departure from the dominant thinking in environmental sociology. Environmental thinking has generally been cautious in its expectations of future resource availability. On this point, Giddens is, to the contrary, unreservedly optimistic: 'We find here the potential for a *post-scarcity system*, coordinated on a global scale.... While some resources are intrinsically scarce, most are not, in the sense that, except for the basic requirements of bodily existence, "scarcity" is relative to socially defined needs and to the demands of specific lifestyles' (1990: 165-166, emphasis in the original).
14. A more explicit statement, perhaps, of this duality comes from Bhaskar, a principal source for Giddens, when he writes: 'Society is not the unconditioned creation of human agency (voluntarism), but neither does it exist independently of it (reification). And individual action neither completely determines (individualism) nor is completely determined by (determinism) social forms. In [this context], unintended consequences, unacknowledged conditions and tacit skills... limit the actor's understanding of the social world, while unacknowledged (unconscious) motivations limits one's understanding of oneself' (Bhaskar, 1982: 286).

15. Beck's intent to supersede Marx could hardly be more explicit: 'The petrification of criticism, which was one meaning of the predominance of Marxian theory among the critical intelligentsia for more than a century is gone. The all-powerful father is dead' (Beck et al., 1994: 12).
16. Seeds of social constructivism were already being scattered in during the Greek Enlightenment of the fifth century BC by the Sophists. Ultimate truth, the Holy Grail of prior philosophical approaches, was little more than an elusive quest: a deceptive fantasy. Any claims of ultimate finality, whether they be philosophical, religious, or anything else could not stand the test of critical argument. 'Truth was relative, not absolute, and differed from culture to culture, from person to person, and from situation to situation.... In the end, they [the Sophists] argued, all understanding is subjective opinion. Genuine objectivity is impossible.... Other than appearances, a deeper stable reality could not be known, not only on account of man's limited faculties, but more fundamentally because such a reality could not be said to exist outside of human conjecture' (Tarnas, 1990: 27-28).

Bibliography

Alexander, J.C. and B. Giesen (1987), From Reduction to Linkage: The Long View of the Micro-Macro Debate, in: J.C. Alexander, B. Giesen, R. Münch and N. Smelser (eds), *The Micro-Macro Link.* Berkeley and Los Angeles: University of California Press, pp. 1-41.

Alexander, J.C., B. Giesen, R. Münch and N. Smelser (eds) (1987), *The Micro-Macro Link.* Berkeley and Los Angeles: University of California Press.

Alexander, J.C., B. Giesen, R. Münch and N. Smelser (eds) (1987), Resisting the Revival of Relativism, *International Sociology* 2: 235-250.

Bhaskar, R. (1982), Emergence, Explanation, and Emancipation, in: P.F. Secord (ed.), *Explaining Human Behavior.* Beverley Hills, CA: Sage.

Beck, U. (1992 [1986]), *Risk Society: Towards a New Modernity.* Translated by M. Ritter. London: Sage Publications.

Beck, U. (1995 [1988]), *Ecological Politics in an Age of Risk.* Cambridge: Polity Press.

Beck, U. (1995 [1991]), *Ecological Enlightenment.* Translated by M.A. Ritter. Atlantic Highlands, NJ: Humanities Press.

Beck, U., A. Giddens and S. Lash (1994), *Reflexive Modernization: Politics, Tradition and Aesthetics in the Modern Social Order.* Stanford, CA: Stanford University Press.

Berger, P.L. and T. Luckman (1966), *The Social Construction of Reality.* New York: Anchor Books.

Burton, I., R. Kates and G. White (1978), *The Environment as Hazard.* New York: Oxford University Press.

Collins, R. (1988), *Theoretical Sociology.* New York: Harcourt, Brace, Jovanovich.

Collins, R. (1994), *Four Sociological Traditions.* New York: Oxford University Press.

Dietz, T., P.C. Stern and R.W. Rycroft (1989), Definitions of Conflict and the Legitimation of Resources: The Case of Environmental Risk, *Sociological Forum* 4: 47-70.

Douglas, M. and A. Wildavsky (1982), *Risk and Culture: The Selection of Technological and Environmental Dangers.* Berkeley, CA: University of California Press.

Durkheim, É. (1965 [1915]), *The Elementary Forms of the Religious Life.* Translated by J. Ward Swain. New York: Free Press.

Durkheim, É. (1982 [1895]), *The Rules of Sociological Method.* New York: Free Press.

Durkheim, É. (1984 [1893]), *The Division of Labor in Society.* Translated by W.D. Halls. New York: Free Press.

Elster, J. (1989), *Nuts and Bolts for the Social Sciences.* New York: Cambridge University Press.

Freese, L. (ed.)(1980), *Theoretical Methods in Sociology: Seven Essays.* Pittsburgh: University of Pittsburgh Press.

Froebel, F., J. Heinrichs and O. Kreye (1980), *The New International Division of Labor.* London: Cambridge University Press.

Giddens, A. (1977), *Studies in Social and Political Theory.* London: Hutchinson.

Giddens, A. (1984), *The Constitution of Society: Outline of the Theory of Structuration.* Berkeley and Los Angeles: University of California Press.

Giddens, A. (1990), *The Consequences of Modernity.* Stanford, CA: Stanford University Press.

Giddens, A. (1991), *Modernity and Self-Identity.* Cambridge: Polity Press.

Gould, S.J. (1997a), Candidate Statement for President-Elect, *Election 97.* Washington, DC: American Association for the Advancement of Science.

Gould, S.J. (1997b), *Questioning the Millennium: A Rationalist's Guide to a Precisely Arbitrary Countdown.* New York: Harmony Books.

Jaeger, C., O. Renn, E.A. Rosa and T. Webler, *Risk, Uncertainty, and Rational Action* (forthcoming).

Kleinhesselink, R.R. and E.A. Rosa (1991), Cognitive Representation of Risk Perceptions: A Comparison of Japan and the United States, *Journal of Cross-Cultural Psychology* 22: 11-28.

Kleinhesselink, R.R. and E.A. Rosa (1994), Nuclear Trees in a Forest of Hazards: A Comparison of Risk Perceptions Between American and Japanese Students, in: T.C. Lowinger and G.W. Hinman (eds), *Nuclear Power at the Crossroads: Challenges and Prospects for the Twenty-First Century*. Boulder, CO: International Research Center for Energy and Economic Development (ICEED), pp. 101-119.

Krimsky, S. and D. Golding (eds) (1992), *Social Theories of Risk*. Westport, CT: Praeger.

Kunreuther, H. (1992), A Conceptual Framework for Managing Low-Probability Events, in: S. Krimsky and D. Golding (eds), *Social Theories of Risk*. Westport, CT: Praeger, pp. 30l-320.

Kunreuther, H., R. Ginsberg, L. Miller, P. Sagi, P. Slovic, B. Borkan and N. Katz (1978), *Disaster Insurance Protection: Public Policy Lessons*. New York: Wiley.

Lichtenstein, S., P. Slovic, B. Fischhoff, M. Layman and B. Combs (1978), Judged Frequency of Lethal Events, *Journal of Experimental Psychology: Human Learning and Memory* 4: 551-578.

Manicas, P.R. (1987), *A History and Philosophy of the Social Sciences*. Oxford, UK: Basil Blackwell.

Merton, R.K. (1967), *On Theoretical Sociology*. New York: Free Press.

National Research Council (1989), *Improving Risk Communication*. Washington DC: National Academy Press.

Nisbet, R. (1982), *Prejudices: A Philosophical Dictionary*. Cambridge, MA: Harvard University Press.

Perrow, C. (1984), *Normal Accidents: Living with High-Risk Technologies*. New York: Basic Books.

Porter, Th.M. (1995), *Trust in Numbers: The Pursuit of Objectivity in Science and Public Life*. Princeton, NJ: Princeton University Press.

Pusey, M. (1987), *Jürgen Habermas*. Chichester, UK: Ellis Harwood.

Renn, O. (1996), *Three Decades of Risk Research: Accomplishments and New Challenges*. Plenary paper, Annual Meetings of the Society for Risk Analysis – Europe. 3-5 June, University of Surrey, Guildford, UK.

Renn, O., Th. Webler and P. Wiedemann (eds) (1995), *Fairness and Competence in Citizen Participation: Evaluating Models for Environmental Discourse*. Dordrecht: Kluwer.

Rosa, E.A. (1994), Mirrors and Lenses: Toward Theoretical Method in the Study of the Nature-Culture Dialectic, in: D. Duclos (ed.), *La Société au Naturel: Functions de la Nature* (Nature's Society: Social Functions of Nature). Paris, France: L'Harmattan.

Rosa, E.A. (1998), Metatheoretical Foundations for Post-Normal Risk, *Journal of Risk Research* (in press).

Rosa, E.A. and D. L. Clark, Jr. (1998), Historical Routes to Technological Gridlock: Nuclear Technology as the Prototypical Vehicle, *Research in Social Problems and Public Policy* (in press).

Ross, R. and K. Trachte (1990) *Global Capitalism: The New Leviathan*. Albany: State University of New York Press.

Roth, G. and W. Schluchter (1979), *Max Weber's Vision of History: Ethics and Methods*. Berkeley: University of California Press.

Schluchter, W. (1981), *The Rise of Western Rationalism: Max Weber's Developmental History*. Berkeley: University of California Press.

Short, J.F., Jr. (1984), The Social Fabric at Risk: Toward the Social Transformation of Risk Analysis, *American Sociological Review* 49: 711-725.
Slovic, P. (1987), Perception of Risk, *Science* 236: 280-285.
Slovic, P. (1992), Perceptions of Risk: Reflections on the Psychometric Paradigm, in: S. Krimsky and D. Golding (eds), *Social Theories of Risk.* Westport, CT: Praeger, pp. 117-152.
Stallings, R.A. (1995), *Promoting Risk: Constructing the Earthquake Threat.* Chicago: University of Chicago Press.
Starr, C. (1969), Social Benefit Versus Technological Risk: What is Our Society Willing to Pay for Safety, *Science* 165: 1232-1238.
Stern, P.C. and H.V. Fineberg (eds) (1996), *Understanding Risk: Informing Decisions in a Democratic Society.* Washington, DC: National Academy Press.
Tarnas, R. (1990), *The Passion of the Western Mind.* New York: Ballantine Books.
Tversky, A. and D. Kahneman (1987), Rational Choice and the Framing of Decisions, in: R.M. Hogarth and M.W. Reder (eds), *Rational Choice: The Contrast between Economics and Psychology.* Chicago: University of Chicago Press, pp. 67-94.
Von Schomberg, R. (ed.)(1995), *Contested Technology: Ethics, Risk and Public Debate.* Tilburg, The Netherlands: International Centre for Human and Public Affairs.
Wolfe, T. (1996), End of the Century and The Spirit of the Age. Brown University Public Lecture. C-Span, Purdue University Public Affairs Video Archives.
Wynne, B. (1992), Risk and Social Learning: Reification to Engagement, in: S. Krimsky and D. Golding (eds), *Social Theories of Risk.* Westport, CT: Praeger, pp. 275-297.
Zeitlin, I.M. (1968), *Ideology and the Development of Sociological Theory.* Englewood Cliffs, NJ: Prentice-Hall.

5

Social Constructions and Social Constrictions: toward analyzing the social construction of 'the naturalized' as well as 'the natural'

William R. Freudenburg

Introduction

The debate on the social construction of environmental problems is in the center of attention in environmental sociology. It seems that more than in any other subdiscipline in environmental sociology social constructionism has found fertile ground as well as fierce criticism. This article will provide a contribution to that debate on the social constructionist approaches to environmental sociology, but not so much by choosing sides. Rather, I want to focus on two main points.

The first point will involve the warning that such discussions, while offering initial opportunities for insights (and potentially for academic self-amusement), have the longer-term potential to become counterproductive, particularly if the early and relatively good-natured exchanges in this debate become transformed into a matter of ongoing obsession. The best antidote to such a potential for obsession, I will argue, lies in devoting energy not so much to 'choosing sides' in the debate, but to comprehending the inherent inseparability and interrelationships between (what we sometimes presume to be) 'strictly' social versus biophysical phenomena.

The second main point will be that the most useful course of action for the future may well lie not so much in having more or less emphasis on social constructionism, but in the development of a fuller and better-balanced constructionism – one that includes systematic attention not just to the social construction of environmental *problems*, but that focuses also instead on what may arguably be both more important and sociologically more interesting in the long run, namely the social construction of environmental *privileges*. As I will spell out in the body of the chapter itself, the issue of 'the Risk Society' offers a particularly important example of the need for sociological analyses to devote greater attention to the socioeconomic inequalities that are embedded within the existing structures of privileges, as well as of the risks, of industrial and post-industrial societies.

Biophysical, social or both?

It would be an oversimplification, albeit only to a moderate degree, to characterize the discussions of the 'social construction' topic in terms of three main positions, each of which includes an 'of course' clause. The first or 'constructivist' position, in simplified terms, would be that, while there can of course be 'real,' underlying problems in the biophysical environment (resource depletion, contamination, global atmospheric changes, etc.), it would be a mistake for sociologists to ignore the important extent to which our understandings (as well as the social consequences) of these problems will tend to be shaped by social construction processes. The 'realist' position, equally simplified, would be that, while social construction processes of course play important roles both in the understanding and in the consequences of various societal uses of the environment, it would be a mistake for sociologists to devote so much attention to social construction processes as to overlook the societal importance of the biophysical changes taking place. The third or 'agnostic' position is that, of course, once the other two positions have been simplified to this degree, it is a relatively simple matter to pronounce that both of them are entirely valid and appropriate, if not entirely complete. What even this third perspective leaves out, however, is an additional problem – one that my colleagues and I have elsewhere described as the risk of becoming 'prisoners of our perspectives' (Freudenburg et al., 1995).

My own argument, in other words begins with but goes beyond agreement with all three of these 'of course' positions. Of course it would be a mistake for sociologists to pay too little attention to the real and tangible significance of social construction processes, and of course it would also be a mistake for sociologists to pay too little attention to the real social significance of physical factors that can continue to exert powerful influences on social life – sometimes even when these physical factors fail to have been socially recognized or understood, let alone to have become the focus of self-conscious social construction efforts. And beyond both points, it would also of course be a mistake for sociologists to become so obsessed with either of these two warnings that we allow ourselves to misunderstand the interactions between the social and physical. It is in the challenge of understanding the interactions between the physical and the socially constructed, however, where I see a need to go beyond the 'traditional' positions that have been spelled out.

One aspect of the interactions is that humans are what my colleague Fred Buttel calls a 'dualistic' species – being subject to many of the same dependencies on biophysical environmental factors as any other species, but at the same time, being uniquely capable of creating (and arguing over) symbols and communication (see e.g. Buttel, 1986). In addition, however, as my colleagues and I have argued (see especially Freudenburg et al., 1995), it is also important to keep in mind that the very distinction between 'the physical' and 'the social' is itself a social construction. The distinction is one that has been created as a matter of analytical convenience, yet if that distinction becomes reified, it can become more than a little 'inconvenient.' In fact, much of what we take to be 'the' biophysical world is inherently and unavoidably a result and a reflection of the way in which we think about it – and at the very same time, many of what we take to be 'strictly social' phenomena inherently involve and respond to the physical realities and constraints that surround us, and affect us, even when we fail to notice the importance of those physical factors.

In practice, in other words, it is arguably most accurate to see the physical and the social as being *conjointly constituted,* or *mutually contingent* – potentially separable, at least conceptually, in the search for analytical convenience, but ultimately impossible to separate completely in practice. Like all such academic distinctions, accordingly, efforts to distinguish social 'versus' physical factors are capable of leading us into the types of efforts to distinguish and prioritize that are initially entertaining, but ultimately counterproductive.

Technology offers a convenient illustration. If one is forced to render a verdict that technology is *either* physical or social, the likelihood of coming up with a sensible answer may be quite small. On the one hand, technology is *inherently* a social product. It is the result of human ingenuity, manipulation, exertion, creativity, blind spots, and other human strengths and weaknesses, and it is often capable of changing what we understand to be 'the' physical limits of a system. On the other hand, technology is *also* inherently physical. It is shaped by physical factors that are sometimes likely to be taken for granted and at other times to be taken as problematic, but that in practice can rarely be ignored with impunity – as illustrated by everything from 'human' flight, to habitable buildings near the north and south poles, to the electronic transfer of documents that physically remain in their initial locations. In the absence of fuel or properly working engines and wings, heavier-than-air transportation devices can and do fall out of the sky; except for traditional dwellings of the Inupiat and other indigenous peoples, habitable buildings in polar regions depend on the importing of energy and insulating materials; and even the electronic transfer of documents can only take place through the physical flows of electronic currents through appropriately designed circuits and across space.

In short, like environmental questions more broadly, technology involves the intersection of the physical and the social; it inherently and inevitably involves, and is shaped by, both. Accordingly, while it is of course still possible to argue whether technology – or environment – should be seen as predominantly 'physical' or 'social,' or even to get caught up in the excitement of the ensuing argument, that is scarcely the approach that appears most likely to lead to improved levels of understanding. An even-handed or dualistic approach – one that sees technology or the environment as being part social and part physical – has the advantage of avoiding the need to invest so much energy in an ultimately distracting debate. An emphasis on mutual contingency or conjoint constitution goes one step further, however, highlighting rather than sidestepping the fact that the attempted allocation into 'social' and 'physical' categories may have the potential to contribute as much to confusion as to understanding. At an analytical level, of course, it is often possible to identify sources of influence that are more nearly social, versus those that are more nearly physical, but ultimately, it can prove to be no more possible to effect a clean and unambiguous separation of the physical and the social than to saw apart the north and south poles of a magnet. Even if the magnet is sawed precisely in half, the net result will be two new magnets, each with a north and south pole. This is far from being a claim that the two are identical; clearly they are not. It is an observation, instead, that the two are perhaps best understood not in terms of their distinctions, but their fundamental interconnectedness – of comprehending not the separations but the inseparability.

Accounts of privilege, privileged accounts?

Once we understand the conjoint constitution of the physical and the social, it is appropriate to return to the issue of social constructionism in our analyses of society-environment relationships. Over the long run, I believe, the more important aspect of this issue will have to do not so much with the *amount* of attention to be devoted to social constructionism, but to the political-economic *balance* or symmetry with which that attention is applied to the understanding of society-environment relationships.

To understand the underlying reasons, it is necessary to recall that what is 'obvious,' at one level, may nevertheless become effectively invisible at another. In one sense, in other words, nothing could be more 'obvious,' particularly to sociologists, than the observation that patterns of power and privilege tend to be socially structured, as well as to be both supported and obscured by strategic, legitimating ideologies. In another sense, however, few of the phenomena that have this degree of importance in society-environment relationships appear to have received less systematic sociological attention than have the ideologies surrounding human uses of the environment. Among the sociologists who have presented or published papers on global environmental change, for example, it is possible to list a number of highly respected scholars who have devoted explicit attention to the social construction of global warming as an environmental problem (see for example Buttel et al., 1990; Fox, 1991; Buttel and Taylor, 1992; Ungar, 1992; Mazur and Lee, 1993; Greider and Garkovich, 1994), yet there is a remarkable paucity of work on what might be called 'the social construction of *non*-problematicity' on the same subject (cf. Bachrach and Baratz, 1970; Lukes, 1974).

For example, while he is clearly aware of what might be called 'the social construction of non-problematicity,' John Hannigan (1995), in his book outlying 'a social constructionist perspective' on environmental sociology, describes his 'chief task' as one of being 'to understand why certain conditions come to be perceived as problematic and how those who register this "claim" command political attention in their quest to do something positive' (Hannigan, 1995: 2-3). Indeed, at least as I peruse the sociological literature more broadly – exempting for the moment the work that my co-authors and I have produced (see e.g., Freudenburg and Pastor, 1992; Freudenburg and Gramling, 1994; Gramling and Freudenburg, 1996) – I can find only a tiny number of works that devote anything like the same kind of systematic analysis to the ways in which certain conditions come to be defined as nonproblematic (see especially the recent work by Krogman, 1996; for other important exceptions, see also Wynne, 1982; Martin, 1989; Draper, 1991).

The antidote, accordingly, may have little to do with arguing over whether or not environmental social problems, like all others, are to an important degree socially constructed. Instead, the need is to correct the asymmetry of awareness within sociological analyses, and to focus substantially increased attention on what

Gaventa's classic analysis of power and powerlessness once called the social creation of 'quiescence' (Gaventa, 1980).

Risk Society risks

The issue of what Beck and others have begun to call 'the Risk Society' offers a particularly important example of an area of inquiry where there is a need for greater sociological attention. In particular, I see a need for greater attention to the socio-economic inequalities that exist within – and have the potential to be obscured rather than highlighted by the approaches we have taken thus far in analyzing – 'the' Risk Society.

At this point, however, I need to pause and to issue a warning about the argument I am intending to make. It would be just as unwise to posit a sharp or black/white contrast between 'American versus European' approaches to the issue of risk as it is to convince ourselves that it is possible to create a clean separation between 'realist versus constructionist' approaches to the environment. At the same time, the simplest and clearest way to illustrate what I see as a greater need for sociological attention to the structure of inequalities and risks *within* societies is to contrast the most prominent of the approaches to issues of risk and the environment that have been taken by European theorists – principally Giddens (1990) and Beck (1992, 1995) – against the approaches that I and a number of my American colleagues have taken (see for example Short, 1984; Erikson, 1990; or the important collection of authors contained in the compilation by Couch and Kroll-Smith, 1991). As a way of underscoring the 'risk' of taking this 'risk contrast' too literally, however, I should note that some of the most interesting and insightful examples of what I might otherwise be tempted to call the 'American' approach to risk society issues have in fact come from some of my European colleagues, such as Renn (1992), Wynne (1992), or Drottz-Sjöberg (1991), and by Europeans and Americans who have worked together on given publications (see e.g. Kasperson et al., 1988; Krimsky and Golding, 1992; Rosa et al., forthcoming). While it is important for these warnings being kept in mind, however, I will spell out the contrasts in relatively stark form in the interest of making my points as clearly as possible.

Despite the attention that Giddens, in particular, places on the forms of trust that are accorded to technical specialists – a key concern for US approaches to risk (see e.g. Short, 1984; Freudenburg, 1988, 1993; Slovic, 1993) – perhaps the key shared focus of their work has to do with the kinds of 'high-consequence risks' for which the societal distribution of concerns 'transcends all values and all exclusionary divisions of power' (Giddens, 1990: 154). Both Giddens and Beck, in other words, have chosen to place much of their emphasis on what Giddens (1990: 124) calls 'truly formidable,' global-scale risks, ranging from nuclear warfare and nuclear winter to 'chemical pollution of the seas sufficient to destroy the phytoplankton that renews much of the oxygen in the atmosphere' (Giddens, 1990: 125). To the degree to which it is possible to discern any consensus among US sociologists about the sociologically significant aspects of the emerging 'Risk Society', by contrast, the emphasis would actually be nearly the opposite of this Giddens/Beck approach, emphasizing precisely the kinds of technological risks for which societal concerns do *not* 'transcend all values and all exclusionary divisions of power.' While the problem has been examined in a growing range of contexts (see for example the edited volume by Couch and Kroll-Smith, 1991), perhaps the clearest example is provided by nuclear risks and technologies.

While public concerns over nuclear power and nuclear waste disposal have often proved to be extremely salient in European contexts, the types of 'nuclear risks' that received the greatest attention from Beck, in particular, were relatively widespread and potentially catastrophic ones – most immediately, the widespread fallout from the Chernobyl nuclear disaster, which truly did affect rich neighborhoods in much the same way as it affected poor ones, just as a nuclear weapon can 'transcend' the social divisions between rich and poor, by effectively obliterating them, in the process of obliterating the rich as well as the poor *people* through its destructive power.

Risks of such a magnitude, clearly, are not to be taken lightly, and Beck, Giddens, and the other sociologists who have begun to devote greater attention to these risks are to be commended for the work they have done thus far. At the same time, however, it would be regrettable indeed if sociological attention to risk, or even simply to nuclear risk, were to stop with a fascination for nuclear fission and the nuclear explosion. There is also a need, instead, for greater attention to the potential for *social* fission – and particularly for what Short (1984), among others, has described as the potential for risks to the social fabric.

Over half a century has now passed since the last use of nuclear weapons in warfare, after all, and yet that half-century has had hundreds of thousands of examples of innocent civilians who have been harmed by 'peaceable' nuclear tests – on both sides of the former Iron Curtain. In addition, we have been learning in recent years that the process of manufacturing the nuclear weapons has been handled in such a hurried and haphazard way that truly phenomenal levels of environmental damage, and risk, were created for the innocent civilian bystanders who happened to live in nearby communities and regions (see e.g. the collection of articles in Freudenburg and Youn, 1993).

On closer examination, in other words, the actual societal experience of risk from nuclear weapons technology proves to have been a good deal more ironic than the approaches of Beck and Giddens would suggest, and in at least two respects. First, at least since 1945, the greatest health damage from nuclear weaponry has been inflicted not on far-away, purported 'enemies,' but on nearby, supposed 'friends.' Second – although this irony ought to be less surprising to sociologists, given that it has unfolded much as Durkheim (1951) would have predicted – at least the net effects of nuclear weapons technologies *within* national societies has been quite paradoxical. The more dramatic and 'scary' risks of nuclear weapons – the specter of shadowy 'enemies,' and the ominous threat of nuclear annihilation – actually did a good deal to *increase* social cohesion within societies. By contrast – but as Short (1984, 1992) and Erikson (1994), among others have helped to explain (see also Freudenburg, 1993, 1996) – the greatest strains to social cohesion, particularly within the US and among the peoples of the former Soviet Union, were created by the 'peaceable' testing and commercialization of the atom, which often did have the effect of putting the social fabric itself at risk.

The socially significant risks, in short, may ultimately prove to have had less to do with the threat of nuclear weapons being used against all of us, by those we see as our enemies, but instead with the less dramatic but far more socially divisive threats that 'ordinary' nuclear technology has posed for some of us, due in large part to the actions of those who proclaim to be our friends. At least to date, in other words, for many of the citizens and societies of the industrialized world, the forms of risk that have had the greatest salience, most of the time, may have been precisely the opposite of those emphasized by Giddens and Beck. The socially salient risks, to repeat, have tended to be precisely those risks that do *not* 'transcend all values and all exclusionary divisions of power.'

Implications and postulates

At least in my view, in case the point is not already clear, the important lesson for sociology from this example has to do with the need to examine the overall risks – and the overall distribution of privileges – more closely. While in one sense, the risks of the nuclear age were created by all of us, or by the larger societies of the nations that have created nuclear weapons, in another sense it is much more accurate to assign the responsibility more narrowly. The average citizen of the US or the USSR – or for that matter of France, Great Britain, and the other members of the nuclear weapons 'club' – actually had almost nothing to do with the creation of nuclear weapons other than to pay the bill in the form of taxes. Equally tellingly, except to the extent to which it is possible to argue that the existence of nuclear weaponry has had the effect of deterring war, the average citizens in the same countries have received little evident benefit from the massive societal investments in weapons that have purportedly been made 'for' those very same citizens. Much smaller proportions of the same societies, however, have indeed received substantial benefits and privileges – most notably in the form of the substantial levels of profit, income and power that have often been associated with the nuclear weapons industry. At least from the perspective of the present paper, accordingly, what is remarkable is the extent to which the problematic and socially structured distribution of privileges and problems – of profits and pains – has escaped sociological attention. While in one sense it is understandable to focus on the overall drama of the kinds of nuclear risks that 'transcend all values and all exclusionary divisions of power,' in another sense it is highly unfortunate for sociology that so little attention, to date, has focused on the kinds of nuclear risks, and rewards, that do not. As a way of attempting to contribute to increased attention of the sort I advocate, accordingly, I would like to propose a pair of postulates that may be useful in offering guidance for future work.

Public property, private benefit

The first postulate involves the observation that, despite the philosophical problematicity of claiming that environmental resources are 'inherently' the property of any individual or small group of individuals – let alone the 'property' of humans as a whole – to the extent to which most natural resources can be seen as the 'property' of humans, the resources are public ones. To the extent to which it is possible to refer to an 'objective' component of environmental social problems and risks, accordingly, *the key factors are likely to involve the social construction of privileged private benefits from public resources.*

Particularly if the matter of privileged access is taken to be an important one at levels of analysis that are more fine-grained and focused than that of the nation-state – or for that matter, of industrial capitalism more broadly – this starting postulate actually points to very different types of analysis than have been found in the better-known analyses of the society-environment relationships to date. To note the examples of the better-known environmental sociologists from the US, for example, Catton (1980) focuses on a collective human 'overshoot' of environmental carrying capacity at a global scale; Dunlap (1980) and Cotgrove (1982) call attention to a pervasive and extensively institutionalized world-view that they term the 'human exemptionlist paradigm' and the 'dominant social paradigm,' respectively; and O'Connor (1988) sees industrial capitalism as having such an extensive resource appetite, in the aggregate, as to be creating 'the second contradiction of capitalism,' adding explicitly Marxian language to Schnaiberg's earlier discussion (1980) of the phenomenon that he termed instead 'the treadmill of production.' Even in more recent work, Schnaiberg and Gould (1994: 94) continue to characterize 'the major argument' of their book as being that there is 'a conflict between economic growth and environmental protection,' at a broad or overall level.

While these points may well all be true, all other factors being equal, the present postulate is that even collectively, such lines of argument are incomplete. In addition, there can be a substantial *disproportion of disruption* – not just at the national or global level (as in the textbook fact that, while the United States contains less than 5 percent of the world's population, it produces over a third of the world's pollution and trash; see e.g. Miller, 1993) – but at the sub-national level, as well. Even within a given society or community, in other words, the disproportion of disruption can be substantial, indeed.

One example I quoted in a recent paper (Freudenburg, 1997a), made possible by the fact that releases of toxic materials in the United States are now subject to at least a degree of reporting and quantification, in the form of a Toxics Release Inventory, or TRI. The TRI data need to be seen as providing only a rough approximation of 'reality,' given a number of reporting and calculation problems (for a discussion and analysis, see for example Szasz, 1994), but a rough approximation is more than sufficient for making the present point. If indeed the key source of environmental problems needs to be seen as 'the economy' as a relatively undifferentiated whole, then it would need to be expected that the production of toxins would need to be at least roughly proportionate across the economy as a whole. So, what proportion of the economy is actually associated with the production of, say, 50 to 60 percent of all the toxic releases in the United States? The answer is not anything close to 60 percent of the economy, but under 6 percent (Freudenburg, 1997a). While it is certainly possible to argue that, at least in certain respects, it is the economic system as a whole that is responsible for such toxic emissions, it is also necessary to recognize that all such emissions, in the end, come from quite specific locations and activities. In addition, the time appears to have come to recognize the growing body of empirical findings indicating that, in aggregate, economic prosperity is not so much helpful as harmed by environmental pollution (for a relatively recent summary of findings, see Repetto, 1995).

Pollution and power

The second postulate is that, all other factors being equal, the greater the level of political-economic inequality in a society, the greater the magnitude of environmental damage that should be expected. This logic builds directly on the observation by the ecological economist, John Boyce (1994), although it is somewhat broader, encompassing imbalances in not merely economic resources, but in effective political power. In cases where one group of people is creating risks or environmental problems for another group, all other factors being equal, the safest bet is that the risks and problems will be created predominately by the powerful, and that the victims will largely be found among the less powerful – not the other way around. The simple fact is that more powerful social actors are more likely to have the ability to create problems that will harm others, but also more likely to have the power to prevent the creation of problems that will harm their own interests. In social contexts where the distribution of power is relatively even, by extension, it may well be that even the most powerful groups in society will have less ability to win the types of disproportionate private access to environmental resources that will create problems for other groups. By this logic, incidentally, it may well be that the reasons for what are generally taken to be more environmentally benign policies in northern European countries, at one extreme, versus high levels of environmental disruption in many so-called Third-World countries, at the other, may have to do not just with the overall level of poverty, but also with higher levels of inequality and with the influence of a *patron* or *comprador* class (cf. Frank, 1967) in many of the developing nations of the South – a point that has recently been argued and illustrated particularly effectively with respect to the developmental history of Pakistan by Niazi (1997).

This is not the place for an extensive discussion of the key factors contributing to what I am here calling 'effective political power,' but there is a need for a brief detour at this point, to note three key respects in which this phenomenon is socially structured, reflecting different considerations than those that might be expected on the basis of extrapolation from individual-level characteristics.

The first consideration is that effective political power involves not so much power *over* as power *to*. It can be largely independent, in other words, of the classic or Weberian conceptualization of power – the ability to get others to behave in a certain manner, even against their will. Instead, it is a more subtle form of power, involving merely the ability 'to' obtain whatever privilege one desires. To a significant degree, this more subtle form of power is similar to what Stone (1980) has termed 'systemic power,' but the conceptualization also draws on other analyses, such as those by Lukes (1974), Bachrach and Baratz (1970), Crenson (1971), and Gaventa (1980). The key point about naturalized legitimation or embedded power is that, if one person or social group is able to obtain privileged access to valued resources without having other persons or groups challenge that privilege – or perhaps even notice it – so much the better.

The second consideration is that the most effective forms of political power tend to be those that are organized. Contrary to what might be expected about electoral power, in other words – namely, that there can be strength in numbers – this second consideration suggests that *concentrated* interests may actually have important strategic advantages over those interests that are relatively diffuse. As pointed out several decades ago by Mancur Olson (1965), the larger a social group becomes, the more difficult it can be to organize, given problems of 'free riders' and the increasing costs of making and implementing decisions when a larger number of actors need to be coordinated. As Olson noted, for example, if there are 10,000 people who each would derive a $2 advantage from some policy, but two actors who would each obtain a $1,000 advantage from the opposite policy, it would be easy to imagine how each of the potential beneficiaries of the second policy might be willing to donate, say, $500 each – legally or illegally – to obtain the benefit, but easier to imagine the difficulties of getting even a tenth of the 10,000 persons to donate $1 apiece.

The third consideration is the one that brings us back to the need to examine the social construction of nonproblematicity. Based in part on the logic spelled out in the initial sections of this paper, the privileged access should be expected to be made possible, in part, by *privileged accounts* – that is, by essentially ideological beliefs that confer differential advantage on one group, often at the expense of other groups, but that in many cases have come to be taken for granted or 'naturalized.' The important ideologies, in other words, include not just the forms of discourse that are socially contested and/or otherwise recognized as being ideological, but also the forms of legitimation that appear to be just the opposite – those that have come to be so widely accepted or unchallenged as to have become 'naturalized' or taken for granted, achieving a form of ideological legitimation that is similar to what Foucault (1977) has described for what he calls 'embedded power.'

Still, even in cases where the privileged access is challenged, powerful actors will often be able to defend their privileges – often by means of techniques that are effective, easy to describe, and frequently used, but that nevertheless have received remarkably little attention from sociologists to date. A particularly important example involves the discursive tactic that Bob Gramling and I have termed 'diversionary reframing.' This tactic is a special form of 'changing the subject' – one that involves *diverting* attention away from a question that powerful actors find uncomfortable (e.g., do a small number of corporations truly 'deserve' to be able to make large profits through practices that largely amount to using up or destroying finite or irreplaceable natural resources?) by *reframing* the debate around other questions – most often, questions that challenge the legitimacy of the privileged actors' critics (e.g., by asking whether the opposition is really 'just' a matter of privileged environmentalists who don't care if they threaten the jobs of working-class people, or for that matter, 'just' a matter of ignorance, irrationality, insufficient patriotism, or the desire to send the entire industrialized world back to the days of caves and candles).

All in all, 'the environment' offers rich opportunities for the constructionist project. As noted above, while it is philosophically problematic to argue that individual, short-lived human beings 'own' a thousand-year-old tree or million-year-old mineral deposit, for example – let alone to 'own' the portions of the planet Earth where those resources happen to be found – such considerations are anything but problematic politically in most industrialized countries. In certain countries, in fact, the situation is very nearly the opposite; in the United States, in particular, the reversal has become so complete that governmental efforts to protect the common good – through measures that range from wide-scale pollution prevention through local-scale land-use planning – have become officially defined as 'takings' of environmental privileges that have become socially defined as 'belonging' to specific individuals.

It is not just in the arena of remarkable legal doctrines, however, where the naturalization of 'rights' to nature would appear to call out for greater attention from social constructionists. Instead, other examples can be drawn from the kinds of language that are normally seen as altogether unremarkable. Given the importance of fossil fuels in the debates over global and environmental change, these materials provide illustrations that are as good as any. What we call the 'production' of oil, for example, is actually nothing of the sort. At least according to geologists, the oil was in fact 'produced' many eons ago, before the species we now know as humans even existed. What we *call* oil 'production' actually involves little more than taking an essentially non-renewable resource out of the ground and burning it up. Yet the ironies could scarcely be said to stop there. Even though the oil was produced almost entirely through natural processes, it is commonly defined as private property; if that oil is subjected to industrial processes, creating fertilizers, pesticides, and a broad range of other petrochemical products, those products are defined as private property, as well. Yet there is an important category of exceptions, involving many of the products that are *not* socially defined as useful. A certain fraction of those other 'products' are potentially dangerous, and at least some of those products will go straight out the smokestack. In contrast to the oil itself, these emissions truly are the 'products' of human activity, insight, and oversight, yet at the millisecond they leave the top of the smokestack – being more fully the result of human 'production' processes than they will be during any other moment in the life cycles of any of the constituent chemicals – they stop being defined as private property, becoming defined instead as part of the common legacy of humanity.

Discussion: ironies and possibilities

The situation, in short, is rich in irony. Some of those ironies have become embedded into our language, as in the case of the examples just noted. Others lurk within our analyses and logic, having to do both with what we see and with what we fail to see. Among the overall effects, however, notable is the fact that so many sociologists whose personal beliefs include strong antipathies toward both environmental problems and societal inequities have nevertheless managed collectively to produce analyses that – in spite or in some cases precisely because of those sociologists' efforts to keep their values out of their analyses – at least hold the potential to contribute both to the worsening of environmental problems and to further the concentrations of privilege and inequality in society, if only through the failure to call the inequities into question. This, however, is a strongly and indeed provocatively stated interpretation – and importantly, even if it does prove to contain more truth than most of us wish to see, what might be true today need not be true tomorrow. The obvious antidote, in other words, is not to argue against the pursuit of social constructionist analyses, but to urge that more of the highly developed analytical skills of sociological constructionists need to be applied to the analysis of environmental *privileges*.

It is within this realm of naturalized yet advantage-conferring belief systems, in other words, where the logic of this paper suggests that systematic analysis of social construction patterns will be both most challenging and most sociologically fruitful. All other factors being equal, after all, it would make sense to expect greater sophistication, closer connections to the underpinnings of power – and ultimately, the potential for greater insights into the dynamics and the consequences of the differential distribution of power in society – not among those who challenge but among those who construct and help to maintain what has often been called 'the dominant social paradigm' (see for example Dunlap and Catton, 1976; Cotgrove, 1982; Dunlap and Van Liere, 1984; Milbrath, 1984).

The process of carrying out more systematic analyses is likely to prove challenging, but important precedents and possibilities exist. Gaventa's classic study of power and powerlessness in Appalachia (1980), for example, provides a number of important methodological considerations for studying the non-emergence of protest. Crenson's still-earlier study of *The Un-Politics of Air Pollution* (1971) showed how it is possible in at least a qualitative yet rigorous way to document a lack of activity/involvement on the part of local officials; also in that earlier time, Molotch's now-classic study of the then-major oil spill at Santa Barbara (1970) offered another, extremely helpful hint, involving the 'accident' – an approach that deserves a bit of additional elaboration.

As I have argued elsewhere (see especially Freudenburg, 1985; Freudenburg and Gramling, 1997; see also Freudenburg, 1997b), while it may make sense to expect that powerful persons and institutions will have a disproportionate ability to shape societal discourse – an ability that not just correlates with but also contributes to the ability of those same actors to enjoy disproportionate advantages in other realms – what is clearly not reasonable is to expect such advantages to be absolute. Instead, it may be particularly among the most powerful actors in society that we should expect to see the importance of *autogenic succession* – a phenomenon that can also be understood as involving a self-induced demise.

The more powerful an actor becomes, in other words – all other factors being equal – the less eagerness we should expect other actors to have toward challenging that actor. Instead, as pointed out by Stone (1980) and Crenson (1971), among others, when an actor comes to be seen as being highly powerful and influential, others may act in anticipation of the powerful actor's wishes, often without the powerful actor being required to initiate the behavioral change, or indeed, even to be aware of it.

Yet while sociologists tend to be well-aware of the potential for such an accumulation of advantages, we should also be aware of the potential for an initial advantage to carry within it the seeds of its own demise. In particular, in at least certain situations where a powerful actor has been able to achieve high levels of success in manipulating the discourse of legitimation, there may be a growing probability that the actor's own behaviors will undermine the position of advantage. In particular, such an actor may tend to become complacent or overconfident – so self-assured about the ability to manipulate ideological systems (or, like the con artist, eventually so convinced by the very arguments being constructed) as to put forth arguments that are ultimately judged to be ludicrous, often in part because they will be 'disconfirmed' by actual experience.

Such experiences bring us back to the classic Molotch definition of an 'accident' (1970): a case where 'a miscalculation leads to a breakdown of the customary order' – or, we might say today, a breakdown of the socially constructed order. In the case of the Santa Barbara oil spill, the oil companies and their allies in federal agencies put forth a strongly and consistently stated belief that, despite the ominous warnings by environmentalists, the risk was perfectly acceptable; oil drilling in the Santa Barbara channel could be carried out in a safe and environmentally responsible manner. The argument proved sufficient for the Federal government to provide permits to the oil companies and for drilling to begin, thus demonstrating something about the power of social construction processes. On the other hand, the same argument did not prove sufficient to keep an oil spill from occurring and fouling many miles of the coast – thus also demonstrating something about the *limits* of social construction processes. As Molotch pointed out, moreover, once the spill led to much closer scrutiny of the leasing arrangement – and induced many previously unconcerned citizens to ask how oil drilling could have been permitted to take place in such an environmentally unsuitable location – 'what leaked out was not just oil but ... a little bit of truth about power' (Molotch, 1970).

The construction of what I call 'privileged accounts' – that is, socially constructed beliefs that help in the appropriation or legitimation of privileged access to natural resources – may well prove to be subject to some of the same potential weaknesses. All other factors being equal, I would expect overt or flagrant accidents – those as visible and as likely to inspire intersubjective agreement as the Santa Barbara oil spill – to be significantly less common than those involving ambiguous (and hence fortuitously 'contestable') evidence. I would also expect that the relatively small groups of social actors enjoying disproportionate or privileged access to natural resources are likely to enjoy the status of being what Galanter (1974) calls 'repeat players,' having been able to obtain the benefits of such privileged access on a relatively consistent basis – a fact that, in its own right, will often confer an increased legitimation for that access. Importantly, however, I would further expect that, partly as a result of their repeat-player status, such actors should have been able to develop considerable skill in establishing 'ground rules' (whether formal or informal), which are likely to have the net effect of permitting them to continue enjoying privileged access to or use of natural resources, even in cases where (and tellingly, *so long as*) the issue remains socially contested.

This last point also deserves more attention than it seems to have received from the sociological community to date. Many of the most pressing risk and environmental issues of our day are to a large degree what Weinberg long ago (1972) called 'trans-scientific' questions. The questions can clearly be asked in what seem to be scientific terms – e.g., what will be the long-term consequences of a 12 percent increase in carbon dioxide emissions over a period of less than half a century, combined with a similar increase in methane emissions in the past twenty years alone? – and yet there is no real way for the questions to be answered in advance of the 'real-world experiment' currently taking place. Particularly in the United States, the social actors who are most heavily implicated in the production of greenhouse gases have been quite effective in arguing against 'premature' government regulations – which, in their view, would impose 'unnecessary' costs on industry, 'before' the scientific questions have been 'adequately' addressed. In such a set of circumstances – and in many others – what can be characterized simply as a prudent call for 'more research' can also be seen as a relatively cynical effort by a privileged few to neutralize an issue that might otherwise call their privilege into question.

For this and many other such questions, accordingly, perhaps the sociologically reasonable approach is to hold four expectations, the first three of which are relatively straightforward. First, even scientific-sounding questions can only be answered in three main ways – essentially, yes, no, and maybe. Second, in any issue that remains contentious, neither a 'yes' nor a 'no' answer is unambiguously appropriate, meaning that the politically effective answer to the scientific question is likely to be a rough equivalent of 'maybe.' Third, a relatively small number of relatively privileged social actors will be likely to have taken steps in advance to assure that the pragmatic consequence of a 'maybe' answer will be that they continue to enjoy privileged private access to earth's collective resources. Fourth, and finally, when or if these special privileges are challenged by 'critics,' those critics are likely to be the targets of diversionary reframing, renaming and blaming – of having their legitimacy challenged by powerful actors who will call into question the genuineness of their concerns, their patriotism, their rationality, their concern for the poor, or almost any other grounds of political-economic 'standing' that happen to be available.

This, however, is clearly just an initial list of the types of expectations that can and should be developed – and the need for that further development is precisely the broader point. The situation, at present, is rich in irony, but it is also rich in possibility. To sum up by stating the matter as bluntly as possible, most of our systematic efforts of examining social construction of environmental issues, to date, have been concentrated in precisely those areas of discourse where the social construction processes have been less sophisticated, less important socially, and less informative about the structure and consequences of differential power and privilege. As a kind of bonus of irony, moreover, to the extent to which our analyses have exerted tangible effects on the practical political discourse of the day, those effects may well have been in the direction of further legitimating the prevailing inequalities in the distributions of power, of access to environmental 'rights' and resources, and of avoidance of responsibility for the creation of environmental destruction.

For those who believe in the potential advantages of reflexivity, however, there is no need to view the existing situation – whatever its degree of irony – as being inevitable or inescapable. The mutual contingencies of society and environment are at least equally rich in the possibilities they offer for insights into the nature and workings of power and privilege, and into the dynamics of discourse. Based on the experiences and skills that have been developed to date, moreover, the sociological community may now be better-prepared than ever to take on that challenge, and to look into the possibilities. The task, fortunately, is more than large enough for all of us, and thus I hope it is one that will ultimately receive focused attention from at least a great many of us.

Bibliography

Bachrach, P. and M.S. Baratz (1970), *Power and Poverty*. New York: Oxford Univ. Press.

Beck, U. (1992), *Risk Society: Toward a New Modernity*. Newburry Park, CA: Sage.

Beck, U. (1995), *Ecological Politics in an Age of Risk*. Cambridge: Policy Press.

Boyce, J.K. (1994), Inequality as a Cause of Environmental Degradation. *Ecological Economics* 11 (Dec.): 169-178.

Burningham, J. and K. Cooper (1998), Misconstructing Constructivism: A Defence of Social Constructionist Approaches to Environmental Problems, in: G. Gijswijt et al. *Social Theory and the Environment*, proceedings of the conference on Sociological Theory and the Environment, Amsterdam: SISWO.

Buttel, F.H. (1986), Sociology and the Environment: The Winding Road toward Human Ecology. *International Social Science Journal* 38: 337-356.

Buttel, F.H., A.P. Hawkins and A.G. Power (1990), From Limits to Growth to Global Change. *Global Environmental Change* 1: 57-66.

Buttel, F.H. and P.J. Taylor (1992), Environmental Sociology and Global Environmental Change: A Critical Assessment. *Society and Natural Resources* 5(3): 211-230.

Catton, W.R., Jr. (1980), *Overshoot: The Ecological Basis of Revolutionary Change*. Urbana: Univ. of Illinois Press.

Cotgrove, S. (1982), *Catastrophe or Cornucopia: The Environment, Politics and the Future*. New York: John Wiley.

Couch, S.R. and J.S. Kroll-Smith (eds) (1991), *Communities at Risk: Collective Responses to Technological Hazards*. New York: Peter Lang.

Crenson, M.A. (1971), *The Un-Politics of Air Pollution: A Study of Non-Decisionmaking in the Cities*. Baltimore: Johns Hopkins Univ. Press.

Draper, E. (1991), *Risky Business: Genetic Testing and Exclusionary Practices in the Hazardous Workplace*. New York: Cambridge.

Drottz-Sjöberg, B.-M. (1991), Risk: Conceptions, Reactions and Communication. *European Management Journal* 9: 88-97.

Dunlap, R.E. (1980) Paradigmatic Change in Social Science: From Human Exemptionalism to an Ecological Paradigm. *American Behavioral Scientist* (Sept/Oct): 5-14.

Dunlap, R.E. and W.R. Catton, Jr. (1976), Environmental Sociology: Why not Human Ecology? Presented at Annual Meeting of American Sociological Association, New York, August.

Dunlap, R.E. and K.D. van Liere (1984), Commitment to the Dominant Social Paradigm and Concern for Environmental Quality. *Social Science Quarterly* 65: 1013-1028.

Durkheim, E. (1951), *Suicide: A Study in Sociology*. John A. Spaulding and George Simpson, Trans. New York: Free Press.

Erikson, K.T. (1990), Toxic Reckoning: Business Faces a New Kind of Fear. *Harvard Business Review* 68 (1, Jan.-Feb.): 119-126.

Erikson, K.T. (1994), *A New Species of Trouble: Explorations in Disaster, Trauma, and Community*. New York: Norton.

Foucault, M. (1977), *Power/Knowledge: Selected Interviews and Other Writings, 1972-77*. New York: Pantheon.

Fox, N. (1991), Green Sociology. *Network* (Newsletter of British Sociological Association) 50 (May): 23-24.

Frank, A.G. (1967), Sociology of Development and Underdevelopment of Sociology. *Catalyst*: 20-73.

Freudenburg, W.R. (1985), Succession and Success: A New Look at an Old Concept. *Sociological Spectrum* 5: 269-289.

Freudenburg, W.R. (1988), Perceived Risk, Real Risk: Social Science and the Art of Probabilistic Risk Assessment. *Science* 242 (October 7): 44-49.

Freudenburg, W.R. (1993), Risk and Recreancy: Weber, the Division of Labor, and the Rationality of Risk Perceptions. *Social Forces* 71 (4, June): 909-932.

Freudenburg, W.R. (1996), Risky Thinking: Re-examining Common Beliefs about Risk, Technology, and Society. *Annals of the American Academy of Political and Social Sciences* 545 (May): 44-53.

Freudenburg, W.R. (1997a), The Double Diversion: Toward a Socially Structured Theory of Resources and Discourses. Paper prepared for annual meeting of American Sociological Association, Toronto, Ontario, August.

Freudenburg, W.R. (1997b), The Crude and the Refined: Sociology, Obscurity, Language, and Oil. *Sociological Spectrum* 17 (1): 1-28.

Freudenburg, W.R., S. Frickel and R. Gramling (1995), Beyond the Nature/Society Divide: Learning to Think about a Mountain. *Sociological Forum* 10 (3): 361-392.

Freudenburg, W.R. and R. Gramling (1994), *Oil in Troubled Waters: Perceptions, Politics, and the Battle over Offshore Oil.* Albany: State University of New York (SUNY) Press.

Freudenburg, W.R. and R. Gramling (1997), How Crude: Advocacy Coalitions, Offshore Oil, and the Self-Negating Belief. Paper presented at annual meeting of American Association for the Advancement of Science, Seattle, February.

Freudenburg, W.R. and S.K. Pastor (1992), Public Responses to Technological Risks: Toward a Sociological Perspective. *Sociological Quarterly* 33 (3, August): 389-412.

Freudenburg, W.R. and T.I.K. Youn (eds) (1993), *Research in Social Problems and Public Policy: A New Perspective on Problems and Policy.* Greenwich, CT: JAI.

Galanter, M. (1974) Why the 'Haves' Come Out Ahead: Speculations on the Limits of Legal Change. *Law and Society Review* 9: 95-160.

Gaventa, J. (1980), *Power and Powerlessness: Quiescence and Rebellion in an Appalachian Valley.* Urbana: Univ. of Illinois Press.

Giddens, A. (1990), *The Consequences of Modernity.* Cambridge: Polity Press.

Gramling, R. and W.R. Freudenburg (1996), Environmental Sociology: Toward a Paradigm for the 21st Century. *Sociological Spectrum* 16 (4, Oct.): 347-370.

Greider, T. and L. Garkovich (1995), The Social Construction of Risk: The Case of Environmental Change. Paper presented to annual meeting of Rural Sociological Society, Washington D.C., August.

Hannigan, J.A. (1995), *Environmental Sociology: A Social Constructionist Perspective.* New York: Routledge.

Kasperson, R.E., O. Renn, P. Slovic et al. (1988), The Social Amplification of Risk: A Conceptual Framework. *Risk Analysis* 8 (2, June): 177-187.

Krimsky, S. and D. Golding (1992), *Social Theories of Risk.* Westport, CT: Greenwood.

Krogman, N. (1996), Frame Disputes in Environmental Controversies: The Case of Wetland Regulation in Louisiana. *Sociological Spectrum* 16: 371-400.

Kroll-Smith, J.S. and S.R. Couch (1990), *The Real Disaster is Above Ground: A Mine Fire and Social Conflict.* Lexington: Univ. Press of Kentucky.

Lukes, S. (1974), *Power: A Radical View.* New York: Macmillan.

Martin, B. (1989), The Sociology of the Fluoridation Controversy: A Re-examination. *The Sociological Quarterly* 30 (1): 59-76.

Mazur, A. and J. Lee (1993), Sounding the Global Alarm: Environmental Issues in the U.S. National News. *Social Studies of Science* 23: 681-720.

Metzner, A. (1997), Constructivism & Realism (Re)Considered, in: G. Gijswijt et al., *Social Theory and the Environment*, proceedings of the conference on Sociological Theory and the Environment, Amsterdam: SISWO.

Milbrath, L.W. (1984), *Environmentalists: Vanguard for a New Society.* Albany: SUNY Press.

Miller, G.T., Jr. (1993), *Environmental Science: Sustaining the Earth* (fourth edition). Belmont, CA: Wadsworth.

Molotch, H. (1970), Oil in Santa Barbara and Power in America. *Sociological Inquiry* 40 (Winter): 131-144.

Netzer, T. (1997), *Ecological Bases of Social Violence in Pakistan.* Madison, WI: Unpublished Ph.D. Dissertation, University of Wisconsin.

Newby, H. (1991), One World, Two Cultures: Sociology and the Environment. *Network* (Newsletter of the British Sociological Association) 50 (May): 1-8.

Niazi, M.T. (1997), *Ecological Bases of Social Violence in Pakistan.* Madison, WI: Unpublished PhD Dissertation, University of Wisconsin.

O'Connor, J.R. (1988), Capitalism, Nature, Socialism: A Theoretical Introduction. *Capitalism, Nature, Socialism* 1 (1, Fall): 11-38.

Olson, M. (1965), *The Logic of Collective Action.* Cambridge, MA: Harvard Univ. Press.

Renn, O. (1992), Concepts of Risk: A Classification, in: Sheldon Krimsky and Dominc Golding (eds), *Social Theories of Risk.* Westport, CT: Praeger, pp. 53-79.

Repetto, R. (1995) *Jobs, Competitiveness, and Environmental Regulation: What are the Real Issues?* Washington, D.C.: World Resources Institute.

Rosa, E.A., O. Renn, C. Jaeger and T. Webler (forthcoming) Risk as a Challenge to Cross-Cultural Dialogue, in: *Selected Papers of the XXXII Congress* of the International Institute of Sociology. Trieste, Italy: International Institute of Sociology.

Schnaiberg, A. (1980), *The Environment: From Surplus to Scarcity.* New York: Oxford Univ. Press.

Schnaiberg, A. and K.A. Gould (1994), *Environment and Society: The Enduring Conflict.* New York: St. Martin's.

Short, J.F. (1984), The Social Fabric at Risk: Toward the Social Transformation of Risk Analysis. *American Sociological Review* 49: 711-725.

Short, J.F. (1992), Defining, Explaining and Managing Risk, in: J.F. Short and L. Clarke (eds), *Organizations, Uncertainties and Risk.* Boulder: Westview, pp. 3-23.

Slovic, P. (1993), Perceived Risk, Trust, and Democracy. *Risk Analysis* 13 (6, Feb.): 675-682.

Stone, C. N. (1980), Systemic Power in Community Decision Making: A Restatement of Stratification Theory. *American Political Science Review* 74: 978-990.

Szasz, A. (1994), *EcoPopulism: Toxic Waste and the Movement for Environmental Justice.* Minneapolis: Univ. Minnesota Press.

Ungar, S. (1992), The Rise and (Relative) Decline of Global Warming as a Social Problem. *Sociological Quarterly* 33: 483-501.

Weinberg, A. (1972), Science and Trans-Science. *Minerva* 10 (2, April): 209-222.

Wynne, B. (1982), *Rationality and Ritual: The Windscale Inquiry and Nuclear Decisions in Britain.* Chalfont St. Giles: British Society for the History of Science.

Wynne, B. (1992), Risk and Social Learning: Reification to Engagement, in: Sheldon Krimsky and Dominic Golding (eds), *Social Theories of Risk.* Westport, CT: Greenwood, pp. 275-297.

Yearley, S. (1997), The Social Construction of Environmental Problems: A Theoretical Review and Some Not-Very-Herculean Labours, in: G. Gijswijt et al., *Social Theory and the Environment*, proceedings of the conference on Sociological Theory and the Environment, Amsterdam: SISWO.

6

Globalization and Environment: between apocalypse-blindness and ecological modernization

Arthur P.J. Mol

Introduction

The notion of globalization has become quite popular in a short time, both in the daily vocabulary of newspapers, business representatives, state officials and non-governmental organizations, and in the social sciences. Before the early 1980s the concept of globalization could hardly be found, neither in academic studies nor in popular newspapers and magazines. Within one decade (cf. Robertson, 1992), however, this notion has reached a firm position in the social sciences and is at the moment one of the leading concepts to analyze and indicate the changing character of the modern world. The popularity of the notion of globalization should, however, not be interpreted as only the latest fashion. Its emergence on the agendas and vocabulary of academics, politicians and commentators forms also an expression of the inadequacy of 'older' concepts such as internationalism, transnationalism or multinationalism as The Group of Lisbon (1994) has adequately argued. This popularity and widespread use of the notion of globalization does not mean, however, that the ideas and definitions of processes and consequences of globalization converge, as we will point out below.

In analyzing globalization processes reference is often made to the environment. Traditionally, environmental deterioration is seen as one of the motors of global awareness, be it primarily in the ecological-material sense (spaceship Earth, Gaia). More recently environmental problems are interpreted as the negative off-spin of tendencies of globalization in basically the economic sphere of production and consumption. And environmental deterioration – especially those problems related to the global commons such as the Greenhouse effect, the loss of biodiversity, the destruction of the rain forests and the pollution of the oceans – is increasingly interpreted as one of the issues around which global action crystallizes.

A more systematic review of the sociological studies that relate globalization processes to environmental deterioration, consciousness and management has to draw upon at least two bodies of sociological literature[1]: globalization studies and environmental sociology. From the references to the environment in the sociological literature on globalization two conclusions can be drawn. First, although references to the environment are often made in globalization literature, they are usually restricted to the notion that the environment is indeed a relevant category when studying globalization processes (e.g. Archer, 1991; Mlinar, 1992; Barber, 1995; Held, 1995; Dalby, 1996; McMichael, 1996). In most cases a more detailed and profound analysis of the relation between processes of globalization in the world economy, in international politics and in global culture on the one hand, and environmental deterioration, environmental consciousness and environmental management on the other, is lacking. Either this linkage is taken for granted, or it is seen as relatively unimportant. And, secondly, in those few cases in which globalization studies deal more extensively with the environment, they emphasize globalization tendencies as a new cause for environmental deterioration rather than stressing the potential benefits of globalization processes for international environmental management (cf. Scholte, 1996). There are a few, though influential, exceptions to these general observations: most notably the work of Giddens (1990, 1991) and – although he also tends to be rather apocalyptic on the environmental consequences of globalization – Ulrich Beck (1986, 1991, 1994a). In focusing on the changing character of modernity they both analyze the interplay between globalization tendencies and environmental challenges as (one of) the main motor(s) behind the transformation of high modernity into late or reflexive modernity. '(..) (I)t has become apparent that questions of ecology are markers of host of other problems that face us' (Giddens, 1994: 189).

In reviewing the literature in environmental sociology (and to a lesser extent environmental sciences in general), the record is slightly better, but still disappointing. The general conclusion should be twofold. First, sociological theories on globalization are insufficiently used in investigating the causes behind global environmental deterioration and change, as well as in focusing on new ways of globalized environmental management. Most contributions focus on empirical phenomena – rather than developing theoretical understanding – such as the environmental consequences of the global economy, often embodied in multinational corporations and institutions such as the World Trade Organization (WTO), the emergence of environmentalism as a global culture, 'materialized' in international NGOs and international opinion polls, and international environmental law and decision-making in organizations such as those of the UN, the EU and NAFTA. Redclift and Benton (1994), Yearley (1996) and Redclift (1996) are more theoretical inspired volumes that form to some extent the exceptions to this rule.[2] And the second conclusion should be that in most of the contributions in environmental sociology and environmental sciences to globalization the emphasis is on the destructive consequences of globalization processes on the natural environment (e.g. Anton, 1995).

Against this background the goal of this paper is twofold. First, we want to connect globalization and environment in a theoretically more profound way. Second, a more balanced evaluation will be made of globalization tendencies for environmental deterioration and reform. In order to do so we start with a selective reading and interpretation of different branches of globalization theories. This rather lengthy part is essential, first to delineate the essence of globalization both in theoretical concepts (especially in relation to the changing character of modernity; section on page 123) and in substantial social transformations and continuities in the distinct domains of economy, politics and culture (section on page 128).

And, secondly, our environment-oriented investigation of globalization can thus be founded in and profit from the insights and conceptualizations developed so far in globalization theories. After this selective overview of globalization theories that only scarcely refers to the environment, we switch to the more substantial questions on the relation between globalization processes and environmental deterioration, consciousness and reform (sections on page 131 and 135). This part is divided in two sections, the first concentrating on the environmental side-effects of globalization, and the latter focusing on the synchronization of globalization and environmental reform. This division is not so much to force an overall assessment of the positive and negative environmental impacts of globalization, but – in contrast – rather to emphasize the dualistic relationship between environment and globalization processes in various forms, to outline the mechanisms of their synchronic and a-synchronic interrelations and to stress the sheer impossibility to give overall assessments. The impacts of globalization processes on the environment are historically shaped and contextual, and will be determined by the outcome of social and political struggles. In that sense, to put it quite bluntly, we will conclude that in itself globalization is neither good nor bad for the environment.

Globalization and the changing character of modernity

Although globalization has moved high upon the agenda of sociological inquiries on the changing character of the modern world, there still exists some debate on three – partly interdependent – points: (i) whether globalization is a tendency that can be identified in the real world, (ii) whether this process of globalization is something new or merely a continuation of a tendency that already started some 400 – or even more – years ago, and (iii) what the core features of globalization are.

On the first point Scholte (1996) provides some clarification. In his analysis of the globalization discourse Scholte identifies three main schools of thought: conservatives who deny that such a trend as globalization exists, liberals (both neo-liberalists and reformists) who celebrate the presumed fruits of globalization and critics (including both historical-materialists and postmodernists and post-structuralists) who decry its alleged disempowering effects. The first group can basically be found in the realists international relations tradition, which insists on the central role of the state and the continuation of sovereignty in social developments around the globe. In stressing that the contemporary debate on globalization basically runs between reformists and critics, Scholte (1996: 54-55) emphasizes the diminishing meaning and importance of the conservative standpoint on globalization. This is in line with Held's (1995: 25) conclusion that 'there is not much evidence to suggest that realism and neo-realism possess a convincing account of the enmeshment of states with the wider global order, of the effects of the global order of states, and of the political implications of all this for the modern democratic state'.

The discussion of whether globalization is a new phenomenon or just a continuation and extension of processes of interdependence and interconnectedness that started with the emergence of capitalism some 400 years ago, has to some extent replaced the discussion on the existence of globalization. While most scholars now do acknowledge that global interconnectedness and interdependence do exist, and that it is fruitful to analyze international economic, political and societal processes in such a way, they disagree on the continuity or discontinuity in these processes. One of the strongest defenders of continuity is perhaps Wallerstein, stressing the continuation of the internationalization of the world economy, parallel to the development of capitalism. Within the political arena it is for instance Gourevitch (1978) who emphasizes the constancy of local and international interdependence and interpenetration from the sixteenth century onwards. Starting from a predominantly cultural analysis of globalization Robertson is in line with the above mentioned authors in stressing the fifteenth century as the starting point of globalization and the 1880s as the acceleration of globalization.[3] In departing from these continuist analyses, we are not denying elements of continuity and acknowledge that most social developments fall short of sudden, radical changes. But at the same time we emphasize that the form and dynamics of interconnectedness and interdependence have changed fundamentally in the last 30 to 40 years. It was especially the new communication technology that made the acceleration in the compression of time and space possible, thereby altering the scope and speed of economic decision-making, enhancing the capacity of the economic system to respond rapidly to fluctuations, but also rendering it more vulnerable through a tendency to enlarge relatively minor disturbances into major crises. Moreover, the new communication technologies not only affected the economic system, but also influenced – as Manuel Castells in his 'Network Society' (1996) has shown – the political and cultural dimensions of modernity. This seems also the interpretation of globalization presented by The Group of Lisbon (1994), who wrote their *Limits to Competition* to analyze the changing dynamics of capitalism and its consequences, in an age marked by globalization.

'Old' and 'new' theories on globalization

In his compact but remarkably complete and informative overview on globalization Waters (1995) makes a useful distinction between old and new theories of globalization. This distinction is also useful in defining the core features of globalization.

The 'old' theories of the third quarter of this century are summarized by Waters under four headings: modernization and convergence theories (focusing on Parsonian functionalists and post-industrialists such as Clark Kerr and Daniel Bell), world capitalism theories (especially those around Wallerstein's World-System, with Sklair's (1991) work as one of the more recent examples), international relation theories (especially those that depart from the realist framework and focus on interdependence, Morse (1976) being one of the first) and those centered around McLuhan's notion of Global Village. These quite diverging studies converge in the underlying idea that something fundamental was changing in the modern world at that time. This group of 'old' theories were developed quite separately from each other, without much exchange, interaction and cross-fertilization, and it is – as McGrew (1992) and Yearley (1996) emphasize – the idea of a prime cause behind processes of globalization that distinguishes these 'old' theories from the 'new' theories that stress the multiplicity of causes.[4] In addition, the theories of the third quarter of this century differ from their successors in that they were often insufficiently able (i) to identify a common denominator of the changing character of modernity in all its institutional clusters, and (ii) to analyze this change as a fundamental transformation or discontinuity in the historical development of capitalism, modernity, the nation-state system or whatever institutional cluster or feature of modernity they focused on.

Robertson – Waters rightly states – should be seen as the grandfather of the 'new' globalization studies that had a major breakthrough by the end of the 1980s and the early 1990s. Around the mid 1980s his first papers began to appear which explicitly focused on the notion of globalization as an 'umbrella' category of social processes that were separately dealt with in the 'old' theories. In moving beyond monolithic theories on the international system of states and economic world-system theories, Robertson aims to give a broader account on what is at stake on the global arena, making especially a point by stressing the relative independence of the cultural sphere. His classical definition of globalization points at both the concrete processes of increasing interdependence and interconnectedness of the world and the growing awareness of it: 'Globalization as a concept refers both to the compression of the world and the intensification of consciousness of the world as a whole' (Robertson, 1992: 8). Giddens is often seen as the second main author on globalization theories. While some identify Giddens as the chief representative of the recent upheaval in globalization studies and include Robertson in his school of thought – e.g. McGrew (1992) – this is at least chronologically, but according to some also substantially, not correct.[5] Moreover, they disagree on one of the central notions in globalization theories, being the relation between globalization and modernity (see below). Nevertheless, Giddens has indeed made a major contribution to globalization theories, especially in his work in the early 1990s (1990, 1991) and – most interestingly for our theme – has connected globalization processes to the environment. 'Ecological problems highlight the new and accelerating interdependence of global systems and bring home to everyone the depth of the connections between personal activity and planetary problems' (Giddens, 1991: 221). In doing so Giddens basically extends his work on the transformation of modern society in analyzing 'the intensification of worldwide social relations which link distant localities in such a way that local happenings are shaped by events occurring many miles away and vice versa' (Giddens, 1990: 64).

Globalization and modernity

In the more recent contributions to globalization theories we witness a return of the discourse on modernity, which has its roots both in the debate on Parsonian modernization theory in the 1960s/1970s and in the postmodernity discourse in the early 1980s. We will briefly elaborate on this relation between globalization and modernity as it seems essential for the environmental dimension of globalization.

In analyzing the consequences of globalization processes various contributions have centered on the question of whether globalization processes had universalizing consequences and would thus result in a homogenization (or even Westernization) of the world. Globalization, through economic (e.g. disorganized capitalism; global chains, networks and flows around manufacturing, trade and finances), cultural (e.g. cultural imperialism, homogeneous cosmopolitans), political (diminishing capacity of the nation state, world polity) and other domains, would then result in a world in which – in general terms – distinct societies become more and more alike. And the standard would be the capitalist-(post)industrial society of the West with its typical cultural and political outlooks. Cohen (1996) provides an example of the new convergency thesis in analyzing the growing equality of major cities in North and South, also with respect to their environmental threats and reforms. And the Dutch government tries to contribute to this homogenization by actively exporting their environmental policy and management models throughout the world, as if they would easily fit in every local context. Various authors (e.g. Robertson, 1992; Nederveen Pieterse, 1995) have criticized this interpretation of globalization as a renewed form of the traditional, Parsonian modernization theories. They claimed that such an evolutionary convergence is not only theoretically inadequate (as has been discussed extensively in the debate on Parsons' functionalism), but it usually neglects the dark side of modernity, so colorfully portrayed by authors such as Zygmunt Bauman. And Bauman explicitly includes (global) environmental threats in the dark side of an ambivalent modernity (Bauman, 1993: 186-222). On the other side of the globalization discourse some have identified globalization as inherently leading to tendencies of heterogenization, especially in the economic field of unequal distribution of goods, money and power, but also related to the unequal access to environmental resources and the unequal distribution of 'environmental space'. Increasingly, however, authors converge in the idea that although there exist strong global economic, political and cultural forces that work in a comparable way in all edges of world society, these have diversifying consequences on a local level. Or as McMichael (1996: 40-41) puts it, 'globalization is ultimately an institutional transformation. It has no single face, as institutions and institutional change vary across the world'. Global cultural messages are received and interpreted differently in various localities, global economic producers tailor their products to local markets and preferences, global politics do not have equal consequences for distinct countries and localities, etc. Or, as Giddens (1994: 188) puts it: 'globalization can no longer be understood as Westernization; developing societies and developed societies mix in culture, economy and politics'. Globalization can result in homogenization, heterogenization or hybridization (Nederveen Pieterse, 1995), depending on the specific structuration of the actors and institutions in the relevant social system or domain. It is too simple to either analyze globalization as a uniform process of Westernization or McDonaldization resulting in an increasingly homogeneous single global system, or interpret globalization as a process causing increasingly diverging and heterogenizing effects in different parts of the world. Notwithstanding this general consensus, there are differences in stressing the homogeneity and heterogeneity of globalization, and we will see that these differences also play a role when linking globalization with the environment.

The debate on homogeneity/heterogeneity relates to a controversy on the relationship between globalization and modernity, especially involving Robertson and Giddens although others have also contributed to this controversy (e.g. Nederveen Pieterse, 1995). Robertson (1995) criticizes Giddens (1990) on the fact that the latter sees globalization as a consequence of modernity, while it should rather be identified as a general condition that facilitated modernity. This depends of course very much on the definition and interpretation of globalization and modernity. Robertson, for instance, seems to have a rather broad conception of globalization, a process that started from the fifteenth century onwards, while Giddens restricts globalization to the last four decades of this century. On the other side, Robertson seems to restrict the notion of modernization to a rather Parsonian one[6] and is therefore able to criticize Giddens on both his emphasis on modernity and his analysis of globalization as the consequence of modernity in a specific era. This reading of Giddens' work on globalization seems not entirely correct, as Giddens' notion of modernity is more analytical and departs from the normative/substantive connotations of Parsonian sociology that have been criticized so strongly by, among others, Giddens himself (Giddens, 1984: 263-274). Giddens links the process of globalization to the development of modern society, and he analyzes it through transformations in four institutional clusters: capitalism, industrialism, surveillance through the nation-state, and the military order. Although all four institutional clusters are relevant for understanding the emergence and persistence of environmental threats in relation to globalization, the most direct link is via industrialism and the global division of labor.[7] Central – or even a prerequisite – to globalization processes in these four institutional clusters are the liberation of time and space, which is closely linked with disembedding, the lifting out of social relations from local contexts of interaction and their recombination across time and space. Symbolic tokens (such as money) and expert systems are the two essential disembedding mechanisms. In analyzing the emergence of globalization since the 1960s Giddens then especially points to the contribution of modern communication technology (being part of the institutional cluster of industrialism) in shrinking the globe. It is especially the growing speed of flows of information (and often connected to that money, capital, culture, images, beliefs, ideas etc.)[8] and the global networks of production and exchange that has changed the modern world qualitatively in the last four decades.

And it is especially this interpretation that links globalization processes with the analysis of the changing character of modernity and the emergence of a new phase in modernity, labeled reflexive modernization. As Kumar (1995) has argued convincingly, globalization studies constitute the last – until further notice – effort in a series of sociological contributions to define the changing constitution of modernity, starting from the post-industrial and information society theories of the 1970s, via the post-fordist theories, to the postmodernist and reflexive modernization theories in all its varieties. Various themes raised in the discourse on globalization find their origin in these earlier theoretical contributions, such as the emergence of new communication technology and the dialectics of the global and the local. This becomes especially clear in the connection of globalization with reflexive modernization. Radicalized or reflexive modernity is analyzed by Giddens (1990, 1991, 1994), among others, as a phase marked by globalization and the end of tradition. It refers to the constant examining and reshaping of social practices in the light of new incoming information about those very practices, and thus denotes the end to the idea that social and natural environments would increasingly be subjected to rational ordering. Within the era of reflexive modernity globalizing processes contribute to experiences of anxiety and uncertainty by lay-actors, while at the institutional level reflexivity involves the routine incorporation of new information and knowledge in social conduct and institutional forms, transforming the institutional order. For instance, and especially relevant for environmental sociology, both science and technology and the bureaucratic decision-making institutions at the political and administrative level shift towards another mode of operation as compared to the era of simple or high modernity. Neither science and technology (or expert systems) nor the national political arrangements will remain unchallenged in confrontation with globalization processes. A similar line of arguing is also put forward by authors such as Scott Lash and Ulrich Beck. And most interestingly, it is especially the latter that makes the most direct connection between globalization and the changing character of modernity on the one hand, and ecological risks on the other. The persistent environmental threats and the chronic environmental anxieties by large segments of society can no longer be counteracted by the institutional mechanisms of simple modernity, and globalization processes are among the most pressing causes for this a-synchronization.

Up until this place we have given a brief overview of globalization theories and positioned globalization in the theoretical debate on the changing character of modernity. It confirmed our introductory starting point: environmental issues are only marginally introduced in globalization theories. Before turning our attention fully to the environment, we have to substantiate this up to now rather abstract overview on globalization theories with the form and outlook of globalization processes in the interdependent economic, political and cultural domains. Only then are we able to link globalization with environmental degradation and reform.

Intermezzo: globalization in economy, politics and culture

It should not surprise us that globalization *avant-la-lettre* found its origin in the economic sphere of production and trade. The development of transportation and communication networks, the rapid growth of trade and the huge flow of capital mainly in the form of direct investment in the late nineteenth century contributed strongly to the internationalization of the capitalist economy. Although international trade started of course earlier, as Wallerstein and Braudel have shown, it should not surprise us that at the eve of the twentieth century Marx emphasized so strongly the international character of capitalism. More recently Dicken (1992) has analyzed in detail what he calls the global character of manufacturing, services and trade, with a shifting division of labor so that a simple categorization in core, semi-periphery and periphery is no longer adequate. Tendencies of universalization, standardization, homogenization and the global product are challenged by the need to diversify, individualize and localize the economic products and processes of global capitalism. In contrast to Dicken, Ruigrok and Van Tulder (1995) and Hoogvelt (1997) relativize the global or footloose character of large corporations and the globalization of production and technology and rather emphasize the concentration of foreign direct investment and multinational enterprises in three economic regions (the triad of NAFTA, EU and South-East Asia/Japan). In comparing the different aspects of economic globalization Waters (1995) gives partially a solution to these diverging studies in concluding that material economic relations (labor) have a tendency to localize, while symbolic relationships (financial markets, ideological arenas such as management concepts of flexibility and increasingly trade and investments in especially services) tend to globalize. It is especially the latter forms of capital circulation via modern communication technologies and along routes of increasing distance and with growing velocity, that lies at the origin of the present phase of disorganized capitalism.[9] In the present phase of 'cultural economy' (Waters, 1995) or 'economies of signs in space' (Lash and Urry, 1994), symbolic markets have moved beyond the (declining) capacity of nation-states to manage them, while the economy becomes subordinate to individual tastes and choices. The leading sectors in this transformation are those that produce symbols themselves: mass media, entertainment industries and the service sectors. This, however, does not mean that manufacturing – often more of direct interest from an environmental point of view – is not transformed by similar processes of liberalization, privatization, deregulation and economic restructuring that are related to processes of globalization in the 'symbolic' sectors.

The most direct and strongest chronological line between globalization studies and the discourse on modernity runs via the typical postmodern emphasis on culture. Although several authors have tried to understand the phenomenon of postmodernity through an analysis of the current and changing condition of capitalism (e.g. Harvey, 1989; Lash and Urry, 1994), it should not surprise us that a large number of contributions to globalization theories still have a strong cultural emphasis (cf. Robertson, Featherstone). The principal issue in most contributions is the changing cultural institutions that are both medium and outcome of processes of globalization. The globalization of culture on a transnational or transsocietal level can take various forms (religion, tourism, mass media, universal norms, eating practices, civil society movements) and is in a number of cases closely interrelated with global economic exchanges. Coming from postmodernism, it was especially these cultural conceptualizations of globalization that moved beyond the ideas of homogenization, universalism and Westernization which dominated the industrialism, capitalism and rationalism brands of modernization theories in the 1960s and 1970s. These culture-oriented studies converge in the idea that the subordination of culture to the structural developments of especially the economic-technological sphere is no longer an adequate conceptualization; rather the diversity, variety and local interpretations of global mass media messages and cosmopolitan cultural phenomena are stressed. In a similar way local, traditional and religious protest movements against global – often seen as western – culture contribute to this diversity. These diversities and local counter movements, however, only exist because of a globalizing culture, and the intensity and rapidity of global cultural flows have contributed to the growing sense that the world is a singular place, be it not everywhere identical.

It has been especially the work of international relation theorists – such as the (neo)realist school of thought and the neo-liberal institutionalists – that have paved the ground for political analyses of globalization. While (neo)realists have hardly payed any attention to environmental issues in internationalization and globalization, institutionalists have recently 'discovered' the environment as an important field of international institution building (cf. Young, 1989 and 1994; Haas, 1990; Haas et al., 1993; Keohane and Levy, 1996). Most of these studies focus on the various state and non-state actors involved in international or global environmental problems, analyzing and identifying the basic constraints and opportunities for effective policy negotiations, institution building and policy implementation. With a few exceptions, these studies hardly concentrate on the changing character of international politics following globalization, but basically extend their line of reasoning to the new field of international environmental policy-making. Of some interest, although only marginally concentrating on environmental issues, has been the work on world polity by Thomas et al. (1987; Boli and Thomas, 1997), that combines cultural theories with political perspectives. These institutionalists – be it not in the strict connotation common in international relation theory but rather in the broader sociological understanding – emphasize the construction of a so-called world polity, relatively independent from and irreducible to the world economic system (Meyer, 1987). In itself the world polity is not reducible to traditional political categories such as states, transnational companies or national forces, but include transnational cultural and institutional frames.[10] In criticizing the inadequacies and shortcomings of a purely or mainly economic analysis of (production and exchange) institutions and relations in the world system, Meyer (1987) and Boli and Thomas (1997) emphasize the importance of incorporating global political and cultural frames in understanding major globalization developments: the transnational and global political and societal institutions, the key political and societal actors at the global level, etc.[11] He and his colleagues interpret the development of the world polity as a process of political rationalization in a Weberian sense (apart from economic rationalization), that develops via conflicts and struggles. In their analyses they rather stress the universalizing tendencies of world culture, resulting in collective identities and interests of firms, societal organizations, states and nations, than emphasizing heterogenizing and diverging developments. Although they conclude that at the moment the world polity can at most be characterized as a world proto-state (as a single authority structure is lacking but it has shared cultural categories and principles of authority), it may eventually result in a world state.

While the world polity analysis concentrates on the universalizing tendencies at the transnational level, another challenging line of argument within the polity-oriented discourse on globalization discusses the changing bureaucratic and political decision-making structures under the notion of declining sovereignty. Although most decisions are still taken by national governments and international organizations are merely a reflexive recognition and reinforcement of national sovereignty, 'in a world of regional and global interconnectedness, there are major questions to be put about the coherence, viability and accountability of national decision-making entities themselves' (Held, 1995). According to Held, sovereignty[12] is undermined by international law, the internationalization of decision-making via regimes and organizations,[13] hegemonic powers and international security structures, national identities and globalization of culture and – not in the last instance – economic globalization. It is the latter that contributes to a process of redefinition of the role of nation-states in relation to (national) economic policy and economic actors, among which are multinational enterprises. This has consequences for the position of the nation-state on both the global and the national level. Although some stick to the idea that sovereignty is still unaffected in the system of nation-states (cf. the vanishing realist school-of-thought) and elites and states often do not want to give up their sovereignty and the powers connected to it, it is no longer the subject of much discussion that (i) 'globalization ruptures the territoriality of conventional international relations' (Williams, 1995), that (ii) the state is less able to dictate the internal developments on economic, environmental and other areas, and that (iii) in global politics the growing contribution of non-state actors undermines the central position of the nation-state. The normative standpoints on the possibilities and desirability of (democratic) governance beyond the nation-state level, however, drive different authors apart. While Held, using environmental problems as one of his arguments (1995: 236), makes a strong case for such a cosmopolitan model of democracy and numerous environmental advocates join him in a request for global environmental management (cf. Esty, 1994), others are less convinced of both the possibility and desirability of global, supranational institutions. And one of the causes for this scepticism is to be found in economic globalization processes, that not only subvert the nation-state but also redefines the relation between state and market. At the national level the consequences of the undermining sovereignty or autonomy lie within the boundaries of national policy-making, where the political room for maneuver as well as the intervention mechanisms of the nation-state are seriously affected, contributing to changing national governance models. This is of course especially true with respect to issue areas that have a clear global outlook: both economic and environmental policy-making are confronted with new models of governance and politics.

If we want to relate globalization processes to the environment we should at least look at two dimensions of this relationship: the way in which the changing institutions of modernity (by processes of globalization) affect environmental deterioration and limit adequate environmental management (next section) and the synchronization of globalization and its institutional transformations on the one hand with processes of ecological modernization and environmental reform on the other (section on page 135).

The globalization of side-effects and apocalypse-blindness[14]

'Students of globalization must surely take seriously the possibility that underlying structures of the modern (now globalized) world order – capitalism, the state, industrialism, nationality, rationalism – as well as the orthodox discourses that sustain them, may be in important respects irreparably destructive' (Scholte, 1996: 55). This quote – which also refers to the ecological consequences of globalization – sets the pace for the critical evaluation of globalizing processes and tendencies which dominates the environmental discourse. Most contributions which try to clarify the relationship between processes of globalization and the environment concentrate on increasing environmental deterioration and diminishing capabilities for environmental management and reform.

In dealing with such environmental side-effects of globalization most of the literature focuses on those environmental threats that have recently been clustered under the heading of global environmental change or the somewhat diverging category of high-consequence risks (Giddens, 1990).[15] We are dealing here with problems which (i) had an agenda setting that ran parallel to the emergence of globalization theories (the late 1980s and early 1990s), (ii) have a truly global outlook in the sense that both the causes and the consequences cannot be restricted to the territory of one nation-state, (iii) are beyond the control of the old institutional arrangements of the nation-state, (iv) give evidence of the diminishing enlightenment character of science and technology as their contribution to both the understanding of the causal mechanisms and the reduction of its threats are heavily disputed, (v) are increasingly triggered by processes of economic globalization in which, for instance, the international division of labor increases the flow of material goods, and (vi) show a heterogeneous distribution of risks that deviates from the traditional economic categories of classes and countries. Notwithstanding these interesting relations between global environmental change (or high consequence risks) and processes of globalization, it would distort our theoretical understanding and models if we would limit our analysis of global change and environmental quality to that specific category or definition of environmental problems. The production, persistence and distribution of 'normal' environmental risks such as water pollution, solid waste disposal, local air emissions or diminishing food quality are also mediated by globalization processes, be it on some points in a different way.

Environmental consequences of globalization processes are usually primarily related to the globalization of economic production and consumption (or, in Giddens' terminology, the industrial dimension of the institutional order), but a more complete analysis can not be restricted to this dimension as will become evident. The global character of chains of production and consumption, where economic relations stretch along large distances and material resources and (intermediary) goods flow over the world economy, results in increasing gaps in time and space between the origins of environmental neglect and the actual environmental consequences and deterioration in specific localities. The most obvious relation between globalization and environmental degradation is that economic globalization processes regarding trade, foreign direct investment, economic decision-making, management concepts, financial markets etc. directly result in an overall decreasing environmental quality. This is occasionally true, for instance where the division of labor connected to processes of economic globalization result in increasing material transport (of natural resources, waste, capital goods, intermediary goods and final consumer products), enhancing both the amount of energy used and greenhouse gases emitted, and the risks of major environmental accidents. The same is true for the enhancing movement of persons around the globe, not only resulting in growing emissions and natural resource depletion, but also causing local nuisance and claims on scarce space. It might also be valid for the loss of biodiversity by accelerating economic development in developing countries, in which the bulk of biodiversity is located. In addition, the global spread of machine technology, both related to production and daily consumption, may in a number of cases harm the environment. These kind of mechanisms of global 'diffusion of industrialism has created "one world" in a more negative and threatening sense than just mentioned – a world in which there are actual or potential changes of a harmful sort that affect everyone on the planet' (Giddens, 1990: 76-77). In addition to such analyses of the environmental effects of more and more rapid circulation of economic goods and persons, the environmental consequences of economic globalization should also include (i) the increasingly global but still unequal (or heterogeneous) distribution of environmental effects, and (ii) the decreasing possibilities of management and control of (global) environmental problems by the institutional arrangements of high modernity, most notably the nation-state and science and technology. At the same time (iii) environmental threats, especially the high consequence risks, contribute to (economic) globalization. We will pay attention to these three points, respectively.

Economic globalization means in a number of cases that environmental consequences are spread around the globe in such a way that it becomes increasingly difficult to escape from them. This is of course true for those polluting substances that freely move around the globe in the ecological system (such as greenhouse gasses, ozone depletion gasses, nuclear fall-out, and persistent pesticides and heavy metals in the oceans). But environmental risks are also transplanted around the globe via more economic mechanisms, such as in the case of pesticides and additives in international food chains or toxic coloring agents in plastics and textiles. It is especially the Risk Society Theory that stresses the truly global character of today's environmental risks, and the inherent apocalypse-blindness of the economic and political institutions. Although no one or no class is able to escape from these threats and often environmental risks rebound on the producers of these risks (e.g. in the pesticides' circle of poison, or Beck's notion of the boomerang curve), this does not mean that everyone is victimized in the same way and that thus the distributional consequences are no longer an item. The diversifying (or heterogenizing) effects of global economic mechanisms and processes appear also in the field of the environment. Goldblatt (1996: 63-64) pays attention to the heterogenizing environmental consequences caused by variations in natural conditions and ecosystems. Homogeneous threats have diverging consequences due to different geographic and environmental circumstances, for instance the case of small and flat islands endangered by potential sea level rise. In addition, the possibilities to protect oneself against, or escape from, these environmental threats are still unequally distributed along primarily economic lines. There exists a spatial distribution of environmental degradation, as not all parts of the world contribute equally to environmental problems and not all parts of the world are confronted with equal levels of environmental destruction and risks. This spatial distribution sometimes runs parallel to economic categories, but at other occasions not. Within nation-states the quest for equal distribution of environmental consequences along spatial and social lines is often articulated in the call for environmental justice or environmental equality (Hurley, 1995); on an international or global scale the equal distribution of 'environmental space' (Spangenberg, 1995) forms a key notion in putting the correlation of economic poverty and disproportionate environmental threats on the public and political agenda. It is in exactly this area that dependencia and neo-Marxist theories are celebrating to some extent a revaluation and revival. Industrial production in developing locations fall relatively short of environmental monitoring, measures and control, and products and technologies used in daily life can have larger ecological consequences due to this. A exemplary case and key mechanism in the economic globalization of manufacturing are the export processing zones, on which Sassen (1994: 19) and Miller (1995: 145) report not only favorable tax regimes but also lenient workplace standards and environmental damage without paying the costs. In relation to this it is also often argued (LeQuesne, 1996: 67ff) that with the rapid growth of foreign direct investments, investments in dangerous production processes and products concentrate in developing countries and regions, where monitoring and the enforcement of costly protective measures are low. In going through the literature it should be concluded, however, that there exists hardly any evidence of systematic international relocation of industrial production caused by differences in environmental regimes (Mol, 2000).

Processes of globalization interfere also with environmental quality in that the former subvert the traditional institutional arrangements for environmental management and control that dominated the phase of high modernity (our second point). In that era it was especially the nation-state and science and technology that played an essential role in both identifying and monitoring environmental change and reducing environmental risks. With respect to the nation-state, it is argued that the state falls short on both global and local issues of environmental management. Diminishing sovereignty via various political and economic mechanisms reduces the capabilities of nation-states to intervene in economic activities that deteriorate its territorial environment. And these limited possibilities for management and control are not only related to those global environmental problems that transcend the national level, such as the greenhouse effect, the ozone layer depletion or the loss of biodiversity. The declining autonomy on, for instance, import restrictions, export bans (of hazardous waste) and the introduction of economic measures to stimulate ecological-sound products also acts on the management of 'normal' environmental problems. At the same time, the growing organizational and technical complexity of production and consumption systems, the growing international competition and requests for environmental standardization and harmonization, the rapid rise and wide-ranging character of social transformations, the fast flows of new incoming information on causes, consequences and interfering factors related to environmental problems, and the increasing request for the fine-tuning of environmental protection towards local physical conditions and social demand, frustrate the traditional governance models of the nation-state. As Lash and Urry (1994) extensively argue in analyzing the changing nature-society relationship in the era of globalization, contemporary nation-states are now too small for the big problems such as high consequence risks and too big for the small problems that need local fine-tuning and close collaboration with the local actors directly involved. In looking for alternative political models that may more successfully deal with environmental threats various ideas are elaborated. Some argue for the development of 'subpolitics' (Beck, 1994a), non-institutional and sub-systemic politics besides and often below the traditional national political institutions of parliament, bureaucracy and political parties,[16] while others (Held, 1995) incline on cosmopolitan governance. These ideas converge, among others, in the fact that they originate from existing initiatives but still have a long way to go before they are put widely into practice.[17]

The undermining of science and technology as institutions that may contribute to the management and control of (global) environmental problems, is related to globalization via the debate on reflexive modernization and the risk society. With global environmental change, and the greenhouse effect as its key example, various authors call upon the fact that science is falling short in drawing any lasting conclusions on the consequences, origins and solutions of environmental impacts. Risk and safety calculations prove to be inadequate in the case of global, complex and perhaps even synergistic interactions. In addition, according to among others Beck, science and technology are heavily involved in the immanent self-endangerment of modern society by producing, preserving and even legitimizing environmental risks, while their contribution to managing these threats enhance ecotechnocracy, as for instance is the case with the IPCC global circulation models and their inbuilt forms of politics (Beck, 1996: 6). As a consequence science and technology (or expert systems as Giddens calls them) become the target of anxiety-induced suspicion and ontological insecurity of individuals. In this perspective the original idea that – in the absence of a global government – science, or epistemic communities as its institutional materialization, could become a kind of supra-national authority that might 'force' nation-states into common action on global environmental problems, is more and more abandoned, both on theoretical grounds and following empirical evidence.

As an answer to these detrimental effects of globalization on environmental quality and environmental management several ways out of this condition are suggested, and some of these have already been mentioned above. The most radical solutions build upon the de-industrialization theories of the 1970s, in arguing for shortening the chains of production and consumption, both on the one hand to limit transport and thus energy consumption and the risk of major accidents, and on the other to increase product quality and control and – sometimes but not automatically related – the trust consumers and lay people have in these products (cf. various contributions in Sachs, 1993). This, however, neglects the dominant forces that carry globalization processes, among which are, paradoxically, environmental threats. Ecological threats are not only caused, sustained, accelerated and mismanaged following globalization processes, but at the same time also contribute to processes of globalization in various domains. For instance, Held (1995: 105-107) argues how concern for our common heritage undermines the traditional Westphalian model of international relations. Environmental problems, both global and local ones, have definitely become an important cornerstone of a global culture and consciousness, and have triggered and/or accelerated global political and economic processes. In that line Strassoldo (1992) identifies the ecological world view as one of the central constituting factors in processes of globalization. Here we touch upon our second dimension of the relationship between globalization and the environment.

Globalization and ecological modernization

Lash and Urry (1994: 3) are right when they – in their own terminology – suggest that 'the sorts of economies of signs and space that became pervasive in the wake of organized capitalism do not just lead to increasing meaninglessness, homogenization, abstraction, anomie and the destruction of the subject'. Nor, we would add, only to environmental destruction. Although a considerable number of authors within the field of globalization theories emphasize – and to some extent with valid arguments as we have pointed out in the former section – the disastrous effects of globalization processes on the environment, there exists also another, contrasting side. The institutional transformations of modernity in the age of globalization are not all detrimental to the environment. It is the perspective or theory of ecological modernization that may serve to focus our analysis on those institutional transformations of modernity that contribute to environmental reform, and supersede the idea of an all pervasive juggernaut of globalization whose side-effects are inevitable and beyond any 'control'.

Ecological modernization

At the basis of the idea, perspective or theory of ecological modernization lie two insights. First, on an analytical level the growing relative independence, 'emancipation', autonomy or differentiation of the ecological sphere and rationality from especially the economic sphere and rationality can be articulated. To understand developments and transformations in institutions and practices, and especially in those related to economic processes of production and consumption, it will be increasingly rewarding to analyze and judge them from both an economic and an ecological point of view. Ecological rationality is slowly catching up with the still dominant economic rationality in the practices and institutions related to production and consumption (Mol, 1996). Second, on a more substantial level, this growing independence of the ecological sphere and ecological rationality becomes evident in (reflexive) environment-induced transformations in the core institutions and practices of modernity, using the 'means' of modernity. The institutions of modernity that have for some time been challenged for their – according to some inherent – ecological destructive qualities, such as modern technology, the capitalist market, industrialism and the nation-state, are (i) increasingly playing a significant role in environmental reform, and (ii) transformed (though not beyond recognition) in order to better fulfil this progressive 'green' role. Needless to say that this process of ecological transformation neither 'unfolds' automatically nor is it to be seen as an evolutionary necessity. Rather, it should be interpreted as the result of daily struggles between different interests that are no longer divided along lines of class and economic position.[18]

The development and growing popularity of the perspective of ecological modernization from the mid 1980s onwards, reflect transformations in the environmental discourse, practices and relevant institutions in a number of OECD countries. Environmental concern and environmental reform have moved beyond their initial stage of window-dressing and marginal measures in the periphery of the environmental crisis in industrial societies. They now approach the core institutions that are held responsible for the environmental treats of modern industrial society. Up till now, almost all interpretations of these environment-induced changes are restricted to the level of the nation-state, or at best to a number of closely related industrial societies (e.g. the EU). But if we take the insights of the globalization discourse serious, the nation-state should no longer be the only or even most essential level or unit of analysis. Global flows of information, capital, ideas and persons, worldwide networks and the development of new institutional arrangements at the global level take a central position in globalization discourses. In using the perspective of ecological modernization for analyzing the synchronization of globalization and ecological reform, the essential innovation is then not so much the scale-up of this perspective from a national to a global scale, but rather to apply a perspective originally developed for industrial societies or nation-states for investigating processes of ecological reform connected with globalization. Section on page 123 makes clear that the essence of this difference is not so much related to scale, but more importantly to the changing character of modernity, in which the nation-state and national institutions are no longer principal units of analysis and (ecological) transformation processes follow a different 'logic'.

In general, the fact that ecological concern, motives and interests are to some extent a driving force behind processes of globalization (in economic, cultural and political domains) indicates that the concern for and the preservation of the environment will in a number of cases reinforce tendencies of globalization (and here we continue the argument made in the last point of section on page 131). In a similar way globalization processes strengthen sometimes environmental reform, as they trigger harmonization of national environmental practices, regimes and standards, construct new institutional arrangements at a supra-national level, transfer environmental technologies, management concepts and organizational models, and accelerate the exchange of environmental information around the globe. In this section we are preoccupied with these kinds of synchronization or mutual reinforcement between globalization and environmental reform by looking at new institutional arrangements related to respectively the globalization of economic manufacturing, the modernization of political arrangements beyond the nation-state level and the emergence of global environmental consciousness and movements.

Global ecological restructuring

The most radical environmental reform was 'promised' in the 1970s discourse on the changing character of modernity and the emergence of globalization. It related to the coming of the post-industrial or information society, in which manufacturing and material consumption would be dematerialized or reduced in favor of services, non-industrial sectors and non-material consumption patterns. This promise did not sustain, as various authors have emphasized (Giddens, 1982; Dicken, 1992; Kumar, 1995). Although the relative contribution of services – compared to manufacturing – to the national economies of most OECD countries increased, it did not result in major environmental improvements by decreasing industrial sectors, dematerialization or changing patterns of consumption into more environmentally friendly directions. At the same time, a new category of New Industrializing Countries (NICs) emerged showing high industrial growth figures and backward environmental institutions that are only slowly catching up with those of OECD countries. Notwithstanding this up till now failing overall dematerialization of industrial societies, environmental considerations have been institutionalized in global economic practices and institutions to some extent and consequently changed those institutions and practices. We will concentrate on multinational enterprises and global economic institutions as the two main global economic mechanisms that propel innovations in institutional arrangements for the environment.

The changing relationship between nation-states and multinational enterprises, and especially the power shift from the former to the latter in the global arena, is often regarded as one of the causes of environmental deterioration, both at the national and the supra-national or global level. But the other side is that especially multinational enterprises, as key actors in the globalization of manufacturing, are not only the driving force behind the harmonization of national environmental regimes (see below), they also push for global environmental regimes and form progressive implementing agencies of Environmental Management and Audit Scheme (EMAS), environmental technologies, new organization principles, environment-oriented economic networks, etc. Notwithstanding the fact that a number of cases still show diverging environmental practices of branches of the same multinational enterprises in distinct local circumstances (e.g. Union Carbide in Bhopal, Shell in Nigeria), clear tendencies of environmental convergence exist, pushed by international standards,[19] regulations, codes of conducts and civil pressure. It is exactly the fact of their global outlook that make these enterprises vulnerable for environmental violations, not only at the location of environmental mismanagement, but at all edges of the world economy where their production takes place and their products are available.[20] Or as Wyn Grant (1993) shortly summarizes: 'the stateless firm has an interest in transnational forms of authoritative decision-making which reduce regulatory divergence between nation-states, yet the existence of the stateless firm gives a further impetus to the extension of the regulatory role of transnational political structures'[21] .....global enterprises have a clear interest in, and thus push towards, homogenization in environmental reforms. Tendencies of economic homogenization and standardization, often evaluated negatively from an environmental point of view, are essential elements of global 'green' firms such as The Body Shop. In general, transnational enterprises cannot be shared among the top polluters, neither in developed countries and certainly not in developing countries. They are usually too visible, vulnerable and 'standardized' to risk environmental confrontation in one locality that may rebound on their global products and production. The larger national oriented firms, primarily confronted with and sometimes even protected by national regimes, are more likely to be among the massive polluters. Environmental reform for these industries show a hybridization rather than homogenization, as global norms and standards mix with distinct national and local policies and interpretations.

Environmental considerations are also increasingly advanced by those global and regional economic organizations that traditionally form the advocates of the global free market, such as the World Bank, the OECD, the European Union, the Economic Commission for Europe, and NAFTA. The World Trade organization (the successor of the GATT) and the IMF – not the least of those mentioned – are the ones that still fall short in combining the advancement of global free market and free trade with environmental reform. Nevertheless, even environment-informed restrictions on import and trade are no longer questioned; the discussion has rather moved to issues of proportionality and unjust use of environmental arguments in protecting national economic interests. These environmental reforms in global economic institutions can no longer be understood or explained by economic reasoning. The environment is becoming a relatively independent factor in these new institutional arrangements, although the strategic question remains whether such institutional reforms of economic organizations should be the main strategy or rather the establishment of global environmental counterpart organizations or institutions (Esty, 1994). Both alternatives, however, will contribute to a tendency of homogenization in environmental reform.

Political modernization beyond the nation-state

One of the most obvious contributions of globalization processes to strengthening environmental reform might be the harmonization or 'homogenization' of national environmental regimes of OECD countries with those in lesser developed nation-states. As Meyer (1987: 48) stresses isomorphism is striking in the present era: 'peripheral societies shift to modern forms of industrial and service activity; to modern state organizations; to modern educational systems; to modern welfare and military systems; in short, to all the institutional apparatus of modern social organization'. And we indeed see in most 'peripheral' states the construction of institutions for environmental reform that are clearly inspired by the OECD ones. Notwithstanding the numerous drawbacks noticed by many scholars,[22] the net effect could very well still be positive, certainly in a long term perspective. But these national developments at 'peripheral' states are not the only political innovations from an environmental perspective that parallel globalization.

Globalization and the declining autonomy of the nation-state – although by some interpreted as disastrous for environmental reform – parallels with new modes of governance and politics, also – or maybe even especially – in the environmental field. Following tendencies of globalization the traditional political institutions that deal with environmental reform at primarily the nation-state level become more and more inadequate and are transformed, modernized, into new institutional arrangements. This can be interpreted as a redefinition of the relation between state and market at the global level, resulting in new institutional arrangements at both the sub-national as well as the supra-national level.

At the sub-national two developments can be seen. A first development entails the reform of traditional command-and-control regulation due to new circumstances of international competition, free circulation of goods, the rapidly varying social context of environmental policy-making and the growing complexity of social relations stretching beyond the national territories. That means that negotiations, cooperation, interaction and consultation with those to be regulated in an early stage of environmental policy-making have become rule rather than exception in environmental politics. We witness similar tendencies of political modernization in all OECD countries, although they are often accommodated to the national policy styles and cultures. Notions of environmental mediation, negotiated rule-making regulatory negotiations, joint environmental policy-making, voluntary agreements all refer to these phenomena.[23] To some extent these innovations in environmental governance can be interpreted as a form of 'sub-politics', although they all center around state environmental policy. 'Sub-politics', as new institutional arrangements following the globalization of environmental destruction, also include politics that move beyond the state politics, although they will always relate to them (see below). A second – partly connected – sub-national development related to changing state-market relations is local diversification. In the era of globalization many regions and communities are faced with evaporating social protections by the national state, through deregulation and decentralization. Diversity, by taking local conditions (both natural and social) more into account and giving more freedom for maneuver for local producers, becomes then a survival strategy and numerous authors (especially in the field of rural and food studies) have argued for the beneficial ecological consequences of local diversification and against the ecological consequences of global food systems.[24] Independent from the question whether that is true, these local diversities can not be interpreted apart from globalization processes, no matter how much the local practitioners or the academic advocates want to abstract from globalization processes.

At the other side, new supra-national political institutions and arrangements on environmental issues can be identified, both at the regional and the global level. To put it shortly, globalization means the end of the primacy of national political arrangements and political modernization in the era of globalization can never be reduced to national political modernization. Globalization in economic processes of production and consumption and the undermining of the state's autonomy go together with (i) a transfer of environmental policy-making capacity from the nation-state to international institutions; (ii) more non-governmental actors on the international stage of environmental politics, such as multinational enterprises and environmental NGOs; (iii) an increase in global or regional environmental regimes formation that trigger tendencies of homogenization. Although the nation-state is still a major actor in (primarily national) policies on for instance solid waste, surface water pollution control, natural reserve protection, and soil pollution clean-up, and states are also major actors in international environmental politics,[25] the modernization of environmental politics moves into the global direction. Supra-national environmental regime formation is of course particularly strong in the European Union. This unique case of supra-national politics is at the moment the subject of heavy criticism requests for an intergovernmental rather than a federal model, defended partly by using environmental arguments. According to these defenders, a strong European Union is not only essential from an economic point of view, but especially to protect the environment in a unifying market. And in the EU of today, supra-national environmental politics do indeed surpass the nation-state although states still have a major say – but in diminishing cases a veto power – in it. Whether this model *status nascendi* will be an example for the construction of new global institutional arrangements can be doubted, although experiences will certainly be valuable in the development of the present state of fragmented global environmental regimes towards more consistent and integrated environmental arrangements on a global level. The necessity for such global arrangements is also pushed by the fact that states, especially the poorer ones, increasingly use environmental resources (both in terms of natural resources to be exploited and emissions to be reduced) in the world struggle for economic resources and power (Miller, 1995).[26] With the articulation of environmental reform at the global level, the environment becomes a powerful resource for nation-states, not only to compensate for the national drawbacks of stringent international environmental regimes, but also to secure other – economic – rewards.[27]

Global discourse coalitions

Global awareness was triggered, among others, by environmental deterioration. It was especially the global environmental risks that forced environmental consciousness in the direction of large scale global (environmental) awareness. Next to global environmental change it is especially the global appearance of the ecological world view that most authors on globalization theories stress when they analyze globalization in relation to environment. 'Globalization has helped to increase ecological consciousness and programs to enhance sustainability', Scholte (1996: 53) writes before he stresses the negative global environmental threats related to globalization. Both global polls (such as the Worldwide Gallop polls on environment) and the growth of global environmental movements such as Greenpeace, Friends of the Earth and WWF are believed to give evidence of the globalization of ecological concern.

Two more theoretical concepts are frequently used to analyze this mutual relation between globalization and what we might call the environmental discourse. In analyzing the changing environmental discourse in some OECD countries Maarten Hajer (1995) uses the concept of discourse coalitions, being the ensemble of a set of story lines, the actors who utter these story lines and the practices in which the discourse is based and (re)produced.[28] The essential element in discourse coalitions is that formerly dispersed groups with varying ideas and belief systems on the environment gather together around a specific discourse, a common political project. Usually the dominant story-line that constructs the cement in a discourse coalition reduces the discursive complexity and at the same time – necessarily – has a multi-interpretable character. Sustainable development, or ecological modernization, can be interpreted as the current notion around which the dominant discourse coalition center. Interpretations and definitions of environmental problems, solutions to these problems and strategies that must be developed accordingly converge more and more around these dominant notions and related story-lines. According to Hajer (1995: 14) it has been especially the processes of globalization and the perception of a global order that have enforced a common environmental story-line among various 'partners' of the environmental discourse coalition. The second notion to analyze the relation between globalization processes and environmental reform is most strongly emphasized by Steven Yearley (1996): the development of universal principles. One of the key issues and one of the major environmental achievements that runs parallel to globalization processes is the development and institutionalization of universal principles in global environmental discourses and strategies. Such universal principles can be founded either in intersubjectivist standards or in scientific knowledge. One of the most important attractions of such universal principles is that they appear to offer a way of speaking with world-wide authority.

The progress of the development put forward in these two notions is that within globalization processes a certain basic – though maybe still low-level – common intersubjective understanding has emerged on the essentiality of protecting environmental values. The so-called universal environmental discourses that are believed to emerge are, however, not that universal as they generally entail diverging sectoral, national/regional or class interests, as has become clear during the 1992 UNCED and the environmental discourse on sustainable development.[29] In a similar way the global environmental polls might measure quite different perceptions and definitions of the environment. The 'universal' ecological world view can be expected to heterogenize in distinct local and cultural contexts. Nevertheless, since the emergence of the green slogan 'think globally, act locally' became popular in the late sixties, two changes can be noted. First, global thinking and analyzing on the environment has expanded to large sections of world society, triggered – among others – by the inadequateness of the ('old') institutions of simple modernity to control high consequence risks, as has been portrayed by Ulrich Beck. Second, actions of the environmental movement have also globalized, be it not without obstacles (Rucht, 1993). The shortcomings of these institutions of simple modernity contributed to some extent to the emergence of global 'sub-politics': new institutional arrangements on a global level that surpass traditional political lines in linking diffuse environmental consciousness within civil society with global decision-making in the economic and political arena.

Epilogue

If this analysis of globalization and the environment leads to one conclusion it should be that there can be no general, overall statement on the consequences of globalization for environmental deterioration and reform. Globalization proofs to be a complex, multi-dimensional phenomenon that materializes in numerous concrete historical processes, and the academic debate on globalization entails, or builds on, quite divers bodies of knowledge: from European integration studies, via studies on cultural hybridization to the debate on economic restructuring of global firms. Consequently, it will not only be an intellectual *tour de force* to draw any lasting conclusions on the environmental consequences of globalization. It also means that quite divers – and interdependent – social mechanisms connect globalization to environmental disruption and reform, of which the net effect will vary and change depending on place, time and the type of environmental problem.

This being said, we as environmental sociologists should not leave it here. A clear need exists to refine this analysis to different segments of the discourse on globalization and the environment. Not all arguments, not all mechanisms and interdependencies stated above, have the same validity for every environmental issue at every location. In the Dutch National Environmental Policy Plan – often quoted for its innovative character, also with respect to its conceptions (Weale, 1992) – a so-called 5-level model was introduced to link distinct environmental problems to different hierarchical levels. Each level – from the local up to the global – has its 'own' environmental problems that are caused and should primarily be solved at that level. This is a rather ecological-material starting point for analyzing internationalization in environmental policy-making. My idea is that environmental sociologists can do better by not limiting such an analysis to the ecological-material dimension, but include the social processes and dynamics of globalization to conceptualizations of and designs for environmental reform. But then sociologists should move beyond both the general 'slogans' that globalization and localization are two sides of the same coin, and an iteration of case studies that have only value by themselves and do not seem to produce generalizing knowledge.

Rather than analyzing the full impact of connecting globalization to environmental reform programs, I will finish with an exemplary analysis of how globalization processes interfere with – converging or diverging – environmental reform in distinct locations, and what political consequences might be drawn from such a refined sociological interpretation. Those environmental issues that are intensively connected to globalization processes will – although they might not threaten the global environment – result in a homogenization of environmental regimes for different locations, and most probably this homogenization will indeed be a westernization. A typical example would be environmental standards for products produced mainly by transnational firms and which are moved around the globe. The dynamics of global firm and of international trade politics will ensure a homogenization in environmental regimes. Environmental issues such as waste collection and treatment will rather heterogenize, as local factors on for instance labor costs, existing recycling practices and the economic value of waste are more important in designing such systems than global processes that might force homogenization. One can expect that such systems in Third World megacities will deviate from those in OECD megacities. But the same will be true for the major part of reform programs to beat the greenhouse effect, in which globalization processes will not always be the dominant factor. This means that in some cases ecological modernization models of OECD countries will – or can – be transferred to developing regions, while in other cases the outlook of ecological reform will vary strongly by location and pressures for convergence will have counterproductive effects.

Notes

1. Hoogvelt (1997: 116) claims that in the confusing and dispersing discourse on globalization, sociologists – rather than economists or international relation scholars – have been at the forefront in trying to give globalization a consistent theoretical status.
2. Redclift (1996), however, is not really concerned with globalization, but basically deals with the distribution of consumption between the North (being the industrial wealthy countries) and the developing countries in the South. As we have noted earlier only a few contributions in Redclift and Benton (1994) are concerned with globalization (Spaargaren and Mol, 1995).
3. The emergence of the idea of the homogeneous nation-state, the increase in the number of international agencies and institutions, the increasing global forms of communications (railways, telephone, post), the acceptance of unified global time, the development of global competition, the emergence of a standard notion of citizenship all contribute, according to Robertson, to this time framing.
4. Recently these 'older' monolithic theories have incorporated a multiplicity of factors that contribute to globalization. Adherents to the World-System Theory school-of-thought have recently dealt with more cultural and ideological factors in processes of globalization and tried to incorporate these into their theory (Wallerstein, 1990).
5. Robertson (1992: 138-145) describes Giddens' work on globalization as a rather late contribution, adding little to the existing theories.
6. Having that in mind, we can understand that Robertson criticizes Giddens' work on globalization as 'an overly abstracted version of the convergence thesis' (1992: 145).
7. Goldblatt (1996: 14-51) criticizes Giddens for restricting environmental degradation only to the institutional cluster of industrialism and the international division of labor, while according to Goldblatt the environmental crisis cannot be understood without taking the capitalist organization of the economy into account. Although Goldblatt is right that capitalism, but also the other two institutional dimensions of modernity, have their relevance for understanding the relation between globalization and environment,

we agree with Giddens on stressing the special relevance of industrialism (cf. Mol and Spaargaren, 1993; Mol, 1995).

8. The emphasis on flows rather than on national institutions return in many recent contributions to globalization theory. Lash and Urry (1994), for instance, identify the present age as one of the disorganization of capitalism, which should be understood not so much as the successor of organized capitalism but rather as a global system that is systematically disorganized, and in which the logic of organizations, especially the nation-state, is replaced by a logic of flows. The logic of flows is also very prominent in the work of Castells (1996).
9. This is in line with the analysis of Sassen (1994), who stresses the large contribution of financial capital and services to the emerging new global economic regimes (especially from the 1970s onward), which nonetheless have major consequences for manufacturing industries. In general, the latter – being most directly relevant for the environmental quality – are declining due to decreasing possibilities for superprofits and thus withdrawal of investment capital from these sectors.
10. A world polity was originally defined as a 'system of rules legitimating the extension and expansion of authority of rationalized nation-states to control and act on behalf of their populations' (Meyer, 1987: 69). Later (Boli and Thomas, 1997) it took a broader perspective by (i) focusing on both political and cultural institutions, and (ii) taking a transnational perspective rather than one in the framework of the system of nation-states.
11. Meyer and his colleagues stress especially the importance of the world system of nation-states as the central political order, and only marginally touch upon the (globalization) tendencies that undermine this system of nation-states.
12. In analyzing the undermining sovereignty Held makes a distinction between internal sovereignty, focusing on the government as the final and absolute authority that exercises supreme command over a particular society, and external sovereignty, which stresses that there is no final and absolute authority above and beyond the state.
13. The diminishing role of the nation-state is emphasized by McMichael (1996: 39): 'internationalization of political authority includes both the centralization of power in multilateral institutions to set global rules and the internalization of those rules in national policy making (...). The definition of an international regime is thereby refined to include the actual determination, or at least implementation, of those rules by global agencies. In other words, the potential global regime is only formally multilateral, as states lose capacity as sovereign rule makers'.
14. Beck (1994b: 180). Apocalypse-blindness refers to a modern society whose institutions have no 'receptors' to notice apocalyptic dangers in an early stage of progression.
15. High consequence risks are defined as risks which are remote from control by individual agents, while at the same time threatening the existence of millions of people and indeed of humanity as a whole. The principle examples given by Giddens are Chernobyl, the depletion of the ozone layer and the greenhouse effect.
16. As Beck explains, '*subpolitics* is distinguished from "politics", first in that agents *outside* the political or corporatist system are allowed to appear on the stage of social design (...), and second, in that not only social and collective agents but individuals as well compete with the latter and each other for the emerging shaping power of the political' (Beck, 1994a: 22; emphasis in the original).
17. As Redclift (1996: 34-35) points out, in relation to environmental threats 'global sovereignty is still largely a rhetorical device'.
18. For a more elaborated analysis of the theory of ecological modernization of production and consumption, see Spaargaren, 1997 and Mol, 1995.

19. The International Standard Organization 14000 series is a special case of global harmonization of environmental practices in manufacturing.
20. The Shell Brent Spar case – in which the dumping of the oil producing platform in the Atlantic was prevented by consumer boycott in numerous countries – has made this especially clear.
21. Grant (1993: 61) defines a stateless company as a firm whose ownership, board of directors and senior management is internationalized. Such firms operate globally and are no longer based in one country nor have loyalty to one country.
22. For instance, it is often emphasized that these institutions hardly function properly, that they have become necessary due to the modernization path of industrialism the countries were 'drawn in', that their construction is to a large extent pushed by and beneficial for western and international consultancies, donor agencies and environmental technology industries, and that these cause rather a weakening of arrangements at other nation-states in bringing about a harmonization at the level of the lowest common denominator.
23. These changing models of environmental governance at the national level are of course not only provoked by globalization processes. Within the field of environmental policy-making the limited successes of traditional command-and-control modes of governance and the tendencies of and requests for deregulation, among others, have also contributed significantly (Mol et al., 1996).
24. Cf. the studies of Marsden et al. (1993), Bonnano et al. (1994) and McMichael (1996) to name but a few.
25. Rowlands (1995), for instance, argues that the UNCED has reinforced the sovereign rights of states and rejected any supra-national authority on monitoring, reporting and inspection, notwithstanding the quest by some for both 'sub-politics' and supra-national politics.
26. Redclift (1996), however, notes that the part of natural resources and primary food production in the export of Third World Countries has decreased considerably from 50% in 1965 to 20% in 1986. In the same period the exports of manufactured goods have increased from 25% to 50% of total exports. A similar trend is reported by Dicken (1992).
27. The dept-for-nature or dept-for-environment swabs and joint implementation are two examples of the use of environmental resources by non-OECD countries to secure economic profits.
28. This notion comes close to Sabatier's concept of advocacy coalitions, although Hajer notes some important differences (1995: 68-72).
29. The Group of Lisbon (1994) rightly observes that the unofficial parallel conference of NGOs to the UNCED also showed that the 'global civil society' is still strongly fragmented, uncoordinated and internally divided along several lines: North vs. South, environmentalist vs. developers, reformists vs. 'revolutionaries', those starting from local interests vs. those arguing from a global analysis, etatists vs. local autonomy defenders, etc.

Bibliography

Anton, D.J. (1995), *Diversity, Globalization and the Ways of Nature*. Ottawa: International Development Research Center.

Archer, M. (1991), Sociology for One World: Unity and Diversity, *International Sociology*, 6 (2): 131-147.

Barber, B.R. (1995), *Jihad vs. McWorld*. New York: Times Books/Random House.

Bauman, Z. (1993), *Postmodern Ethics*. Oxford/Cambridge MA: Basil Blackwell.

Beck, U. (1986), *Risikogesellschaft. Auf dem Weg in eine andere Moderne*. Frankfurt am Main: Suhrkamp.

Beck, U. (1991), *Politik in der Risikogesellschaft*. Frankfurt am Main: Suhrkamp.

Beck, U. (1994a), The Reinvention of Politics: Towards a Theory of Reflexive Modernization, in: U. Beck, A. Giddens and S. Lash, *Reflexive Modernization. Politics, Tradition and Aesthetics in the Modern Social Order*. Cambridge: Polity Press, pp. 1-55.

Beck, U. (1994b), Replies and Critiques. Self-Dissolution and Self-Endangerment of Industrial Society: What Does This Mean, in: U. Beck, A. Giddens and S. Lash, *Reflexive Modernization. Politics, Tradition and Aesthetics in the Modern Social Order*. Cambridge: Polity Press, pp. 174-183.

Beck, U. (1996), World Risk Society as Cosmopolitan Society? Ecological Questions in a Framework of Manufactured Uncertainties. *Theory, Culture & Society* 13 (4): 1-32.

Boli, J. and G. M. Thomas (1997), World Culture in the World Polity: A Century of International Non-governmental Organization. *American Sociological Review*, 62, (April): 171-190.

Bonnano, A., L. Busch, W.H. Friedland et al. (eds) (1994), *From Columbus to ConAgra. The Globalization of Agriculture and Food*. Lawrence, KS: University Press of Kansas.

Castells, M. (1996), *The Rise of the Network Society*. Vol. 1. Oxford: Blackwell.

Cohen, M.A. (1996), The Hypothesis of Urban Convergence: Are Cities in the North and South Becoming More Alike in an Age of Globalization?, in: M.A. Cohen, B.A. Ruble, J.S. Tulchin and A.M. Garland (eds), *Preparing for the Urban Future. Global Pressures and Local Forces*. Washington D.C.: The Woodrow Wilson Center Press.

Dalby, S. (1996), Crossing Disciplinary Boundaries: Political Geography and International Relations after the Cold War, in: E. Kofman and G. Youngs (eds), *Globalization: Theory and Practice*. London/New York: Pinter, pp. 29-42.

Dicken, P. (1992), *Global Shift. The Internationalisation of Economic Activity*. London: Paul Chapman (2nd revised edition).

Esty, D.C. (1994), *Greening the GATT. Trade, Environment and the Future*. Washington D.C.: Institute for International Economics.

Giddens, A. (1982), *Profiles and Critiques in Social Theory*. London and Basingstoke: MacMillan.

Giddens, A. (1984), *The Constitution of Society*. Cambridge: Polity Press.

Giddens, A. (1990), *The Consequences of Modernity*. Cambridge: Polity Press.

Giddens, A. (1991), *Modernity and Self-Identity. Self and Society in the Late Modern Age*. Cambridge: Polity Press.

Giddens, A. (1994), Replies and Critiques. Risk, trust, reflexivity, in: U. Beck, A. Giddens and S. Lash, *Reflexive Modernization. Politics, Tradition and Aesthetics in the Modern Social Order*. Cambridge: Polity Press, pp.184-197.

Goldblatt, D. (1996), *Social Theory and the Environment*. Cambridge: Polity Press.

Gourevitch, P. (1978), The Second Image Reversed: the International Sources of Domestic Politics. *International Organization*, 32 (4): 21-68.

Grant, W. (1993), Transnational Companies and Environmental Policy Making: the Trend of Globalisation, in: J.D. Liefferink, P.D. Lowe and A.P.J. Mol (eds), *European Integration and Environmental Policy*. London and New York: Belhaven, pp. 59-74.

Haas, P.M. (1990), *Saving the Mediterranean: The Politics of International Environmental Cooperation*. New York: Columbia University Press.

Haas, P.M., R.O. Keohane and M.A. Levy (1993), *Institutions for the Earth: Sources of Effective Environmental Protection*. Cambridge, MA: MIT Press.

Hajer, M.A. (1995), *The Politics of Environmental Discourse. Ecological Modernization and the Policy Process*. Oxford: Clarendon Press.

Harvey, D. (1989), *The Condition of Postmodernity*. Oxford: Basil Blackwell.

Held, D. (1995), *Democracy and the Global Order. From the Modern State to Cosmopolitan Governance*. Cambridge: Polity Press.

Hoogvelt, A. (1997), *Globalisation and the Postcolonial World. The New Political Economy of Development*. Houndsmills and London: MacMillan Press.

Hurley, A. (1995), *Environmental Inequalities. Class, Race and Industrial Pollution in Gary, Indiana, 1945-1980*. Chapel Hill and London: The University of North Carolina Press.

Keohane, R.O. and M.A. Levy (1996), *Institutions for Environmental Aid. Pitfalls and Promise*. Cambridge, MA and London: MIT Press.

Kumar, K. (1995), *From Post-Industrial to Post-Modern Society. New Theories of the Contemporary World*. Oxford (UK) and Cambridge (USA): Blackwell.

Lash, S. and J. Urry (1994), *Economies of Signs and Space*. London: Sage.

LeQuesne, C. (1996), *Reforming World Trade: the Social and Environmental Priorities*. Oxford: Oxfam.

Marsden, T., J. Murdoch, P. Lowe et al. (1993), *Constructing the Countryside*. London: University College London Press.

McGrew, A. (1992), A Global Society?, in: S. Hall, D. Held and T. McGrew (eds), *Modernity and its Futures*. Cambridge: Polity Press, pp. 61-102.

McMichael, P. (1996), Globalization: Myths and Realities. *Rural Sociology* 61 (1): 25-55.

Meyer, J.W. (1987), The World Polity and the Authority of the Nation-State, in: G.M. Thomas, W. Meyer, F.O. Ramiorez and J. Boli, *Institutional Structure. Constituting State, Society, and the Individual*. Newbury Park: Sage, pp. 41-70.

Miller, M.A.L. (1995), *The Third World in Global Environmental Politics*. Buckingham: Open University Press.

Mlinar, Z. (1992), Individuation and Globalization: The Transformation of Territorial Social Organization, in: Z. Mlinar (ed.), *Globalization and Territorial Identities*. Aldershot: Avebury, pp. 15-34.

Mol, A.P.J. (1995), *The Refinement of Production. Ecological Modernization Theory and the Chemical Industry*. Utrecht: Jan van Arkel/ International Books.

Mol, A.P.J. (1996), Ecological Modernisation and Institutional Reflexivity. Environmental Reform in the Late Modern Age. *Environmental Politics*, 5 (2): 302-323.

Mol, A.P.J. (2000), Globalization and Changing Patterns of Industrial Pollution and Control, in: S. Herculano (ed.), *Environmental Risks and the Quality of Life*. Rio de Janeiro: UFF.

Mol, A.P.J. and G. Spaargaren (1993), Environment, Modernity and the Risk-Society: The Apocalyptic Horizon of Environmental Reform. *International Sociology*, 8 (4): 431-459.

Mol, A.P.J., V. Lauber, M. Enevoldsen and J. Landman (1996), Joint environmental policy-making in comparative perspective. Paper to the 'Greening of Industry' Conference, Heidelberg, Germany.

Morse, E.L. (1976), *Modernization and the Transformation of International Relations.* New York/London: The Free Press.

Nederveen Pieterse, J. (1995), Globalisation as Hybridisation, in: M. Featherstone, S. Lash and R. Robertson (eds), *Global Modernities.* London: Sage, pp. 45-68.

Redclift, M. (1996), *Wasted. Counting the Costs of Global Consumption.* London: Earthscan.

Redclift, M. and T. Benton (eds) (1994), *Social Theory and the Global Environment.* London and New York: Routledge.

Robertson, R. (1992), *Globalization: Social Theory and Global Culture.* London: Sage.

Robertson, R. (1995), Globalization: Time-space and Homogeneity-heterogeneity, in: M. Featherstone, S. Lash and R. Robertson (eds), *Global Modernities.* London: Sage, pp. 25-44.

Rowlands, I.H. (1995), *The Politics of Atmospheric Change.* Manchester and New York: MUP.

Rucht, D. (1993), 'Think globally, act locally'? Needs, Forms and Problems of Cross-national Cooperation among Environmental Groups, in: J.D. Liefferink, P.D. Lowe and A.P.J. Mol (eds), *European Integration and Environmental Policy.* London and New York: Belhaven, pp. 75-95.

Ruigrok, W. and R. van Tulder (1995), *The Logic of International Restructuring.* London and New York: Routledge.

Sachs, W. (ed.) (1993), *Global Ecology. A New Arena for Political Conflict.* London/New Jersey: Zed Books.

Sassen, S. (1994), *Cities in a World Economy.* Thousand Oaks/London/New Delhi: Pine Forge Press.

Scholte, J.A. (1996), Beyond the Buzzword: Towards a Critical Theory of Globalization, in: E. Kofman and G. Youngs (eds), *Globalization: Theory and Practice.* London/New York: Pinter, pp. 43-57.

Sklair, L. (1991), *Sociology of the Global System.* London: Harvester and Baltimore: John Hopkins University Press.

Spaargaren, G. (1997), *The Ecological Modernization of Production and Consumption. Essays in Environmental Sociology.* Wageningen: Wageningen Agricultural University (dissertation).

Spaargaren, G. and A.P.J. Mol (1995), Book Review of Redclift and Benton, Social Theory and the Global Environment. *Society and Natural Resources*, 8 (6): 578-581.

Spangenberg, J. (1995), *Towards Sustainable Europe.* Wuppertal: Wuppertal Institute/FOEI.

Strassoldo, R. (1992), Globalism and Localism: Theoretical Reflections and Some Evidence, in: Z. Mlinar (ed.), *Globalization and Territorial Identities.* Aldershot: Avebury, pp. 35-59.

The Group of Lisbon (1994), *Limits to Competition.* Cambridge, MA: MIT Press.

Thomas, G.M., J.W. Meyer, F.O. Ramirez and J. Boli (1987), *Institutional Structure. Constituting State, Society, and the Individual.* Newbury Park: Sage.

Wallerstein, I. (1990), Culture as the Ideological Battleground of the Modern World-System. *Theory, Culture & Society*, 7: 31-55.

Waters, M. (1995), *Globalization.* London/New York: Routledge.

Weale, A. (1992), *The New Politics of Pollution.* Manchester: Manchester University Press.

Williams, M. (1995), Rethinking Sovereignty, in: E. Kofman and G. Youngs (eds), *Globalization: Theory and Practice.* London/New York: Pinter, pp. 109-122.

Yearley, S. (1996), *Sociology, Environmentalism, Globalization.* London: Sage.

Young, O. (1989), *International Cooperation: Building Regimes for Natural Resources and the Environment.* Ithaca, NY: Cornell University Press.

Young, O. (1994), *International Governance: Protecting the Environment in a Stateless Society*. Ithaca, NY: Cornell University Press.

7

Environmental Social Theory for a Globalizing World Economy

Michael Redclift

Introduction

Environmental social theory needs to address both the phenomenon of globalization, and the environmental problems associated with it. Globalization is ubiquitous in the late twentieth century; marking both a series of observable, empirically defined issues, and the way in which we have come to understand them. In this contribution I want to examine some of the links between globalization and the environment, and to identify the principal issues with which a new body of social theory will need to be concerned.

The article starts with exploring the links between global economic integration and the unsustainable use of natural resources. It argues that global integration at the level of markets and material processes, carries different implications for the global environment, distinguishing between the diffusion of sources for environmental problems, and that of impacts. But these economic and material processes carry additional implications for sociological analysis, notably the resistance to economic ideology that is perceived as 'Northern' or occidental. This resistance takes the form of a fragmentation of individual and culturally specific perspectives, grounded in quite distinctive experiences of livelihoods and localized environments. The nature of cultural resistance to monolithic globalization suggests a new point of departure for social theory. In place of one 'integrative' social theory this analysis suggests the need for alternative theoretical paradigms, to help explain both changes in the observable 'real world', and the historical and cultural processes through which this world is understood and managed.

Finally, and as a consequence of this discussion, I will raise questions about our failure to develop a global environmental management commensurate with the scale and urgency of the problems we face. The ineffectiveness of global attempts at environmental management takes us back to an issue which was of primary concern for sociology since its inception: how are social impulses internalized and communicated? The article concludes that environmental sociologists may be in a unique position to address this problem, in a postmodern world, by building a concern with sustainability into our understanding of global society. In the search for an explanatory social theory to help explain global environmental problems we may thus be contributing to the greatest challenge on our doorstep: how might our knowledge (and ignorance) of the global environment enable us to refashion social theory?

Economic growth and sustainable resource use

In considering the viability of the generic model of economic globalization, from the perspective of resources and ecological sustainability, three views of the relationship between economic growth and sustainability can be distinguished. The first and most common view is held by most governments, and most conventional economists. This is that sustainability and economic growth are more or less compatible, provided that we recognize the need for minimal international regulation, and make efforts to protect endangered ecosystems and species. The second view is that they are totally incompatible. As Herman Daly has expressed it '...sustainable growth is an oxymoron...' (Daly, 1992). On this reading, the pursuit of economic growth implies increased throughput of energy and materials in an economy, and this in turn serves to undermine the sustainability of the environment. The third perspective is somewhat different. It asserts that whether or not economic growth is compatible with sustainability depends on prior definition of a number of concepts, notably 'wealth', the interests of 'future generations' and the nature of 'economic efficiency'. In the view of its protagonists this third view requires a prioritization of sustainability, as a goal rather than a set of *ex post* management tools. Given a political commitment to consider sustainability as a goal of international politics, economic growth would itself become redefined, and the reduction of waste and pollution would constitute an objective of policy, together with the eradication of world poverty. It is a measure of the cynicism that pervades many international policy fora, that such sentiments should appear utopian, rather than practical and necessary. But before considering ways of grounding these objectives in practical policy measures, it is worth stepping back and considering the global balance sheet of environmental and resource degradation.

The discussion of climate change has concentrated our attention on shifts in the climate system which are difficult to understand and predict and we will return to that below. But at the same time severe problems of pollution, spread around the globe, lie near at hand. In the rural and peri-urban areas 'indoor' pollution, from burning wood, and poor ventilation harms over 400 million people worldwide, and contributes to acute respiratory infections, from which four million children die annually. Household sewage is another 'global' problem which starts in the family. Sewage is the major cause of water contamination, and poor sanitation for over one and a half billion people worldwide contributes to even larger numbers of infant deaths. In the developing world almost 95 per cent of sewage is untreated. The provision of clean drinking water would enable two million fewer children to die from diarrhoea each year; currently thirteen and a half million children die as the result of the combined effects of poor diet and domestic sanitation (Ekins, 1996). Perhaps more alarmingly, there is evidence that the numbers of people without adequate sanitation actually increased during the 1980s, and measures to mount preventative campaigns are jeopardized by growing water scarcity, in many parts of the world. Forty per cent of the global population experiences periodic droughts but, at the same time, rising per capita water consumption (by fifty per cent since 1950) has largely been for irrigated agriculture, and at the expense of domestic water provision.

In the world's cities, the problem is frequently the quality of the air, as well as the water. Today almost one and a half billion people live in cities with air quality below minimum standards set by the World Health Organization (WHO), producing respiratory and other ailments from which half a million people die prematurely, every year (Ekins, 1996). Even the environmental problems that we usually regard as remote from these problems of primary health and livelihoods, such as the production of industrial wastes and the extinction of species, are inextricably linked not only to levels of consumption ('ecological footprints') but also of resource degradation. The losses to our global environment through the pursuit of unsustainable economic growth, in the form of increased pollution and resource degradation, put individual livelihoods at risk throughout the developing world. They are survival issues for the poor, as well as the subject of informed speculation for policy analysis.

Globalization does not only interfere with those global environmental problems commonly referred to as global environmental change, but more importantly also with issues of direct survival of the poor. It goes one step too far to say that economic globalization was the originator of these processes, but it has definitely intensified them. In the neo-liberal view global technology and markets provide efficiency gains. Transnational actors, particularly large corporations, pursue their interests even where they conflict with those of the national state or supranational institutions. This now constitutes the 'natural order'; it follows the grain of history. The creation of the World Trade Organization (WTO), for example, marked a departure from the GATT in several important respects. The management of trade was expanded to include trade in Intellectual Property Rights, and services. There are now calls for the WTO to expand still further to consider other issues such as labor standards and environmental issues. This is in line with the prevailing view that global welfare is best served by a liberal economic order, however substantial the evidence of social and ecological dislocation. The intellectual near-hegemony of this consensus has meant that alternative approaches to understanding the relationship between global trade and social or environmental issues have almost been lost by default. Globalization means more uniformity in economic goals, as well as more competition to achieve them. It carries important implications for international political action and, particularly, the survival of local, frequently oppositional, cultures.

Globalization, the environment and the poor

The notion of globalization, as a means of social and economic ordering within a highly integrated capitalist world economy, has a vital bearing on our understanding of change within the global environment. In a more systematic way two variants to global environmental problems can be distinguished, each derived from the 'development decades' of the 1950s, 1960s and 1970s. First, we can identify the *diffusion of sources*, through which wastes and pollution are relocated globally.

Examples are the transfer of industrial agriculture to new settings, and its sourcing from new locations. Similarly the spread of polluting industries and energy and transport technologies, predominantly from the North to the South, are other cases in point. The problems that ensue – such as the dispersal of hazardous waste, and the threat posed by radioactive leakages and fallout – are global in the sense that economic globalization implies global relocation.

The second form of global environmental problems occurs through the *diffusion of impacts*. What we observe in this case are the effects of systemic changes in natural resource systems – through the medium of water, soils and the atmosphere. Many of the problems that are regarded as characteristically 'global', and most of those contained in the texts of international negotiations, such as global warming, ozone depletion and acid deposition, are of this second variety. The existence of this range of problems reminds us that far from being 'autonomous' (in Robertson's view) global environmental problems are a consequence of the development of the global economic system, and deeply embedded in conventionally designated societal and socio-cultural processes. The parlous state of much of the global environment is a direct reflection of our habits of 'getting and spending', in which the biophysical system is employed as a source of materials and as a sink to assimilate wastes and pollution.

Taking Giddens' definition of globalization as 'the intensification of worldwide social relations which link distant localities in such a way that local happenings are shaped by events occurring many miles away and vice versa' (Giddens, 1990: 64), it follows that the site of environmental degradation may be far removed from its agent of causation. It is central to our understanding of globalization that the location of these material sources, the sites where these materials are transformed and where they are consumed, and the location of the waste sinks need not – and increasingly do not – coincide. It is apparent that such situations result from interdependent patterns of development involving the mobility of capital and the relocation of industrial processes that deepen the international division of labor while reflecting comparative advantage and regional patterns of specialization. Globalization simply provides an interpretive device in which environmental change arising from the use and transformation of sources, sinks and resources can be traced and attributed to a set of structured practices and processes guided by the underlying dynamic of material accumulation. This dynamic, argues Saurin (1996), concentrates wealth in certain locales and amongst certain social groups largely by extracting and dispossessing from other locales and social groups. As the penetration of the market displaces the production of use values by that of exchange values, the co-evolutionary basis by which local environmental resources and ecological processes provide a sustenance base for rural people is broken (Norgaard, 1994). Incorporation into the world economy effectively diminishes the capacity of local producers to exercise control over their choice of production systems and the way resources are to be managed. Instead, a web of decisions made many miles away, that might involve the imposition of externally derived macro-economic goals and market incentives, can exert increasing influence over local production systems and the local environment.

One sector which perfectly illustrates this development is the food industry which exercises an extraordinary level of influence over land use in the South. The Netherlands, for example, appropriates the production capabilities of 24 million hectares of land, ten times its own area of cropland, pasture and forest (Postel, 1996). Part of these 'ghost acres' support cassava production in Thailand and Indonesia which enters the feed industry for intensive pig farming in the Netherlands. Indonesia has been anxious to maintain or even increase its share of the European Union's cassava quota and its vigorous promotion of exports has seen domestic prices rise.

> In response, farmers are switching from more sustainable and less erosive mixed-cropping and perennial crop-farming systems to mono-cropping cassava. They are even removing terracing and other soil and water conservation structures to increase the area of cassava cultivation. ...(T)he on-site productivity costs and off-site erosion impacts of the recent price distortions may have already impaired the prospects for secure livelihoods for many upland farmers, and for the sustainable management of upper watersheds as a whole. (Conway and Barbier, 1990: 76)

This example of the 'cassava connection' shows how a distortionary trading structure, combined with short-term policy objectives, can lead to overall economically and environmentally unsustainable outcomes with the costs borne in Indonesia, principally by small farmers whose livelihoods are undermined. While many observers might attribute such unsustainable outcomes to local agents who may be thought to be acting out of ignorance or willful self interest, a structural analysis informed by the principles of political ecology (Peet and Watts, 1996) serves to challenge this received wisdom as one based on simplified and inherently localized models of linear causation (Leach and Mearns, 1996). Such challenges are vital, not least to counter tendencies toward a form of global environmental management which would strengthen the hand of transnational and multilateral institutions in the name of the common good.

Buttel and Taylor (1994) have argued that packaging multiple environmental problems and concerns within a common rubric conveys a scientific legitimacy and the political rationale for responding urgently. In the case of global environmental management such a basis can provide the opportunity for the powerful to exert control over the resources of others in the name of planetary health and sustainability (Peet and Watts, 1996). The emergence of a paradigm of global environmental management, with its curative rather than preventative approach to environmental problems, rests upon the rise to dominance during the past 25 years of a discourse in which the metaphor of Spaceship/Planet Earth has played an important role. Arturo Escobar argues that the visualization of Earth as a 'fragile ball' offers a narrative for managerialism best exemplified by *Scientific American's* September 1989 Special Issue on Managing Planet Earth which asked: 'What kind of planet do we want? What kind of planet can we get?' (Escobar, 1996: 50). While Escobar in turn enquires as to the identity of this 'we' who knows what is best for the world as a whole, the answer to his question is self-evident. The existing international political economy is managed by a small number of multilateral institutions (principally the World Bank, the International Monetary Fund and the World Trade Organization) with policies determined by the richest industrialized countries (the Group of Seven – G7) who control over 60 per cent of world economic output and over 75 per cent of world trade.

In reflecting on the process of globalization, then, we can make a number of tentative concluding observations. Globalization is often, *but not always*, territorial in nature. Where globalization operates territorially it relocates environmental functions, the sourcing of economic growth and the disposal of waste. The metabolic imprint of human metabolism on the global landscape has an increasing spatial dimension, through both concentration and dispersal. The environmental effects of economic globalization are very uneven in their impact, both spatially and socially. Diffusion of sources are determined by the process of economic development, its policy mechanisms (investment, trade etc.) and its main actors (transnational companies, multilateral institutions, and the richest industrial nation-states). At the same time diffusion of system impacts are more diffuse, usually irreversible and likely to be passed on to future generations.

Cultural constructions of environmental problems

We should acknowledge that globalization consists of the way we see things, as well as the things themselves. Some commentators, such as Roland Robertson, have argued that the '(...) autonomy of the globalization process, operates (...) in *relative* independence of conventionally designated societal and socio-cultural processes' (Featherstone, 1990: 5). This intellectual position can be challenged, not because it is wrong, but rather because it fails to identify the contribution of global economic and social restructuring to the development of global culture, and the tensions that ensue when cultural changes arrive 'on the back', as it were, of economic restructuring. Thus globalization involves representations of the global not because global culture is 'autonomous', in some way disconnected from lived experience, but because cultural change reflects changes in economic and social restructuring, it reflects 'lived experience', as well as reflecting upon it.

It is equally important to emphasize that environmental 'problems' (including global ones such as climate change) are not defined in a cultural vacuum. The way that environmental issues are represented reflects social and cultural perspectives. Although most of the mass media is devoid of explicit environmental messages, it clearly communicates a great deal about the environment. Cinema and television depict environmental values in the form of buried messages about consumption, nature and the world of goods. The values associated with media messages may not be readily assimilated into existing cultural categories. The tension between Western (primarily Northern) values, and those of traditional Hinduism, or Islam, serve to underline the difficulties posed by 'globalization'. If, as we have illustrated above, clean drinking water is the principal environmental issue for millions of people in the South (and its absence contributes to the premature death of thirteen and a half million children each year) it is hardly surprising that global climate predictions appear even more remote in the South than they do in the developed world.

Equity and sustainability

The underlying social commitments and practices of everyday life, constitute the 'filter' through which people, and their governments, perceive 'global' environmental problems as well as the importance and definition of equity. In the poorest countries the 'environment' consists of problems associated with health, shelter and food availability. In the newly industrializing countries the 'environment' is bound up with the short and medium-term costs of pursuing very rapid economic growth, such as high levels of air pollution in cities. In the privileged, developed world, the 'environment' increasingly involves exposure to largely invisible, and unforeseen, risks, such as levels of radon, or beef contaminated by BSE. Issues of equity appear less important, in dealing with these problems, than freedom of information, or civil rights.

However, from a sociological perspective equity is important for global environmental change not only because of measurable differences in the flow of energy and materials, levels of personal consumption, or the difficulty in arriving at international agreements, but also for ideological reasons. Ideology shapes the discussion of global environmental change in several important ways. First, as we have seen, the perceptions of environmental issues are subject to ideology, and vary markedly in different societies and cultures. Second, ideological assumptions govern the trust that can be placed in the behavior of others over the environment. One of the most persistent debates within the global environmental change literature concerns the so-called 'free rider' question: the extent to which countries that do not sign up to global environmental agreements will be able to benefit from them. Similarly, the loss of 'sovereignty' means quite different things in different contexts. Sovereignty itself is an ideological construct which needs to be investigated in order to establish how rights and responsibilities to the environment vary. In addition, many of the structural processes at work in the international economic system have ideological components, which we should not ignore. For example, this is the case with the view taken of environmental regulation (or deregulation), and with the value contributed to trade liberalization and the market economy, rather than command economies or economic protection. The political economy of the environment is governed by opposing ideological precepts, as well as disparate economic strengths and weaknesses.

Perhaps the best example of ideological factors influencing the debate, is the discussion of 'sustainable development' or 'green development' itself, which is often polarized between the respective connotations of techno-economic transformations (ecological modernization) and more thoroughgoing cultural changes (deep green solutions). Some writers claim sustainability for small-scale societies, or earlier civilizations. Others argue that sustainable development needs to totally embrace global technology and global means of communications – the Internet, satellites and the media. In addition and linked to sustainability, there are two dimensions to international equity: intra-generational (or spatial) and intergenerational (or temporal). The concern of most commentators in the North has been with future generations of 'our people', although this is usually implicit, while the concern of those in the South has been with the current generation. The importance of intra-generational equity, however, is such that unless we can successfully resolve differences between North and South, the prospects for future generations in the developed world (too) is bleak.

These differences in both ideology and related material interests complicate successful global environmental policies. Consequently, the conclusion to this paper examines the obstacles to a global compact that prevents the successful 'management' of global environmental problems, using the example of climate change.

The 'ownership' of global environmental problems: the case of climate change

The importance of ideology, and specifically 'Northern' constructions of global environmental problems, is well illustrated by the discourse surrounding climate change. From the viewpoint of a dispassionate observer, greenhouse gas emissions are so much lower in the South, on a per capita basis, that developing countries should have more room to increase their emissions, while developed countries reduce theirs. At the same time, the technological transition to lower energy intensity (the amount required to produce a corresponding increase in Gross Domestic Product) is not as advanced in the South. Thus the potential for economic growth through energy saving and technological improvements is greatest in the developing countries. Logically speaking, and from a 'global' environmental perspective, the industrialized countries could therefore optimize their financial assistance to the South by providing incentives to the adoption of cleaner, more efficient technology, and help the South to effectively 'leapfrog' the stage of dirty, labor intensive industrialization where it is not yet fully underway.

Unfortunately, this dispassionate analysis holds several serious difficulties. First, calculations of atmospheric emissions based on per capita contributions spell disaster for the global situation, since the knowledge that developing countries are not the main contributors to 'the problem' provides little inducement for these countries to pursue more efficient use of energy. Developing countries have little incentive to endorse global objectives above meeting the more immediate needs of their own people. Second, the transition to cleaner technology in the South is not encouraged by most major economic agencies of the Northern, industrialized economies, whose efforts (in so far as they are geared to 'ecological modernization') are focused on gaining for themselves the market advantages conferred by higher environmental standards in tradable products. They have an interest, in other words, in *not* transferring advanced cleaner, more energy-efficient technologies to the South. Acting in the global environmental interest is secondary, for most transnational companies, to their pursuit of profit, until such time as profits reflect the internalization of environmental values. Third, the labor-displacing effect of 'high-tech' industries recommends them to countries where labor costs are high; but this provides fewer obvious advantages for economies where labor costs are low (hence World Bank official Lawrence Summers' celebrated remark, prior to UNCED in 1992, that it was more efficient to direct pollution to poorer countries). For most developing countries the incentives to achieve lower energy intensities pale into insignificance beside the economic benefits of providing dirty (and frequently unsafe, and unhealthy) employment. In addition, the need to reduce levels of some Greenhouse gas emissions, such as methane, is confused by the fact that goods traded in the world economy take no account of emission levels (Agarwal, 1996).

From the perspective of a developing country, then, the real costs of reducing emission levels, in the 'global' interest, are very high. It is by no means clear that 'sustainable development' should be given precedence over achieving increased economic growth. Environmental gains can even be measured in terms of economic growth foregone, and jobs that would have been created. Posed as a conflict between intra-generational equity, and intergenerational equity, most developing countries are more likely to choose to reduce the inequalities in the present global economic system, rather than make sacrifices to achieve gains for future affluent generations (in the North).

Much of the discussion of climate change policy has followed a completely different trajectory. Taking the idea of 'carbon budgets', for example, it has sought to identify principles for allocating reduced emissions among countries. It begins, then, with what is assumed to be a shared problem, and proceeds to allocate national reductions in emissions, as measures for resolving it (Krause et al., 1992). Within the international relations community, and especially among environmental economists, attention has been given to mechanisms which might ease the differentials (or inequalities) between countries, that currently make them unwilling to act for the 'greater good'. International environmental taxes, tradable permits, and joint implementation, are all methods for seeking to overcome current opposition to collective action. Nevertheless, there has been a marked reluctance, even on the part of most of the industrialized countries, to agree to implement these environmental policy measures. This is largely because they place their own short-term economic advantages above longer-term, global benefits. In practical terms, most developed countries are unwilling to endorse the precautionary principle as a guide to resolving the contradictions between equity and efficiency at the international level, or to face the severe political problems which would be likely to accompany a new world economic order.

Distributive problems lie at the heart of the failure to take global action, as Grubb and others (1993) have observed. First, there are differences of opinion over the extent to which historical contributions to climate change should be considered. The problem of anthropogenic Global Warming is essentially one which the industrialized countries have induced through their own industrialization. Since developed countries have a larger responsibility for current Greenhouse gas concentrations, than for current emissions, the question arises as to whether they have a responsibility for their earlier contribution to the problem. Before energy intensities began to decline in the North emissions were linked to unbridled pollution, for which we are all paying the costs.

Second, aggregating emissions is not the same as equating them. Some commentators, such as Anil Agarwal (Agarwal and Narain, 1991) have argued that methane emissions from the cultivation of rice paddies, or ruminant animals, are essentially 'livelihood emissions', rather than 'lifestyle emissions', like those of the North that are based on high levels of personal consumption. In addition, methane emissions are extremely difficult to quantify and assess, and the data very unreliable. Similarly, CFCs, which are overwhelmingly produced in the North, have no known sinks, and should not, therefore, be equated with carbon dioxide or methane, gases for which sinks exist.

Third, it is not clear that we all share an equal stake in the global commons, whether it is common property or open access resources that are being considered. The global 'commons' are being privatized through the patenting of nature itself. The inconclusive deliberations over the Biodiversity agreement at UNCED demonstrated that, for most countries in the North, the wish to preserve tropical forests was linked to the preservation of patent rights in nature. The resources of developing countries needed to be preserved, so that they could be 'discovered', or utilized as a 'laboratory' for commercial ends on a 'first come, first served' basis.

The key question, as Martinez-Alier (1993) has suggested, is the economic value we place on the functions of natural ecosystems. At present the forests and oceans are global sinks, but they do not attract payments to developing countries, even those containing large amounts of forested land, for their global conservation value. Acknowledging this principle would immediately upset the existing consensus surrounding global environmental change, and serve to undermine the current basis for global environmental management. Similarly, it is clear that we cannot begin by agreeing to share 'emission reserves', on an equitable basis, without undermining the very nature of the global economic system on which they, ultimately, depend. If we have all allocated pollution quotas, or carbon budgets, the problems posed by global environmental change are not resolved, they are merely deepened. Would such quotas still be available if the poor become richer? What would be their price? Which international institutions would collect the receipts on quotas, and to what ends would they then be applied? It is clear that the use of the concept of stakeholder, to imply that everybody has a 'stake' in the global environment, is ambiguous. If the countries of the South are disadvantaged through being unequal stakeholders in the global environmental problem, how can they benefit from being stakeholders in global environmental solutions that take their start in this inequality?

A globalizing world economy and the search for social theory

It is the contention of this contribution that '...the great names of classical social theory offer us very little ...(instruction in how to understand global environmental changes)' (Goldblatt, 1996). They also offer us no more than a few insights into the relatively new world of global environmental policy-making. Rather than seeking to uncover a rich vein in existing social theory, we might ask: since globalization is so ubiquitous economically, as well as culturally, what makes global *environmental* management so difficult to achieve? This may not lead us to a discrete body of theory, but it will help us to identify the criteria against which we should begin to measure our theories. This article has sought to examine both the material forms of globalization, and their cultural construction, including cultural opposition to globalization. This results in at least three important conclusions.

Environmental issues, especially at the 'global' level, are addressed through different, and frequently competing, policy frameworks, such as market and regulatory instruments. Global problems are approached from within different paradigms, leading to a number of apparently irresoluble contradictions. For example, economic convergence through global market integration frequently undermines the social and cultural autonomy of distinct groups, their 'being different'. This process of undermining is intensified by exposure to global culture. Globalized culture tends to be Northern culture, and produces its own antithesis – in Commandante Zero in Chiapas, in the occupation by Peruvian guerrillas of the Japanese Embassy in Lima, in the resurgence of fundamentalist Islam throughout much of North Africa, Asia and the Middle East. Environmental problems are also contextualized by these cultural frames, but globalization frequently removes them from their context. They become like a cultureless language, an *esperanto* of the mind, that speaks to everybody and therefore, ultimately, to nobody.

Secondly, the opening up of global markets, and global consumption, and the penetration of 'remote' ecological systems by global capital, leads to conflict with the 'real world' economics that characterizes peoples' everyday lives. There is increasing tension between compliance with neo-liberal economic policies and the struggle to establish sustainable, long-term livelihoods, which depend in turn on sustainable environments. This tension can be eased by policies to arrest the degradation of the physical environment, but global environmental 'management' remains elusive, precisely because it contradicts the integrity of culture and place. It has been argued that attempts to manage the global environment come 'on the back' of global economic restructuring, and are consequently resisted by many people in the South for the same reasons as structural adjustment policies have been resisted. In developing countries 'global' environmental management is as likely to breed resistance as compliance. It frequently represents a narrowing of policy options, and the erosion of local self-determination.

Third, political consequences follow social and environmental dislocation. Addressing global environmental problems requires legitimate political action, and a domestic constituency in support of global environmental management. But many national state governments have been weakened by neo-liberal policies, especially in the South, while global institutions, particularly those of the United Nations system, are increasingly beholden to fewer national governments, notably that of the United States (Hurrell and Woods, 1995). Questions are raised, in the corridors of power, if not across the negotiating table, about the legitimacy of both environmental conditionality, and the coercion used by global institutions seeking to engineer management solutions. Global environmental management is either toothless (and ineffective) or it is coercive (and also ineffective). Especially in the later cases the moral imperative to ensure the integrity of the biosphere is used to justify the creation of a new regulatory world order.

And in that sense the problem of globalization and the environment brings us back to the origins of social theory, be it in a new constellation. It is instructive that the global environment faces us with both a crisis of authority, and an absence of social and political legitimacy: the very processes which social theory in the nineteenth century sought to understand.

Bibliography

Agarwal, A. (1996), *Down to Earth: The State of the Environment*. Delhi: Center for Science and Environment.

Agarwal, A. and S. Narain (1991), *Global Warming in an Unequal World: A case of environmental colonialism*. New Delhi: Center for Science and Environment.

Buttel, F. and P. Taylor (1994), Environmental Sociology and Global Environmental Change, in: M. Redclift and T. Benton (eds), *Social Theory and the Global Environment*. London: Routledge, pp. 228-255.

Conway, G. and E. Barbier (1990), *After the Green Revolution: Sustainable agriculture for development*. London: Earthscan.

Daly, H. (1992), *Steady-State Economics*. London: Earthscan.

Ekins, P. (1996), *The Relationship between Economic Growth, Human Welfare and Environmental Sustainability* (Unpublished Ph.D. Thesis). London: Department of Economics, Birkbeck College, University of London.

Escobar, A. (1996), Constructing Nature: Elements for a poststructural political ecology, in: R. Peet and M. Watts (eds), *Liberation Ecologies: Environment, development, social movements*. London: Routledge.

Featherstone, M. (ed.) (1990), *Global Culture: Nationalism, globalization and modernity*. London: Sage.

Giddens, A. (1990), *The Consequences of Modernity*. Cambridge: Polity Press.

Goldblatt, D. (1996), *Social Theory and the Environment*. Cambridge: Polity Press.

Grubb, M., M. Koch, A. Munson et al. (1993), *The Earth Summit Agreements: A guide and assessment*. London: Royal Institute of International Affairs/Earthscan.

Hurrell, A. and N. Woods (1995), Globalization and Inequality, *Millennium* 24: 3.

Krause, F., W. Back and J. Koomey (1992), *Energy Policy in the Greenhouse: From warming fate to warming limit*. London: Earthscan.

Leach, M. and R. Means (eds) (1996), *The Lie of the Land: Challenging received wisdom on the African environment*. London: James Currey.

Martinez-Alier, J. (1993), Distributional Obstacles to International Environmental Policy: The failures at Rio and prospects after Rio. *Environmental Values* 2: 97-124.

Norgaard, R. (1994), *Development Betrayed: The end of progress and a coevolutionary revisioning of the future*. London: Routledge.

Peet, R. and M. Watts (1996), Liberation Ecology: development, sustainability, and environment in an age of market triumphalism, in: R. Peet and M. Watts (eds.), *Liberation Ecologies: Environment, development, social movements*. London: Routledge.

Postel, S. (1996), *Dividing the Waters: Food security, ecosystem health, and the new politics of scarcity*. Washington: Worldwatch Institute.

Redclift, M. (1996), *Wasted: Counting the cost of global consumption*. London: Earthscan.

Saurin, J. (1996), International Relations, Social Ecology and the Globalization of Environmental Change, in: J. Vogler and M. Imber (eds), *The Environment and International Relations*. London: Routledge, pp. 77-98.

8

The Ideology of Ecological Modernization in 'Double-Risk' Societies: a case study of Lithuanian environmental policy

Leonardas Rinkevicius

Introduction

A growing number of scholars in the field of environmental sociology and policy argue that since the mid-1980s there has occurred a shift from 'revolutionary' environmentalism towards reformist environmentally informed social and institutional change epitomized by the concept of *ecological modernization* (Weale, 1992; Hajer, 1995; Mol, 1995). This co-evolutionary change of modern institutions of economy, science and technology, and public administration is contrasted with the environmental counter-movement of the 1970s which is labeled so due to its 'totalkritik', its antagonist attitudes towards other actors, particularly industrialists and state authorities, and its promulgation of 'alternative' values and life styles (Hajer, 1995; Spaargaren, 1997). 'Ecologization of economy' and 'economization of ecology' (Mol, 1995; Spaargaren, 1997) are the core metaphors of an eco-modernist ideology of the 1990s, indicating the belief in viability of reconciliation of ecology and economy in the course of societal change without 'leaving the path of modernization'.

Based on a set of beliefs and expectations, epitomized by the slogan 'pollution prevention pays', eco-modernist ideology opens up a new perspective for looking at the greening of industry, environmental policy reforms and other processes of environmentally informed institutional and social change. Those ideas, beliefs and practices are increasingly promulgated by different actors and networks, mainly rooted in industrialized countries. This ideology encompasses such ideas, principles and beliefs as precautionary principle, pollution prevention, shared responsibility, public right-to-know, shift from command-and-control environmental policy to consensus-oriented soft steering regimes based on economic and informative instruments, and belief in the ability of institutional learning. In the course of 'compression of the Globe' (Yearley, 1996) one can observe not only the trans-national transfer of pollution, but also a trans-national transfer of eco-modernist ideas and beliefs to other regions, for example, Eastern Europe.

This paper aims at characterizing ecological modernization as an *ideology* – a set of shared ideas, beliefs, values and claims regarding environmentally informed social and institutional change of modern society. It also aims at exploring attitudes and beliefs of different categories of actors in Lithuanian society and interpreting them vis-à-vis eco-modernist ideology. In doing so we apply an interpretation of ecological modernization as 'institutional learning' as opposed to a 'technocratic project' (Hajer, 1996). We want to argue and aim to empirically verify the claim that attitudes, beliefs and respective cultural tensions between various categories of actors shaping an eco-modernist discourse and environmental policy correlate with or stem from intrinsically conflicting doctrines, principles and ethos characteristic of different societal domains (structures) – bureaucratic, economic, academic, and civic. Institutional learning is thereby interpreted as the gradual alleviation of those cultural or institutional tensions, leading to social consensus.

Reflexive modernization and the 'double-risk society'

According to Beck (1992) the issue of distribution of 'bads', first of all globalizing ecological risks, becomes a dominant characteristic of the 'risk-society' which contrasts with the society of 'simple' modernity where the main issue has been the fair distribution of goods. An important trait of this shift towards the 'risk-society' is the change of social values and beliefs reflected in the demystification of a positivist advancement of science and technology, the disillusion in linear modernization. Members of the 'risk-society' are losing faith in science and technology-based anticipatory control of side-effects of economic growth. This disillusion in modern science and technology has to do with overly globalizing and intensifying risks and hazards, particularly the environmental ones. There is no social class or geographic region which could escape a potential catastrophe. Industrial society is shifting towards reflexive modernization (Beck et al., 1994), because the institutions of 'simple' modernity are not capable of controlling globalizing risks. Reflexivity is becoming an immanent characteristic of a 'risk-society'. Reflexivity marks 'the decline of the (Enlightenment) idea that social and natural environments would increasingly be subjected to rational ordering. It includes the reflection upon the nature of reflection, (...) resulting in an institutionalization of doubt' (Mol, 1995: 17).

The distinction between the two phases of societal development based on the issues of social distribution of 'goods' and 'bads' as well as the disillusion in science and technology opens up new perspectives for analyzing not only Western societies, but also comparing, for instance, Lithuanian and other societies in transition vis-à-vis industrialized nations. In contrast to the latter ones, the current state of development in many countries of the world, including Lithuania, is characterized by the mixed importance of both issues: the distribution of goods as well as the distribution of 'bads'. Because of the acuteness of both types of issues social anxiety in many developing and transitional countries, we argue, is more complex than in industrialized societies which have already reached a certain level of material welfare. Because of the complex social anxiety over the distribution of goods and 'bads', developing countries and the ones in transition might be termed *'double-risk' societies* as compared with affluent societies which have supposedly already passed the phase of 'simple' modernity.

Due to this 'double-risk' character, many social controversies in transitional society are more painful and complex as compared to wealthy industrialized nations. An illustration of this might be the dichotomy between the need to continue thc generation of cheap electricity in the insecure Chernobyl-type nuclear power plant of Ignalina in Lithuania, because of the necessity to fulfill daily needs for millions of people while compromising the very basis of the physical survival of individuals as well as the entire nation in the case of nuclear disaster. Social-economic choices reflect that environmental awareness of the globalizing 'risk-society' is not yet a decisive factor shaping people's attitudes and beliefs in such countries as Lithuania. Transitional bifurcation and still pervasive unaffordability of what Beck calls 'goods' make people in 'double-risk' societies marginalize (temporarily) attention to ecological risks although social anxiety about them is apparent. This is part of the context in which the transition to a market economy and civic society in Central and Eastern European (CEE) countries takes place.

The pursued modernization – transition to a market economy and democratic society – reflects certain faith in (or mimetic isomorphism of) values and institutions dominant in industrialized societies. However, this 'learning-by-doing' in Eastern Europe is inevitably intertwined with numerous disillusions regarding institutions of market economy, democratic representation and social justice that were promised or believed-in at the out-set of transition. Different countries ranging from Poland and Hungary to Bosnia and Albania show different pace and success of transition to a market economy and to a democratization of society and imply different levels of social anxiety and potential disillusion, and hence different scenarios for (reflexive) modernization. The risks associated with the processes of globalization make this process even more complex. The trans-national transfer of pollution, such as the trade in hazardous waste and re-allocation of polluting industries to 'double-risk societies' still takes place in spite of the growing awareness, as it is stressed by 'risk-society' theory, that there are no social or geographical boundaries and ways to escape globalizing ecological threats. The exploitation of cheap labor and overwhelming supply of imported goods at dumping prices still takes place in CEE societies in spite of promises by Western democracies to create equal social-economic opportunities if Eastern Europe would follow the free-trade institutional path. All those mismatches between initial beliefs and social reality add to a certain disillusion and anxiety regarding the institutions of market economy and democratic governance. Those dimensions are to be added to the general analytical framework when analyzing 'risk-society' and reflexive modernization in transitional societies.

A framework for analyzing eco-modernist ideology

Following the eco-modernist literature, particularly Weale (1992), Hajer (1995), Mol (1995), Spaargaren (1997) and my own interpretation, the main beliefs and value-orientations shared by adherents of this stream in contemporary environmentalism could be summarized as follows (Figure 1).

- 'Pollution prevention pays' and self-regulation among actors rooted in the economic domain
- Shift from the command-and-control to the soft, flexible regulatory regimes based on economic and informative policy instruments
- Diffusion of precautionary principle
- Shift of research from identifying environmental problems towards developing cleaner technologies and products based on life-cycle assessment
- Polluter pays
- Public right-to-know, shared responsibility and subsidiarity emphasizing delegation of decision-making to the lowest possible levels in the hierarchy of public administration

Figure 1 *The main beliefs and value-orientations shared by the eco-modernist ideology*

Eco-modernist beliefs and values can be interpreted either as a process of *institutional learning* or as a *technocratic project*. According to Hajer (1996) there are different structuring principles which cause differentiation in beliefs and attitudes among actors.

> The structuring principle of institutional learning interpretation of ecological modernization is that nature is 'out of control'. The central assumption of this paradigm is that dominant institutions can learn and that their learning can produce meaningful change (...) The interpretation of ecological modernization as a technocratic project holds that the ecological crisis requires more than social learning by existing social organizations. Its structuring principle is that not nature but technology is out of control. (...) It holds that ecological modernization is propelled by an elite of policy makers, experts and scientists that imposes its definition of problems and solutions on the debate.

According to Hajer, those who regard ecological modernization as a technocratic project see the ecological modernization doctrine rather as a 'repressive answer to radical environmental discourse than its product' (Hajer, 1996: 251-3). Ecological modernization can therefore be interpreted as *cultural politics* – a process of competition between various actors and groups (Hajer calls them 'discourse coalitions') whereby certain ideas are defended, dominant approaches are socially constructed, beliefs, norms, organizational structures and practices are institutionalized (Hajer, 1996; Szerszynski et al., 1996).

This conceptualization opens up a new perspective for sociological inquiry into the process of environmentally driven change of modern society. It also opens up opportunities for new typologies of actors depending on their identity with regard to eco-modernist ideology. Attempts to contrast eco-modernist attitudes and beliefs in line with 'institutional learning' as opposed to the 'technocratic project' more or less parallel earlier attempts of other theorists to contrast the so-called 'Arcadian' and 'Imperialist' approaches to man-nature relations and traditions (Table 1).

Table 1 *'Arcadian' and 'Imperialist' environmental traditions*[1], *Worster (1977)*

'Arcadian'	'Imperialist'
Back to nature	Manage nature
Localist-experiential	Systemic-experimental
Preservation	Rational use
Natural history/'thick description'	Classification/modelling
'Harmony'	'Domination'

Ecological modernization ideology can in various respects be seen in line with an 'imperialist' approach stemming from the Enlightenment belief in the 'Empire of Reason', as opposed to an idealistic and romantic 'Arcadian' world view. Although the eco-modernist ideology emphasizes belief in the ability of modern institutions to adjust, learn, accommodate to natural constraints, it still does not depart from utilitarian (hence 'imperialist') approaches aiming at rational use of natural resources within the boundaries of nature's carrying capacity, the 'global commons', 'limits to growth', 'sustenance base', 'ecological space' or whatever concept is used. The very belief in the ability of globalizing society to scientifically define such boundaries and reshape institutions accordingly so that growth of production and consumption would not exceed the predefined sustenance limits is very central to what Worster calls 'Imperialist' tradition.

However, eco-modernists do not claim the free intervention in nature, rather they accentuate accommodation, adoption and learning. In this sense, an 'Imperialist' tradition in environmentalism could be subdivided. Here we come close to O'Riordan's (1981) classification of 'eco-centric' and 'techno-centric' trends in contemporary environmentalism. Eco-modernist ideas, beliefs and expectations regarding environmentally informed societal change are in various ways similar to what O'Riordan and Pepper (1993, 1994) call an 'environmental management' approach within the techno-centric trend of contemporary environmentalism.

We raise the hypothesis that differences in the way various actors perceive and accept the ideology of ecological modernization depend on the institutional spheres or social domains those actors are rooted in or associated with. At the same time, human agency is crucial in shaping the dominant ethos, doctrines, principles and steering mechanisms in such social domains and thus in fostering or inhibiting the diffusion of eco-modernist ideology. Therefore, it seems relevant to look at the various social domains that shape the public space of ecological modernization, allowing us to project possible implications of changes in different domains for overall changes in the environmental discourse and environmental policy-making.

Jamison and Baark (1990: 32) in their work regarding science and technology policy have distinguished three policy cultures which may be classified as bureaucratic, entrepreneurial and academic.

> While in practice, they often become intertwined in the process of policy-making, for analytical purposes it is useful to separate them as 'ideal types'. They exist primarily as interest lobbies, or institutional networks, and as such exercise significant influence over practical policy-making.

Table 2 *An analytical framework: cultural tensions between different institutional domains (Jamison and Baark, 1990; Jamison, 1997)*

Domain ⇨	Bureaucratic	Economic	Academic	Civic
Doctrine	*Order*	*Growth*	*Enlightenment*	*Democracy*
Steering mechanism	*Planning*	*Commercial*	*Peer review*	*Assessment*
Ethos	*Formalistic*	*Entrepreneurial*	*Scientific*	*Participatory*

This analytical framework has been recently expanded by adding a fourth 'ideal type' policy, institutional or social domain, namely the public or civic domain (see Table 2).[2] Each policy domain is characterized by different traits. On the most general, ideological, level those four domains are distinguished on the basis of different doctrines. For instance, the doctrine of order dominates the bureaucratic domain, in contrast to the doctrine of growth prevailing among actors rooted in the economic domain. In terms of steering mechanism, i.e. institutionalized practices or the way of working, those domains manifest different traits as well. For instance, such practices among those rooted in the bureaucratic domain are based on planning in contrast to the commercial transactions institutionalized as routine practice among economically oriented actors. Steering mechanisms are closely associated with certain predominant value orientations and beliefs inherent in particular institutional environments. For instance, most of the civic activity is based on faith in democratic participation. This participatory ethos is quite different from the ethos of formalism prevailing in the bureaucratic domain. The latter can be contrasted with the ethos of entrepreneurship characteristic of and institutionalized among economic actors.

Following some of the sociological literature (e.g. Eisenstadt, 1968) those policy domains might be more generally described as social institutions or institutional spheres, because these structural entities imply a conjunction of prevailing guiding principles, beliefs and expectations as well as routinized practices and social organization or networks. More important, the structural entities seem to encompass two ways of 'patterning' social relations and social phenomena. To put it in Giddens' (1984: 16-17) words, as 'understood by functionalists and, indeed, by the vast majority of social analysts (...) 'structure' here appears as 'external' to human action, as a source of constraint on the free initiative of the independently constituted subject. (...) on the other hand (...) it is characteristically thought of not as a patterning of presences but as an intersection of presence and absence; underlying codes have to be inferred from surface manifestations'.

This (post) structuralist approach as well as the distinction made between 'ideal' policy domains echoes the distinction which Max Weber has made in his book, *Economy and Society* (1930), between politically and economically oriented behavior. His sociology of politics is based on a distinction between the essence of economy and polity which is conditioned by the subjective meaning of human behavior. According to Weber, economically oriented behavior is basically shaped by the pursuit to satisfy personal needs and wishes associated with benefit and profit-making, whereas politically oriented behavior is characterized by efforts of a person or group to rule other people. Economic behavior might at times be oriented towards using power, hence acquiring political profile. Similarly, the ruling or extension of power upon other actors often needs an acquisition or possession of certain economic means to fulfill certain political ends. Thus, there is a political economy and economic polity. As it is difficult to segregate economic and political dimension in practice, those distinctions are analytical, or 'ideal', rather than real (Aron, 1967).

Approaching ecological modernization as cultural politics, the distinctions between the political and economic institutional sphere, or politically and economically oriented behavior, seems a relevant guiding concept in looking for regular differences in perceptions, beliefs and attitudes of different categories of actors. Conceptualizing political, economic and other institutional spheres in such a way, we hope to enrich and extend the approach of policy and economic networks – focused primarily on social organization – as developed by Mol (1995) in his analysis of ecological modernization. In our perspective, ecological modernization can be analyzed and interpreted as a culturally and institutionally controversial process whereby various actors gradually learn about beliefs, doctrines, codes of conduct (norms and ethos) shared by actors rooted in other institutional spheres, and learn to adjust their behavior and beliefs with respect to them.

For instance, politically oriented actors are expected to learn that their aims at planning, setting standards and targets of pollution reduction within the period of their political governance and formally imposing those norms upon industrialists are often inadequate given the entrepreneurial ethos of the latter. They are expected to learn that economically oriented actors welcome regulatory regimes that favor flexibility and predictable environmental improvements in the perspective of a long-term business development. More generally, an eco-modernist rhetoric of 'institutional learning' and a shift from command-and-control policy to one based on trust, tolerance and dialogue can be viewed as a belief in humanization and harmonization of regulatory regimes whereby actors rooted in the economic or bureaucratic sphere learn to adequately respond to, for instance, the search of economic actors for entrepreneurship or pursuit of the common citizens for democratic participation. The recent empirical evidence of the latter type of 'institutional learning' is epitomized by slogans and emerging practices of corporate environmental accountability, transparency, and public-right-to-know.

Similarly, 'institutional learning' might be seen as the diffusion of certain entrepreneurial skills and ethics in regulatory agencies or environmental NGOs. The development of various regulatory innovations – pollution charge waivers, emission credits ('bubbles'), revolving environmental investment funds, etc. – are the empirical examples from various countries indicating this type of change or learning process in the bureaucratic domain. The fact that WWF or Greenpeace promulgates 'cleaner production' based on the belief that pollution prevention pays epitomizes similar types of 'institutional learning' in the civic domain. Moreover, there are cases that can be interpreted as empirical examples of more complex 'institutional learning' processes. For example, actors rooted in the civic domain, such as WWF, involve the Duke of Edinburgh (symbolizing the political order) to fund environmental campaigns so that resources could be mobilized to involve highly paid academic experts believing that 'there is a great deal that the WWF and its experts can do, if they are given our support' (see Jamison, 1997: 236). The recent mushrooming in various countries of research and teaching departments promoting pollution prevention, cleaner technology and industrial ecology is yet another empirical example of an entrepreneurial ethos being diffused among actors rooted in the academic domain.

In the remaining – empirical – part of this article, we will use the eco-modernist framework in analyzing and discussing some important aspects of environmental policy transformation and the related beliefs and attitudes of two important categories of actors – industrialists and environmental authorities. We will take Lithuania as a case study country and consequently start with outlining the specific place of ecological modernization in this country.

Ecological modernization in Lithuania

Some ecological modernization theorists suggest that this theory 'has more practical relevance for Central European countries such as the former GDR, Poland, Hungary and Czech Republic, and less for countries like Bulgaria, Albania and several CIS republics' (Mol, 1995: 55). Rather than comparing various countries of Central and Eastern Europe (CEE) in terms of their potential for ecological modernization, we just argue here that the eco-modernist ideology as a dominant framework originating in industrialized societies is likely to diffuse in most of CEE, including our case study country Lithuania. It is beyond the scope of this paper to describe the complex multi-dimensional development of Lithuanian society and its (late) modernization. In short, its degree of industrialization and urbanization, the high educational level of its population, the development of modern state institutions, particularly those of science and technology, and other institutional traits enable us to hold to the opinion that Lithuanian society reached the level of social and institutional development to be essentially entitled as modern industrial society (Melnikas, 1995).Consequently, we argue that ecological modernization theory is relevant for analyzing and interpreting environment-informed social and institutional change in this country.

The legacy of Lithuania's five decades under the Soviet rule makes it doubtful whether (neo)Marxist ideology and hence its theoretical approach to environmentalism would be in line with the social values, beliefs, and expectations prevailing among the members of Lithuanian society in the period of transition. The rapid shift towards private ownership, (regulated) market economy and parliamentary democracy provides sufficient empirical evidence to support such a claim. The Marxist ideology accentuates a capitalist mode and private ownership of the means of production as underlying structural causes of social and ecological degradation and this would hardly gain broad social acceptance in Lithuania, nor would the de-industrialization or industrial dismantling ideological approach.

This is not to deny the social premises for reflexivity in Lithuanian society, which are likely to support the ideological claim on social and ecological counter-productivity of linear economic growth. Moreover, precaution and social-ecological questioning of industrialization and urbanization are important characteristics of traditional conservatism in Lithuanian society which is reflected by the metaphor of 'the first generation beyond ploughshare' (Sliogeris, 1993). The latter epitomizes the fact that the social departure from traditional rural peasantry communities in Lithuania emerged only since the Interbellum, the period between the two World Wars in the twentieth century. Due to deportations and emigration (of upper and best educated pro-modernist classes) after the Second World War, the departure of Lithuanian society from 'ploughshare' took a new turn in the 1950-60s. Such features of Lithuanian society, however, are not sufficient to argue that 'industrial disarmament' or institutional dismantling of the 'treadmill of production' would be favored by its members in the period of transition.

Quite on the contrary, a growing demand for material welfare, and hence further industrial-economic growth, is observable, particularly in terms of consumerism among the younger generation (Luobikiene, 1995). This is not surprising after long decades of scarcity and deficit of material welfare in the Soviet system. These points are to strengthen our earlier argument that the ideology of ecological modernization, emphasizing consensus, institutional learning and reconciliation of ecology and economy without leaving the path of modernization, is most likely to diffuse in Lithuanian transitional society, especially compared to the other contrasting ideologies. In addition, the return of Lithuania and other CEE countries to the international political, economic, scientific-technological community and the globalizing exchange of consumer goods, technologies, know-how and respective learning is another strong premise to expect that eco-modernist ideology is likely to diffuse in this region.

Therefore, ecological modernization theory as a set of claims on environment-induced societal change seems relevant for describing and interpreting changes in Lithuania. The issues of 'double-risk society' make this process more complex and multi-dimensional as compared to transformations in advanced (post)industrial societies.

The relation between eco-modernist ideology and environmental policy reform

At first sight, policy changes such as the revision of a pollution charge system can be understood as a solely economic behaviorist project aiming at enforcing environmental standards with the help of economic instruments, and 'optimizing' charges so that they penalize polluters seriously enough while keeping companies at a reasonably competitive level. In doing so, the 'bureaucratic domain' furthers its formal planning procedures and steering regime in order to reduce pollution and collect state revenues from the 'economic domain' in a *predictable way*. From an eco-modernist standpoint, however, pollution charges are important as a way to discourage companies to simply rely upon the state authorities as institutions responsible for mitigating and curing environmental illnesses, and to transfer more social-environmental responsibility to economic actors.

It can be argued that 'properly designed' pollution charges are supposed to create new stimuli for the greening of companies in a cost-effective way. From this rational-choice perspective, pollution charges and fines are hoped to encourage *entrepreneurship* of industrialists, to provide more incentives for pollution prevention at source, hence to reduce the conflict between ecological and economic motives. Such a policy approach is not new to environmental economists and policy analysts. For environmental sociologists, however, this issue is especially important because it illuminates one of the core beliefs of an ideology of ecological modernization: the faith (expectation) that pollution charges can stimulate companies to get *continuously* greener, i.e. that this is a process of institutional learning rather than technocratic cosmetics. It ultimately leads to a cultural change rather than simple compliance with prescribed emission standards enforced with the help of charges and penalties. Therefore, it seems meaningful to explore who believes in such an eco-modernist scenario of change ('institutional learning'); and *why* they do so; and, on the other side, which actors view it as a technocratic project that merely creates an illusion of deeper transformations while in fact only inducing an incremental change.

From this vantage point the reform of the pollution charge system is more than a two-way behaviorist traffic – reaction or anticipation on the side of economic actors, and the regulatory push from the side of state bureaucrats. To move beyond this point a broader institutional outlook is essential, understanding this institutional reform in the context of changes in other institutional spheres and analyzing the ways in which it shapes, and is shaped by, attitudes, expectations and actions of various actors.

An eco-modernist concept of environmental charges, or the 'polluter pays' principle *per se*, has been known and partially implemented in Lithuania long before the revival of the national state in the late 1980s and even long before *Perestroika*, i.e. before the period of intensifying global exchange of eco-modernist ideas, approaches and techniques got its momentum in the former Soviet empire. Lithuanian companies were levied pollution charges in the Soviet times since the mid 1970s. This policy was of little relevance, because in the state-owned, centrally planned economy it meant re-allocation of the same money from one state 'pocket' to another (see Rinkevicius, 1992). Such pollution charges did not induce considerable institutional learning or closure among industrialists and environmental authorities as well as other institutional spheres. Nevertheless, it is important to emphasize that the Soviet bloc, although rather isolated ideologically, was not deprived from Western eco-modernist concepts and principles such as the 'polluter pays' or 'low-waste' technologies. Pollution charges are just one of numerous examples indicating that important dimensions of an ideology of ecological modernization formed part of the environmental discourse in Lithuania long before the end of the Soviet era (for further discussion on the greening of Lithuanian industry in the Soviet times, see Rinkevicius, 1997).

Social and institutional traits of the conventional regulatory regime

The current Lithuanian regulatory system in the environmental field is characterized by bureaucratic domination. Command-and-control, namely pollution emission standards and permits, makes the core of the regulatory regime. However, some authors (see Semeniene et al., 1996) argue that such a regime might be described as a mixed one, because it encompasses not only administrative regulation but also economic instruments: pollution charges for those complying with pollution standards, and environmental fines for those exceeding permitted levels. The latter are defined in two ways: the so-called Greatest Permissible Pollution (GPP) and Temporarily Permissible Pollution (TPP) norms. In the cases when GPP levels are not attainable for economic, technological or other reasons, environmental authorities negotiate the levels of temporarily permissible pollution with industrialists. Enterprises are supposed to develop plans containing environmental measures according to which they are expected to transit from TPP to GPP levels, whereas authorities would be committed to reduce temporarily the regulatory pressure. According to regulations, it is the responsibility of economic actors to collect all relevant information, to calculate potential levels of pollution emissions and assess potential environmental impacts before applying to environmental authorities for a pollution permit. Thus, economic actors themselves are supposed to be actively involved in defining/negotiating pollution discharge limits which are later formally prescribed to them by regulatory agencies. In principle, plans for emission control and environmental measures are legally binding, but practice often differs from theory. Many companies do have environmental plans formally, but their implementation is often hindered due to social, economic and technical difficulties. Such plans are continuously modified or their implementation is postponed, particularly during recent years of economic recession. Such changes and postponements are negotiated between industrialists and regulators based on various arguments, including such issues as employment and company's contribution to the infrastructure of a particular city.

In general, there is little in-depth research conducted examining how effective pollution permitting in combination with a charge system was in stimulating environmentally informed transformations of Lithuanian industry. Our interviews with representatives of industrial enterprises as well as statistical research (Semeniene et al., 1996) indicate that the general burden of pollution charges is too low for many companies to encourage pollution reduction at source (Bluffstone, 1995). Instead of investing in environmental measures many industrialists prefer to simply pay charges to the state, thereby transferring responsibility of environmental protection to the state authorities. Such a behavior can probably be best explained by rational choice theory.

Besides formal regulatory pressure via pollution permits and levies, there are some indications that regulators are willing to promote environmental entrepreneurship of industrialists, although in formally predefined ways. The Law on Pollution Charges of the Republic of Lithuania (1991) provides economic incentives to companies which implement pollution reduction measures. According to this Law, any company which reduces emissions of certain pollutants by at least 25% is allowed to retain its pollution charge up to three years in order to compensate the costs of the implemented measures. However, only about 3% of the enterprises have enjoyed pollution charge waivers provided by this law (Semeniene et al., 1996). Again, rational choice might provide an explanation: low pollution charge burden does encourage formal compliance of industrialists rather than stimulating industrial innovation aiming at pollution reduction at source. Environmental permitting in combination with a charge system does not provide any impetus for diffusion of an eco-modernist belief that pollution prevention pays, and does not encourage deployment of entrepreneurial skills of industrialists leaving the state bureaucrats as the main actors in charge of environmental protection.

This is one stream of the critique which led to the current reform of Lithuanian pollution charge system. Besides the low charge burden, a second shortcoming of

Lithuanian environmental permitting and charge system is that it does not encourage companies to seek continuous pollution reduction beyond certain predefined standards of greatest permissible pollution. Yet another important gap hindering implementation of the 'polluter pays' principle is the lack of public transparency and accountability due to an underdeveloped environmental monitoring system. It is an eco-modernist belief that modern societies are turning towards greater public accountability and transparency of business enterprises, and that this is a prerequisite for softer regulatory regimes. Such a shift can be interpreted as a learning process of actors rooted in the bureaucratic and economic domain to harmonize principles and ethics guiding their activities with characteristics of civic society, such as the right-to-know and democratic participation.

According to the Law on Pollution Charges, the total number of taxed (and hence, supposedly monitored) pollutants is rather high – 151 pollutants. There are 800 air and 4000 water quality standards that remain applicable to polluting enterprises since the Soviet era (Semeniene et al., 1996). In spite of the vast number of pollutants which are supposed to be regularly scrutinized, the monitoring system in Lithuania does not allow accomplishment of this goal in practice. Capacity of environmental authorities to continuously and systematically monitor emissions of particular polluters is rather limited, both technically and in terms of human resources. This is reflected in the results of a survey of Lithuanian companies representing various industrial sectors and geographic regions (see Table 3).[3]

As indicated in Table 3, environmental protection agencies of the Ministry of Environmental Protection are checking companies on average once every six months, whereas 25% of the respondents claim that their companies are checked only once a year. On the other hand, more than 11% of the respondents maintain that EPA is monitoring their emissions once a month. Our analysis did not show any significant correlation between frequency of companies' control vis-à-vis branch of industry, size, geographic location of companies, or any other variable.

A peculiar aspect is that a large proportion of the respondents did not answer this questionnaire at all. One way to interpret this fact is that those companies are monitored by public authorities too seldom to be worth noting. On the other hand, it might indicate lack of tradition of public disclosure, or certain shadow-relations (and possible corruption) in the relationships between controlling agencies and industrialists.

Table 3 *Frequency of environmental monitoring of Lithuanian industrial enterprises (n= 463) by public authorities*

Frequency of monitoring (once per *N* months)	Number of companies checked by respective public authorities					
	Environmental protection agency		Municipal water supply companies		Public health authorities	
	N	%	*N*	%	*N*	%
1	52	11.2	155	33.5	39	8.4
2	13	2.8	5	1.1	8	1.7
3	74	16.0	49	10.6	44	9.5
4	10	2.2	8	1.7	5	1.1
5	3	0.6	0	0	1	0.2
6	76	16.4	14	3.0	48	10.4
8	2	0.4	0	0	0	0
12	110	23.8	34	7.3	90	19.4
18	0	0	0	0	1	0.2
24	1	0.2	0	0	2	0.4
36	0	0	0	0	2	0.4
Missing answers	122	26.3	198	42.8	223	48.2

Pollution charge system reform and conflicting attitudes and beliefs

The shortcomings of the conventional environmental charge system which we discussed in the section above have led to a reform of this system in Lithuania, starting around 1994. It is not our aim to discuss the entire process of policy change. We use this case in order to gather empirical evidence on a possible structural determination of attitudes and beliefs of actors to this policy change, and to interpret revealed attitudes and beliefs vis-à-vis the ideology of ecological modernization. The usage of pollution charge revenues is a good starting point to search for structural reasons for cultural/institutional tensions between various social domains and for conflicts between beliefs and value orientations of various actors. The question of how to reallocate pollution charge revenues is not only an economic one, but a political and social one at the same time. In outlining several options for the use of these revenues, Bluffstone (1995: 13) highlights some important controversial areas:

- use them as a revenue for the state budget, perhaps substituting for other tax instruments used;
- use them to finance environmental investments by commercial enterprises and/or the state and/or municipalities;

– use them as a general revenue source for municipalities;
– return them to the polluters using some methodology to avoid hampering economic growth;
– distribute to the general population in the form of tax cuts or payments;
– distribute to vulnerable populations and/or industries.

The listed ideas and approaches to allocate pollution charge revenues exemplify different principles and ethos implied by possible policy changes. Some reforms would strengthen the bureaucratic search for order, predictability and planning, whereas other shifts in the regulatory regime would be more coherent with the entrepreneurial ethos shared by industrialists. Yet other options would assure greater social equity while running into conflict with the principles and the ethos of businesses and/or state bureaucracy.

According to the current system in Lithuania, 70% of the pollution charge revenues go to the respective municipalities in which polluting industries are located, and 30% go to the state budget. Municipalities are to use those funds only for environmental measures. However, there are no prescriptions what kinds of environmental measures can be funded. In practice, municipalities often use a large proportion of these charge revenues for various infrastructure and public health projects, e.g. planting trees, cultivating public parks, installing garbage collection containers, cleaning water reservoirs, etc. At the same time, pollution charge revenues are seldom allocated back to the polluting companies which paid those charges in the first place. Hence, funds are seldom tailored at eliminating the *sources of pollution.* In general, it might be argued that the allocation of pollution charge revenues by the state bureaucrats in Lithuania works in the opposite direction of what is implied by the metaphor of 'ecologization of economy'. It deepens alienation and mistrust between polluting companies and state authorities, instead of encouraging polluters to take on environmental responsibility. Rather than thinking of possible ways for pollution reduction at source, industrialists tend to blame public authorities and claim that it is the 'function of the Government' to provide subsidies to polluters in order to encourage on-site pollution reduction.

Industrialists vis-à-vis state authority

This controversy is more visible when we contrast the different attitudes of various actors regarding the issue of charge waiver, or tax relief, which essentially reflects belief in or rejection of the (neo-liberal) eco-modernist idea that environmental protection based on personal initiative and social responsibility of business entrepreneurs will be more fruitful for the ecological revival of society than state intervention and domination. An opinion survey was used in our research to contrast attitudes of Lithuanian industrialists and environmental authorities with regard to this controversial issue (see Tables 4-7).

Table 4 *Preferences of industrialists regarding different mechanisms for financing environmental investments (scores from 1 (unimportant) to 5 (very important); n=463)*

Mechanisms for environmental investments	Mean score	Rank
Permit to retain part of profit taxes to be used for environmental investment	3.80	1
Permit to retain part of pollution charges to be used for environmental investment	3.64	2
Soft loan from a specialized revolving Environmental Investment Fund	3.23	3
Waiver of companies 'transitional' debts to the state (e.g. for energy resources) if they would invest equivalent sum in environmental measures	2.87	4
Soft loan from a regular bank	2.83	5

Among the five mechanisms for environmental investment, Lithuanian industrialists favor, albeit not so strongly, the most liberal option – profit tax relief – under condition that an equivalent sum will be invested in environmental measures. Such attitudes and beliefs are coherent with an (theoretically conceived) entrepreneurial ethos and commercial steering mechanism, both dominant among actors rooted in the economic domain/institutional sphere. Taking into account that current pollution charge burden is marginal for most companies, it is surprising from a rational-choice standpoint that the waiver of pollution charge was ranked as the second most favorable mechanism of environmental investment. Such beliefs might be better explained by interpreting expectations of industrialists on the most important stimuli fostering the greening of Lithuanian industry (see Table 5).

As indicated in Table 5, nearly 46% of the industrialists expect environmental regulations, and hence pollution charges, to significantly increase in the future. This might explain their second-ranked preference (see Table 4) of pollution charge waiver which is also coherent with theoretically conceived traits of the economic domain, namely entrepreneurship, flexibility and commerce. This attitude, similar to the first-ranked profit tax relief, implies preference among Lithuanian industrialists for a market-oriented as opposed to a centralized environmental regulatory regime. Their attitudes towards seemingly economic issues – pollution charge waiver and profit tax relief – also reflect their more general value orientations: they are in favor of environmental protection that encourages private initiative and responsibility thereby transferring part of the responsibility from the state bureaucracy to the market forces. This is all in line with eco-modernist beliefs.

Table 5 *Incentives for the greening of Lithuanian industry: opinions of Lithuanian industrialists and environmental authorities*

Incentives	Environmental authorities (n = 321)		Industrialists (n = 463)	
	%	Rank	%	Rank
Escalation of prices for energy and raw materials	57.0	3	73.2	1
Environmental charges and fines	73.5	1	-	-
Strict administrative control by environmental authorities	70.7	2	-	-
Strict control by environmental authorities, high environmental charges and fines	-	-	66.7	2
Expectations that environmental control and standards will strengthen	15.9	5	45.4	4
Pressure by the greens, other NGOs and local communities	11.2	6	6.0	7
Seek to reduce negative impact on nature and humans	5.3	9	51.4	3
Environmental requirements by owners and shareholders	6.5	8	3.4	9
Requirements by business partners (investors, contractors)	9.7	7	4.1	8
Seek to catch up with other peer companies	2.8	10	8.8	6
Requirements by Lithuanian banks and insurance companies	1.9	11	2.4	10
Environmental standards to sell products in foreign markets	38.0	4	14.7	5
Other factors	0.9	12	-	-
Not responded	6.5		24	

On the other hand, as indicated in Table 5, Lithuanian industrialists foresee or expect that administrative control by the state environmental regulatory agencies will strengthen significantly in the future. This is perceived as one of the main stimuli for the greening of Lithuanian industrial enterprises. This expectation can be seen as diverging from both a scenario of institutional change/learning and a command-and-control regulation. Thus, such empirical results indicate a controversy between beliefs and expectations associated with state administrative control and liberal trust of entrepreneurship and responsibility among business actors in the field of environmental protection. On the one hand, from the eco-modernist perspective it might be interpreted as an initial phase of 'institutional learning' and expected to change towards greater preference of market forces over state domination in the future. On the other hand, it might be regarded as an area which is theoretically under-discussed in ecological modernization theory. We will return to these issues below.

The results of the opinion survey of the employees of Lithuanian environmental authorities regarding their attitudes to a pollution charge waiver for industrialists is reflected in Table 6.

Table 6 *Pollution charges relief following environmental investments: opinions of Lithuanian environmental authorities (n = 321)*

Proportion of pollution charge relief waiver	Number of supporting answers	%
100%	67	20.9
50%	176	54.8
25%	57	17.8
0%	15	4.7
Missing answers	6	1.9

Nearly 76% of the state environmental officers are in favor of allowing industrialists to retain at least 50% of pollution charge revenues if this money were invested by respective companies in pollution reduction measures negotiated with and approved by environmental authorities. It should be noted that the latter phrase 'approved by environmental authorities' was included in the question purposefully. The call for charge waiver might be interpreted as coherent with entrepreneurial ethics, whereas the call for 'approval by authorities' matches with a doctrine of order, a mechanism of planning and predictability, and an ethos of formalism characteristic of the bureaucratic domain. In such a way charge waiver as a policy instrument recombines principles and steering mechanisms dominant in two important domains – the bureaucratic and the economic. Implementation of such a policy is based on increasing respect to entrepreneurial ethos among environmental authorities (charge waiver symbolizes it), and might be similar to what in the ecological modernization theory has been called 'institutional learning'.

Positive attitudes of environmental officers and industrialists to regulatory change implementing pollution charge waivers might be viewed as an indication that mutual learning, that is an eco-modernist change towards shared responsibility among industrialists and environmental authorities, is relevant and feasible in Lithuania. The preference of environmental policy change in an 'eco-modernist direction' is also indicated in the responses of Lithuanian environmental authorities to the other, more general, question (see Table 7).

As the survey results indicate, the vast majority (76%) of environmental administrators and policy-makers associate the strongest expectations with flexible economic instruments: soft loans and charge waivers which would enable industrialists to implement cleaner technologies. This is again coherent with an eco-modernist belief that pollution prevention pays. Attitudes of public authorities favoring flexible economic policy instruments can be interpreted as supporting our earlier theoretical point and empirical finding: state environmental bureaucrats are gradually learning to respect entrepreneurial ethos and adjust their policy to the steering mechanism of commerce prevailing among industrialists. The shared eco-modernist ideology is also reflected in policies emphasizing environmental education (ranked second) and provision of information to industrialists about cleaner technologies (ranked fourth).

Table 7 *Attitudes of environmental authorities to different policy instruments (n = 321)*

Environmental policy instruments	*N*	%	Rank
Environmental education and raising people's environmental awareness	225	70.0	2
Strengthening environmental standards and their enforcement	136	42.4	3
Substantial increase of pollution charges and taxes	53	16.5	6
Information about environmentally more benign cleaner technologies	133	41.4	4
Information and training about industrial environmental management	31	9.7	7
Economic instruments (soft loans, pollution charge waivers) to enable companies to invest in waste and pollution reduction measures	244	76.0	1
Dialogue between public authorities and industrialists, encouraging mutual understanding and seeking compromises in order to fulfil business interests and protect the environment	106	33.0	5
Not responded	35	-	-

On the other hand, the policy of 'strengthening environmental standards and their enforcement' is ranked third among other alternatives (42%), and hence the claim about eco-modernist tendency in beliefs and expectations of Lithuanian environmental authorities becomes quite controversial. Earlier we noted a similar controversy reflected in the expectations of industrialists that environmental regulations will become more rigid in the future. This controversy, however, might not only have to do with the responses of the two categories of actors, but also with the theory. The theory of ecological modernizaton emphasizes the shift from administrative to consensual regulation, however, the proportions of command-and-control versus economic/educational instruments and, more normatively, the expectations regarding the time-scale of a shift between the two regulatory regimes remains under-discussed or unclear in this theory. The belief in a mixed-type economic and administrative steering regime is reflected not only in the empirical study of attitudes of Lithuanian industrialists and environmental authorities, but also in the texts of the theorists of ecological modernization: 'Enlarging the steering capacity of the state is regarded as a necessity because the steering potential of markets and market-actors is structurally weak' (Spaargaren, 1997: 14).

This ideological controversy and institutional tensions, primarily between the bureaucratic and economic domain, manifest themselves more strongly in the actual process of environmental policy reform in Lithuania. To illustrate this, we will consider an example of the development of the Lithuanian Environmental Investment Fund (LEIF).[4] One of the main objectives of this fund, according to its statutes, is to provide soft loans to industrial and other enterprises in order to foster implementation of cost-effective waste and pollution reduction measures. Similar revolving funds have been set up earlier in Poland, Hungary and other CEE countries. In Lithuania, this fund was established with strong financial and intellectual support from the US (USAID, Harvard Institute for International Development and particular American consulting enterprises) and the PHARE program of the European Union. The model of LEIF was to a large extent borrowed by reflexively drawing on experiences of similar institutions in other CEE countries, hence undergoing certain process of institutional learning as well.

The fact that the Lithuanian Ministry of Environmental Protection established such a fund implies that the belief in 'pollution prevention pays' is diffusing among state environmental administrators, and they gradually learn to adjust environmental policies and to make them more coherent with the entrepreneurial ethos dominant among economic actors. However, there are some controversial aspects of institutional/cultural tensions in the actual process of developing this fund, indicating that institutional learning is quite fragmented so far. Although LEIF was established following the eco-modernist belief in economic viability of pollution prevention, other steps in environmental policy reform indicate that the search for order, planning and formalistic ethos is still prevailing among ministerial officers. The draft Law on Pollution Charges (as of 1996 and 1997) provides empirical evidence to support this statement. It is envisaged in this new draft law that 70% of pollution charge revenues will be channeled through the state budget to the newly established Environmental Investment Fund (LEIF), whereas the remaining 30% will be channeled to municipalities where the respective polluting enterprises are located. Thus, ministerial officers who drafted this law, albeit believing that pollution prevention pays, are still convinced that centralized reallocation of funds (through LEIF) is most cost-effective and 'optimal' as opposed to the provision in the current law to reallocate 70% of pollution charge revenues back to municipalities. Thus, strong tendencies towards order, planning and predictability are still clearly visible in the actions of the actors rooted in the bureaucratic domain. Paradoxically, the Ministry of Environmental Protection is grounding the centralized pollution charge system reform upon the new Lithuanian Environmental Strategy (1996) which was published by the same ministry, and which – on the contrary – supports several important decentralization-oriented environmental policy principles such as social partnership, shared responsibility, and subsidiarity. The latter inter alia says 'Only the problems which cannot be locally resolved should be addressed at a higher level' (*Lithuanian Environmental Strategy*, 1996: 26).

Thus, the institutional tensions between the orientation towards entrepreneurship versus order and planning, de-centralization versus centralized and formalized redistribution of environmental funding (hence power and responsibility) is more clearly visible in the course of practical environmental policy reforms. According to the earlier-mentioned survey results, the majority of Lithuanian environmental administrators share the opposite attitudes and values associating highest expectations with flexible and entrepreneurial policy instruments. However, the short case study of LEIF indicates that the deeds of (some) environmental bureaucrats differ from the sermon. This, as suggested by Hajer (1996) can be interpreted as a cultural politics between those who see ecological modernization as an institutional learning, and those who regard it as a technocratic project.

Conclusions

Ecological modernization can be viewed as an ideology – a set of values, beliefs and expectations regarding environmentally informed processes of social and institutional change of modern society – shared by various actors. It opens up opportunities for a new typology of actors depending on their self-identity with regard to various aspects of an eco-modernist ideology, distinguishing between those who believe in institutional learning and those who regard it as merely a technocratic cosmetic adjustment and societal illusion. It can also be described as a process of competition between relevant actors and groups whereby certain dominant approaches are socially constructed, beliefs and expectations institutionalized. This conceptualization opens up new space for empirical sociological inquiry into the process of environmentally driven change of modern society.

When drawing analytical lines along particular institutional spheres/domains – political, economic, academic, civic – it is possible to contrast actors and to look for correlation between their attitudes, beliefs and value orientations vis-à-vis their institutional roots. In this paper, we have only in a very fragmented manner considered actors rooted in the bureaucratic and economic domain, and tried to contrast their attitudes, beliefs, expectations, and actual behavior. We also tried to show that 'institutional learning' can be regarded as a gradual adjustment between the institutional spheres, alleviating arising cultural tensions and reaching closure between conflicting discourses, principles, doctrines, steering mechanisms and the ethos of actors associated with different domains. Some empirical evidence from the transitional society in Lithuania shows that this approach might be fruitful both for sociological inquiry into ecological modernization ideology and actual processes of environmental policy reform in various countries.

One of the issues which in our view remains under-discussed in ecological modernization theory is the relation between rational choice 'economizing of ecology' vis-à-vis the moral imperative as an impetus for environmentally friendlier production and consumption. This aspect is particularly important when looking at changes in countries such as Lithuania which have kept certain 'Arcadian' beliefs and values regarding society-nature relations. Such a controversy of beliefs and values is reflected, for example, in the fact that Lithuanian industrialists rank the pursuit to reduce harm to nature and humans third among the most important stimuli for the greening of Lithuanian industry (see Table 5). By contrast, environmental administrators rank this factor lowest which indicates tensions in beliefs and expectations regarding the moral imperative as opposed to eco-modernist rationalization of environmentalism. Ecological modernization theorists accept that the issue of nature's intrinsic value remains underscored in this theory, and that it 'narrows the concept of nature to the sustenance base' (Spaargaren, 1997: 85).

In spite of those drawbacks we would argue in favor of ecological modernization theory as a framework to carry out further theoretical and empirical research into environmentally informed social and institutional change of modern societies – including Lithuania.

Notes

1. This table has been developed by Andrew Jamison for the course on Environment, Technology and Society taught at the universities of Lund, Sweden, and Kaunas, Lithuania.
2. Such an analytical framework is applied in the international comparative project PESTO (Public Participation and Environmental Science and Technology Policy Options), involving six European countries (UK, Italy, Netherlands, Sweden, Norway and Lithuania) (Jamison, 1997).
3. The survey was conducted by Harvard Institute of International Development (HIID) in the framework of Central and Eastern European Environmental Economics and Policy project, supported by USAID, the US Agency for International Development. The questionnaire was designed by Leonardas Rinkevicius, Kaunas University of Technology, in collaboration with Randall Bluffstone, HIID.
4. Established by the *Decree of the Minister of Environmental Protection of Lithuania, 11.16.1991.*

Bibliography

Aron, R. (1967), *Les Etapes de la Pensee Sociologique*. (Russian translation, *Etapy Razvitya Sociologiczeskoy Mysli* (1993).) Moscow: Progress.

Beck, U. (1992), *Risk Society. Towards a New Modernity*. London: Sage Publications.

Beck, U., A. Giddens and S. Lash (1994), *Reflexive Modernization. Politics, Tradition and Aesthetics in the Modern Social Order*. Cambridge: Polity Press.

Bluffstone, R. (1995), *Achieving Environmental and Fiscal Goals in Lithuania Using Pollution Charges*. Working paper. Harvard Institute of International Development, February.

DiMaggio, P.J. and W.W. Powell (1991), 'The Iron Cage Revisited: Institutional Isomorphism and Collective Rationality in Organizational Fields', in W.W. Pawell and P.J. DiMaggio (eds), *The New Institutionalism in Organizational Analysis*. Chicago and London: The University of Chicago Press.

Eder, K. (1996), 'The Institutionalization of Environmentalism: Ecological Discourse and the Second Transformation of the Public Sphere', in S. Lash, B. Szerszynski and B. Wynne (eds), *Environment, Risk and Modernity*. London: Sage Publications.

Eisenstadt, S.N. (1968), 'Social Institutions: the Concept' and 'Social Institutions: comparative study', in D.L. Sills (ed.) *International Encyclopedia of Social Sciences*, vol. 14. New York: Macmillan and Free Press, pp. 409-429.

Fisher, K. and J. Schot (1993), 'The Greening of the Industrial Firm', in K. Fisher and J. Schot (eds), *Environmental Strategies for Industry*. Washington D.C.: Island Press.

Giddens, A. (1984), *The Constitution of Society*. Cambridge: Polity Press.

Giddens, A. (1989), *Sociology*. Cambridge: Polity Press.

Hajer, M.A. (1995), *The Politics of Environmental Discourse. Ecological Modernization and the Policy Process*. Oxford: Clarendon Press.

Hajer, M.A. (1996), 'Ecological Modernization as Cultural Politics', in S. Lash, B. Szerszynski and B. Wynne (eds), *Environment, Risk and Modernity*. London: Sage Publications.

Jamison, A. and E. Baark (1990), *Technological Innovation and Environmental Concern: Contending Policy Models in China and Vietnam*. Discussion paper Nr. 187. Research Policy Institute, University of Lund.

Jamison, A. (1997), 'Introduction', in A. Jamison and P. Ostby (eds), *Public Participation and Sustainable Development: Comparing European Experiences. Pesto Papers 1*. Aalborg: Aalborg Universitetsforlag.

Janicke, M. (1985), *Preventive Environmental Policy as Ecological Modernization and Structural Policy*. Berlin: WZD (IIUG paper 85-2).

Law on Environmental Protection of the Republic of Lithuania. Vilnius: Parliament of the Lithuanian Republic, Decree Nr. 5-75, 21.01.1992.

Law on Pollution Charges of the Republic of Lithuania. Vilnius: Parliament of the Lithuanian Republic, Decree Nr. 11-274, 21. 03. 1991.

Lithuanian Environmental Strategy and Action Programme. Vilnius: Parliament of the Lithuanian Republic, Decree Nr. I-1550, 25.09.1996.

Luobikiene, I. (1995), *Jaunimo Vartotojisku Orientaciju Kaita (Change of the Consumerist Orientations Among Young People)*. Doctoral dissertation. Kaunas: Technologija.

Melnikas, B. (1995), 'Transformacijos Rytu Europoje: Valdymo Procesai ir Raidos Mechanizmai' (Transformations in Eastern Europe: Processes of Management and Mechanisms of Change), in *Socialiniai mokslai: vadyba* Nr. 1(2). Kaunas: Technologija.

Mol, A.P.J. (1995), *The Refinement of Production. Ecological Modernization Theory and the Chemical Industry*. Utrecht: Van Arkel.

O'Riordan, T. (1981), *Environmentalism*. Second edition. London: Pion.

Pepper, D. (1993), *Eco-socialism. From Deep Ecology to Social Justice*. London: Routledge.

Pepper, D. (1994), *The Roots of Modern Environmentalism*. London: Routledge (first published by London: Croom Helm, 1984).

Rinkevicius, L. (1992), 'Environmental Responsibility in the Period of Transition to a Market Economy: a Case of Lithuania', in *Extended Producer Responsibility*. Lund: UNEP IE/PAC and Lund University.

Rinkevicius, L. (1997), 'The Greening of Lithuanian Industry: Past and Present', in R. Hillary (ed.), *Environmental Management Systems and Cleaner Production*. London: John Wiley and Sons.

Semeniene, D., R. Bluffstone, L. Cekanavicius (1996), *The Lithuanian Pollution Charge System: Evaluation and Prospects for the Future*. Working Paper. Harvard Institute of International Development, January

Simmons, P. and B. Wynne (1993), 'Responsible Care: Trust, Credibility, and Environmental Management', in K. Fisher and J. Schot (eds), *Environmental Strategies for Industry*. Washington D.C.: Island Press.

Sliogeris, A. (1993), 'Kulturine Situacija: Vakarai ir Lietuva' (Cultural Situation: the West and Lithuania), in *Metmenys* Nr. 64, Vilnius.

Spaargaren, G. and A.P.J. Mol (1992), 'Sociology, Environment and Modernity. Ecological Modernization as a Theory of Social Change', in *Society and Natural Resources*. Nr.5, pp. 323-344.

Spaargaren, G. (1997), *The Ecological Modernization of Production and Consumption. Essays in Environmental Sociology*. Wageningen: Landbouw Universiteit.

Szerszynski, B., S. Lash, B. Wynne (1996), 'Introduction: Ecology, Realism, and the Social Sciences', in S. Lash, B. Szerszynski and B. Wynne (eds), *Risk, Environment and Modernity: Towards a New Ecology*. London: Sage Publications.

Weale, A. (1992), *The New Politics of Pollution*. Manchester/New York: Manchester University Press.

Weber, M. (1930, 1978), *Economy and Society: An Outline of Interpretive Sociology*. Berkeley: University of California Press.

Worster, D. (1977), *Nature's Economy: A History of Ecological Ideas*. Cambridge: Cambridge University Press.

Yearley, S. (1996), *Sociology, Environmentalism, Globalization*. London: Sage.

9

Political Modernization Theory and Environmental Politics

Pieter Leroy
Jan van Tatenhove

Introduction

Since the 1970s environmental issues have widely been recognized as important political issues, a growing concern paralleled by the gradual institutionalization of environmental politics. During these years there was a great optimism to solve (urgent) environmental problems with technical measures and legislation. To clean up the most threatening environmental problems, such as water pollution, air pollution, waste and noise, end-of-pipe technologies were used. The policy response to this upsurge involved legislation and the creation of ministries and other administrative bodies dealing with environmental problems. A common feature of substantive environmental policy in Western European countries in the 1970s was the use of traditional administrative regulatory strategies. Policy making relied to a large extent on direct regulation by legal rule: general standards were set at the national level and formed the basis of permits for individual plants or activities.

The renewal of environmental policies and the development of new policy arrangements in the 1980s and 1990s arose to counter drawbacks of the 1970s, such as the 'implementation deficit', the 'fragmented institutionalization' and problems of cross-media transfers. The renewal of environmental policies and the development of new policy arrangements was influenced by two developments: a general discussion of the role of the state in Western European welfare states on the one hand, and discussions within both the environmental movements and the environmental sciences about the directions of modern societies with regard to the solution of the environmental crisis on the other hand.

Firstly, the discussions about the role of the state. As a result of the economic crisis in the 1970s, the idea that the state could blueprint societal developments gradually disappeared. The belief in planning and detailed regulations came under ideological, political and scientific attack, leading to a political discussion about the scope and the level of detail of governmental regulation. As we shall see discussions about political modernization had their effects on the designing and the actual organization of environmental politics.

Secondly, the formation of new policy arrangements was the result of new social, political and scientific constructions of the environmental crisis by the environmental movement and concerned scientists. More generally, to renew environmental policies, lessons were derived from the anti- or de-modernization debate and later the reflexive modernization debate. Characteristic for theories of de-industrialization and counter-productivity was their pessimism about the possibilities to change modern capitalistic society and modern technology. The analysis resulted in a (anti-modernist) rejection of modern society and technology. From the late 1980s these theories were followed up by analysis of the environmental issue in terms of risk society and the more optimistic perspective of ecological modernization. Instead of de-modernization and risk society, the ecological modernization theory emphasizes the benefits of transformations in technological trajectories for environmental reform (Mol, 1995). Moreover the ecological modernization approach is considered as a specific theory of political modernization, with its focus on new forms of political intervention e.g., the changing role of the state (Spaargaren, 1997: 15).

In this chapter we concentrate on the changes in the organization of environmental politics induced by its changing societal context, especially the interrelation between political and ecological modernization, and effects on the (re)production of traditional and new arrangements in environmental politics.

We firstly clarify the perspective of political modernization and its central theses, emphasizing thereby its organizational implications, in casu its consequences for the interorganizational relations between the state, civil society and the market (see next section). Secondly we pay attention to the theory of ecological modernization which, at least in Western Europe, has been presented and developed as a promising perspective for the analysis of recent and coming changes in the organization of environmental politics. We discuss the ecological modernization theory and its central concepts, emphasizing its theoretical and empirical biases. Our criticism concentrates on the (partly implicit) suggestions and assumptions of the ecological modernization theory about the state-market-civil society interrelations (see below). We argue that the ecological modernization theory does not do justice to the multiplurity of processes in contemporary society in general and in environmental politics in particular. New developments in environmental politics (see below), especially the new political arrangements that have been set up, firstly illustrate the actual multiplurity in processes of policy making and policy implementation. Secondly, they indicate the emergence of a political modernization of environmental policies in a way the ecological modernization theory so far does not take into account sufficiently.

Political modernization: changing relations between state, market and civil society

Political modernization is about changing relations between state, market and civil society, and about the way governance and policy making are affected by these changing relations (van Tatenhove, 1996). Inspired by Alexander (1995) we distinguish four phases of political modernization: early, anti, post and reflexive political modernization. Each of these ideal-typical forms of political modernization accentuates specific interrelations between state, civil society and market, implying thereby also specific forms of policy making. To understand the political modernization perspective we firstly discuss some characteristics of the modernization debate, since each of the phases of political modernization refers to a specific theoretical-cum-ideological period in Western post-war social thought.

The modernization debate in general

Early- and anti-modernization The changes in contemporary society are indicated by many authors as a transition from early and anti-modernity to post- and/or reflexive modernity. The modernity-reflexive/postmodernity debate is complex and contradictionary. Different interpretations of early, anti, post and reflexive modernity are given. On a general level modernity refers to 'modes of social life or organization which emerged in Europe from about the seventeenth century onwards and which subsequently became more or less world-wide in their influence' (Giddens, 1990: 1). Modernity is not only the product of the American and French Revolutions, it is in itself a permanent revolution of ideas and institutions. Characteristic for the 'grand narratives' of modernity are History and Progress; Truth and Freedom; Reason and Revolution; Science and Industrialism. These elements of the 'grand narratives' reached a point of crystallization in the great social theories of the eighteenth and nineteenth centuries: in the Age of Reason (Kumar, 1995: 84).

The 1970s showed both a renewal of theories of modernization as radical alternatives such as theories of anti-modernization. Anti-modernization is both a societal and scientific reaction to the unsolved 'reality problems' in the modernization model. Theories of anti-modernization were proposed as more valid explanations of these problems in the 1970s. Scientific expressions varied from inequality-, conflict- and state-centered political theories on the one hand, to counter-productivity or de-industrialization theories on the other hand. The accent was on the emancipation of suppressed groups in society, neglected political issues (for example external effects) and on the introduction of ideas of small-scaleness, autonomy and self-development. A societal expression was the upheaval of new social movements. New social movements played a crucial role in the attack on modernization theory, by defining social reality in terms of conflict, revolution and collective emancipation, more generally speaking, in terms of the emancipation of (bourgeois) society from the state (Offe, 1986).

Post- and reflexive modernization In postmodernity the contents of 'the modern' are completely changed. Not only contemporary society is considered to be fragmented, pluralistic and individualistic, also the dividing lines between the different realms of society: political, economic, social and cultural are said to be broken down. Although

> the irreducible pluralism and diversity of contemporary society is not denied (...) that pluralism is not ordered and integrated according to any discernible principle. There is not, or at least no longer, any controlling and directing force to give it shape and meaning (...). There is simply a more or less random, directionless flux

> across all sectors of society. The boundaries between them are dissolved, leading however not to a neo-primitivist wholeness but to a post-modern condition of fragmentation. (Kumar, 1995: 102-103)

Accentuating fragmentation, pluralism and the absence of a 'centralizing force' has lead to a 'dissolution of the social', not in the sense of denying society as such, but in denying its power as an embodied collectivity (Kumar, 1995: 132).

Reflexive modernization takes up a middle position between early modernity on the one hand and postmodernity on the other hand. According to Giddens (1990, 1994) and Beck (1994, 1996b) contemporary society is in the phase of 'high' or 'radicalized' modernity, which can be characterized by a high degree of reflexivity. Reflexivity not only means reflection, but also self-confrontation (Beck, 1994: 5). According to Beck the project of modernity resulted in the risk society. The risk society is understood as a phase of development of modern society in which the social, political, ecological and individual risks created by the momentum of innovation increasingly elude the control and protective institutions of industrial society. Essential is the unintentional and unseen transition from modern industrial society to the risk society. In other words, risk society is not an option which could be chosen or rejected in the course of political debate (Beck, 1996a: 27-28). Reflexive modernization means for Beck 'self-confrontation with the effects of risk society that cannot be dealt with and assimilated in the system of industrial society – as measured by the latter's institutionalized standards' (Beck, 1994: 6). Although Giddens' definition of reflexivity differs from Beck's definition (Giddens, 1994), both take up a middle position in the modern-postmodern debate. According to Giddens reflexivity of modern life exists 'in the fact that social practices are constantly examined and reformed in the light of incoming information about those very practices, thus constitutively altering their character' (Giddens, 1990: 38). In its essence Giddens understands reflexive modernization as a radicalized form of modernity, by which the premises of the modern industrial society are discussed in order to make new forms of modernity possible.

Political modernization as changing relations between state, market and civil society

Early and anti-political modernization Characteristic for early political modernization is the relative insulation of the subsystems state, market and civil society within the nation-state model. Nation-states are conceived as coherently organized systems whose subsystems are closely interdependent, and each have distinctive rationalities (respectively bureaucracy, competition and solidarity). This insulation of societal subsystems, however, means not a fixed position in relation to each other. The specific interrelations between state, civil society and market and the extent of encroachment depend upon ideological and political preferences, and are giving rise to different institutional arrangements. Examples of different preferences and arrangements, which also specify conceptions of governance, governability and policy making, are neo-corporatism, neo-liberalism and neo-étatism.

In *neo-corporatist theories and policy making* the emphasis is on exchange relations between state and sections of civil society and market. Neo-corporatism is 'a socio-political structure of interest articulation and policy formation, in which functional interest organizations possess a representational monopoly, cooperating between each other and with the state on the basis of political-economic consensus at the top. The participating organizations are granted privileged influence on public policy-making in exchange for disciplining their constituency and restraining their demands' (Frouws and van Tatenhove, 1993: 223). Essential in neo-corporatist policy making are exchange relations, such as public and political prestige, authority, cooperation, disciplining and legitimation, between the state and functional interest organizations.

In *liberal democratic theory and pluralist models of policy making* the emphasis is on the relation between 'the state, as an independent authority with supreme right to declare and administer law over a given territory, and the individual, with a right and interest to determine the nature and limits of the state's authority' (Held, 1989: 48). The concept of the individual refers both to the sphere of citizenship and family life (civil society) as to the sphere of producers and consumers (market). Central problem for liberal democratic theorists is to overcome the dilemma of (state) authority versus (individual) liberty. In other words: how can individuals with social and political freedom of choice be protected against the state as a monopoly of coercive power?

In *neo-étatist theory* the Welfare State model is elaborated in the concept of the *'Keynesian state'*. In this concept the state is seen as a compromise between the liberalization of the international capitalist economy and the subjection of the market in national policies by social democrats. It is a combination of neo-corporatism and parliamentary government. The individual citizen is not only associated with the state by the right to vote, but also by a complex of decision making mechanisms, which constitute the interrelations between the state and categories of interest groups (Stuurman, 1985). In the Keynesian state the emphasis is on the intermingling of state, civil society and market.

A special line of criticism on early political modernization was formulated by anti-modernization political and social theorists. We deal with this kind of criticism not only because it influenced to a great extent the range of thought of the environmental movement, but also because ecological modernization theorists were inspired by the criticism of counter-productivity theory on science and technology. The normative and ideological criticism of anti-modernist theorists concerned the role of science and technology and the role of the state. According to anti-modernist theorists modern science and technology are one-sided, one-dimensional and reductionist. These are seen as the main causes for environmental problems, for alienation and the colonizing of the 'life-world' (Marcuse, 1964; Habermas, 1968). According to counter-productivity theorists and (some) neo-Marxists, the state is suspect for it is conceived as an extension of civil society, reinforcing the social order for the enhancement of particular interests (Held, 1995). As a reaction, forms of more or less radical democratization and self-government are formulated. This process of emancipation of state influence requires a politicization of the institutions of society. Politicization and emancipation are ideological arguments used by both scientists and representatives of new social movements. In policy making this type of political modernization emerges in the call for politicized conceptions of participation and self-government.

Post- and reflexive political modernization In post- and reflexive political modernization not only are the relations between state, civil society and market less clear, but these relations are also analyzed from the perspective of global interconnectedness. The nation-state model is more or less exceeded, and both societal transformations and processes of policy making are analyzed in the context of regional and global interconnectedness. In general globalization implies firstly that many political, economic and social activities are becoming world-wide in scope, and secondly it suggests that there has been an intensification of levels of interaction and interconnectedness within and between states and societies (Held, 1995: 21). Policy making will be highly influenced by homogeneous and heterogeneous tendencies of globalization (compare Robertson, 1992 and 1995; Kumar, 1995; Beck, 1996b): on the one hand central functions of nation-states such as legitimacy and governance seem to be less relevant in the context of global governance, whereas on the other hand new local (inter)regional and supranational policy arenas and arrangements are formed both within and outside the nation-state. Examples are the cooperation of localities, the competition between cities and regions, the clustering of nation-states (EU), and the emergence of global cities as new economic power structures (Sassen, 1994).

Postmodern theorists describe society as a network of loosely connected communities, inventing their own forms of life and finding their own means to express them. The result is a renewed interest for the local, sub-national and regional culture. More permanent institutions and organizations are no longer encased within the framework of the nation-state. Essentially we are dealing with a revaluation of (fragmented) civil societies. Instead of 'grand narratives', such as the emancipation of the rational, the liberation of the exploited, or the belief in progress (Lyotard, 1984), the accent is on a diversity (multiplurity) of local cultures and specific narratives. Dividing lines between state, civil society and markets fade away, and these subsystems merge into fragmented collectivities or networks of calculating and negotiating actors, with different social definitions of reality. Problems defined as policy problems are treated from different perspectives and attempts are made to be in an 'perpetual dialogue' (De Wit, 1995). Policy making takes place in networks, which are constructed around problems with different coalitions of public and private actors. Those networks are governed by local narratives and rules, instead of universal meta-languages or rules. We have to acknowledge that 'any consensus on the rules defining a game and the "moves" playable within it *must* be local, in other words agreed on by its present players and subject to eventual cancellation' (Lyotard cited by Kumar, 1995: 136).

In reflexive political modernization there is not only a relativization of the insulation of subsystems (and a relativization of the role of the nation-state), but also a revaluation of the (global) civil society. For a good understanding of changing relations between especially the state and civil society we use Beck's distinction between 'rule-directed politics' and 'rule-altering politics'. Rule-directed politics functions within the rule-system of the nation-state. Rule-altering politics on the other hand concerns altering the rules of the game. In reflexive political modernization 'rule-directed' and 'rule-altering' politics are mingled, overlapping and interfering with each other. This results in a metamorphose of the nation-state and a re-organization of governmental tasks. 'The state must practice self-restraint and self abnegation, give up some monopolies and conquer others temporarily and so forth' (Beck, 1994: 41). According to Beck the essence of contemporary politics is (reflexive) sub-politicization of society.

> Sub-politics (...) means shaping society *from below*. Viewed from above, this results in the loss of implementation power, the shrinkage and minimization of politics. In the wake of sub-politicization, there are growing opportunities to have a voice and a share in the arrangement of society for groups hitherto uninvolved in the substantive technification and industrialization process: citizens, the public sphere, social movements, expert groups, working people on site; there are even opportunities for courageous individuals to 'move mountains' in the nerve centers of development. Politicization thus implies a decrease of the central rule approach; it means that processes which had heretofore always run friction-free fizzle out in the resistance of contradictory objectives. (Beck, 1994: 23)

More generally sub-politics refers to politics outside and beyond the representative institutions of the political system of nation-states. Essentially this form of 'direct' politics refers to the constitution of a global civil society of ad hoc 'coalitions of opposites', in opposition to modern political institutions.[1]

Intermezzo So far we have discussed phases of political modernization and their impact on the interrelations between state, civil society and market and thereby, on patterns of policy making. An important difference between early political modernization on the one hand and post- and reflexive political modernization on the other, is the context of policy making and the relations between public and private actors. In early political modernization the accent is on rule-directed arrangements. This means that the initiative of policy making is with public actors, who define both the rules of policy making processes and the types of interaction between public actors and private actors from civil society and market. Another characteristic is that policy making takes place within or is guided from the context of the nation-state. In post- and reflexive modernization on the contrary the accent is on rule-altering policy arrangements and the intermingling with rule-directed arrangements. In rule-altering policy arrangements private and public actors themselves define the rules of policy making processes. These arrangements often exceed the nation-state. New coalitions are formed crossing the classical divisions and traditional boundaries of individual nation-states.

However, policy making, in particular environmental politics, is not only affected by changing relations between state, civil society and market, i.e. by political modernization itself. Environmental policy making is also affected by the way environmental problems are perceived and constructed, by the way political arrangements to tackle them are set up, and by the way (environmental) policy discourses are legitimizing those constructions and arrangements. According to Jänicke (1993) reflexive (political) modernization is both the medium and outcome of the ecological crisis. On the one hand environmental policy making is influenced by reflexive political modernization in general. On the other hand, however, innovations in environmental politics have been provoked more particularly by the nature of environmental problems and the obvious ineffectiveness of environmental policies so far.

The concept of ecological modernization deals with these latter issues. It has been put forward as both a theoretical concept for the analysis of environmental (policy) problems, and as a political program to direct environmental policy (Spaargaren and Mol, 1997). In the next section we will discuss the concept of ecological modernization and the kind of political modernization it presupposes.

Ecological modernization

The concept of ecological modernization

Since it has been brought forward, the concept of ecological modernization has been described in many different ways. Its basic idea – the emergence of an ecological sphere, introducing and institutionalizing an ecological rationality – however, is very much related to the classical sociological understanding of processes of modernization. Modern social theory has been interpreting modernization as a process of differentiation of a 'holistic', 'one sphere'-society into different and, subsequently, relatively insulated sub-spheres: the economic sphere, the political sphere, the personal life sphere etc. Each of these spheres was developing its own rationality, its own guiding principles and institutions. On a similar basis society was regarded as differentiated into subsystems such as the state, the market and civil society, ruled respectively, as we stated before, by bureaucracy, competition and solidarity. These modernization and differentiation processes are said to be induced by a modern, capitalist economy, by modern techniques, and to be reinforced by and reinforcing in turn cultural changes such as secularization and the dissemination of (economic) rationality in all spheres.

Though modernization in general has been regarded as positive, from the early 1970s especially, attention was drawn to the side-effects of these comprehensive processes of modernization, in particular to its impacts on the physical environment. They gave rise to, among others, anti-modernist or de-modernist ideologies, theories and utopias, which played a central role in the phrasing and articulating of environmental concern (Berger et al., 1974). We have been discussing the basic characteristics of this anti-modernist ideology in the previous section. For a reconstruction and interpretation of the rise of environmental concern among (social) scientists and the subsequent emergence of a sociology of the environment, see among others Buttel et al. (1990), Dunlap and Catton (1992), Leroy (1996) and Spaargaren (1997). Though in a different way, both Luhmann (1989) and Giddens (1990) interpreted the 'ecological crisis' as a series of specific consequences of the process of modernization, especially as the 'disembedding' of society out of its physical and natural context.

Ecological modernization, therefore, was conceived primarily as a process of re-embedding of society within its physical context and constraints, as a necessary next step in modernization, induced not by economic but by ecological rationality. Some authors suppose this ecological rationality to be necessary for mere politico-normative reasons (Dryzek, 1987), whereas others see ecological modernization as the emergence and emancipation of a new societal sub-sphere, the ecological one, not only as a result of growing environmental concern, but as the institutionalization of ecological rationality in society. Huber, the founding father of ecological modernization theory, was among the latter.

From this point on, one should make a difference between two basically different concepts of ecological modernization: (a) its use, function and value as an analytical concept and a theory of social change on the one hand, and (b) its use and functioning as a political program on the other (Spaargaren and Mol, 1997). In the latter, the concept of ecological modernization is primarily used on a practical level to orient and direct environmental policy. The two are narrowly related and interwoven in both scientific literature and political praxis (Weale, 1992), whereby the analytical value of the concept is not always paralleled and furthered by its political ambitions. In this section we do pay attention mainly to the analytical concept, regarding ecological modernization primarily as a theory of (supposed or actual?) social change.

This social change can, as said above, be characterized as the emergence of ecological rationality vis-à-vis the predominant economic rationality, and as the emancipation and institutionalization of the former into an ecological sphere. Ecological modernization theory thus, as most classical modernization theories did, suggests an ongoing process of societal differentiation into newly emerging sub-spheres. Apart from the economic, the political and the socio-ideological sub-spheres, the ecological sub-sphere arises. These sub-spheres should of course not be conceived as material worlds, but as analytical distinctions. It is striking that the ecological modernization theory, in its basic assumptions of sub-sphere differentiation and even in its terminology, fully unintentionally, comes close to Parsonian schemes of functional differentiation, used by Luhmann (1989) as a framework for his 'ecological communication' thesis.

The process of differentiation and emancipation of an ecological sphere is not meant to end up in the abolition of economic rationality (as anti- or de-modernist theories and ideologies in the 1970s suggested), but in a new kind of equilibrium between economic and ecologic rationality (Mol, 1995). This is what Huber calls the 'ecologization of economy' on one hand, and the 'economization of ecology' on the other. The 'ecologization of economy' refers to the physical and organizational changes in production and consumption processes. The 'economization of ecology' refers to the economic valuation of environment and nature, which are recognized to be the third force of production (apart from labor and capital).[2]

The ecological modernization theorists differ, among other things, on the question as to who initiates (or, is expected to initiate) the processes of 'ecologization of economy' and 'economization of ecology'. As Spaargaren (1997) states rightly, Jänicke, at least in his earlier work, stressed the responsibility of the nation-state. State intervention, instrumented by a new industrial policy, a new energy policy etc., should be the motor of ecological modernization. Huber on the contrary regards state intervention as an obstacle to environmental reform. Unlike both modernist and anti-modernist theories and ideologies, he sees no other role for the state but to formulate general standards and to create favorable conditions. Apart from that, he believes ecological modernization primarily to be the responsibility of 'self-regulation' on the market, implying a transfer of responsibilities from the state to the economic sphere of production and consumption, and based upon technological and organizational innovations initiated by these economic actors, especially by the industrial sector itself. In his empirical research based upon ecological modernization theory, Mol (1995) has been looking at recent processes of restructuring within Dutch chemical industries (paint, plastic and pesticide industries). He assesses both the analytical and the explanatory power of ecological modernization theory and concludes that its main strength is its integration of theoretical concepts and insights from distinct areas (technology innovation and diffusion theories, economic theory, environmental management and business strategy, political theories etc.). While it is true that his research reveals the different (power) relations and networks around each of the industry branches mentioned, the differences between them do not allow a single and univocal conclusion on the empirical validity and reliability of ecological modernization theory, on the initiating mechanisms and motives of industrial restructuring etc.

The theoretical biases of ecological modernization theory

Despite the efforts made by Jänicke (1993) and Mol (1995), there is still little empirical evidence for the ecological modernization theory – conceived as a theory of social change, especially as far as the basic initiators and mechanisms of this change are concerned. This doesn't mean however that ecological modernization theory is merely rhetoric. As Hajer (1996: 249-250) states, the concept, even if it might be regarded as 'greenspeak', has been very influential, both on the conceptualization of environmental problems and politics by environmentalists and industrialists, by politicians and scientists, and on the actual development of contemporary environmental politics in for example the Netherlands and Germany. But most of this is related to ecological modernization as a concept for praxis, as a political program.

Though the analytical (added) value of ecological modernization theory seems probable, as Mol has illustrated, questions arise when looking at its value as a theory of social change. Partly based on arguments brought forward by other authors, we would like to criticize some specific biases and weaknesses of the ecological modernization theory. This criticism, both internal and external, affects its theoretical validity and reliability, especially when the ecological modernization theory is related to (theories about) political modernization.

As Spaargaren (1997) has stated rightly one of the biases of ecological modernization theory so far has been its unilateral emphasis on (the innovation of) industrial production. Moreover, the concepts and sub-theories involved are focused upon the larger, both economically and ecologically important branches and sectors with a high potential of technological process and/or product innovation and a positive cost-benefit or investment-impact balance (chemical industry, automobile industry on one hand; bio-technology on the other). Much less attention has been paid to more traditional (energy production, steel, metal) or smaller (housing, furniture) industrial sectors, or to the agricultural sector, the retail business etc. This empirical focus or bias is directly related to what Mol (1995) has been describing as the typical Western scope of the ecological modernization theory. We would argue that the theory is intrinsically based upon the reality of only a few Western, highly industrialized countries with, one should add, a high level of environmental concern, an already institutionalized environmental policy and a well developed, politically accepted and professionally organized environmental movement. As one can easily enumerate those and other societal and political conditions for 'ecological modernization' to be likely, it is very questionable whether the theory is applicable in either politically, economically or culturally different situations. Ecological modernization seems to presuppose (a certain degree and a certain kind of) political modernization. We will come back to this point later.

Spaargaren adds another bias: the 'productivist' orientation of the ecological modernization theory, or its neglect of the consumption aspect of ecological modernization in general and of the actual position and possible role of consumers in the supposed processes of social change more specifically. He has been trying to overcome this neglect in combining elements of the ecological modernization theory with elements of a sociology of consumption (Spaargaren, 1997: 161-205). Though a fair attempt, his suggestions and amendments only partly succeed in restoring the balance in the ecological modernization theory for two main reasons. One is that the (mainly socio-cultural) starting points of his sociology of consumption do not fit easily into the (mainly techno-economically) basis of the ecological modernization theory. The second is that, where his suggestions are plausible on a theoretical level, they definitely need empirical testing and – as the ecological modernization theory as a whole – empirical evidence.

This leads us to what probably is the most important criticism about the ecological modernization theory: its optimism about ecological modernization to occur, as if it was a very probable, if not 'natural' process. As nobody seems to oppose it – and as the theory neglects power relations, there hardly is any room for opposition at all (Hannigan, 1995 and Blowers, 1997), ecological modernization is assumed to occur almost automatically. Or as Hannigan (1995: 184) states: 'all that is needed (...) is to fast forward from the polluting industrial society of the past to the new superindustrialized era of the future'.

In the previous section, we made a clear distinction between ecological modernization used either as a theoretical concept or as a political program. But this is the point where the latter affects and even infects the former. It is true that the ecological modernization concept, when launched in the mid-1980s, meant a break with some discourses, predominant till then, in the conceptualization of environmental problems and environmental policy. The plausibility of these discourses was disputable indeed. The ecological modernization concept, in its 'Huberian' not in its 'Jänickean' variant for instance, challenged the idea that the state had – and was able – to play an initial and steering role in the design and implementation of environmental policies. As we have been arguing earlier, ecological modernization in 'Huberian' view implied the transfer of responsibilities from the state to the economic actors, especially to the industrial production sectors themselves. By doing so, the ecological modernization theory fitted well with the neo-liberal political offensive of the 1980s. But what is the plausibility of these discourses (on 'less state', 'self-regulation' etc.)? In a similar way, the ecological modernization theory broke with the idea that environmental regulation could ever provoke technological innovation within the industry. But is self-responsibility succeeding to do so? The empirical evidence brought forward so far about the effectiveness of covenants and self-discipline as instruments for environmental policy is not very encouraging. In a similar way, ecological modernization was expecting growing environmental concern, sensitivity for the business image and competition over environmental soundness to induce both process and product innovations, while the market situation in many industrial sectors looks quite different and does not make ecological modernization very likely to occur (see below).

In short, ecological modernization theory in our view does pay too less attention to the institutional (pre)conditions for ecological modernization. It does so by insulating the ecological sphere from the other spheres: the political, the economic, the socio-ideological (or cultural), hardly giving attention to the developments in the latter ones, and suggesting that the ecological sphere, as if it were Baron van Munchausen, could in itself bring about a social change from a situation of environmental deterioration into a situation of environmental soundness.

The (hidden) assumptions of ecological modernization theory

From the argument above, it is clear that the ecological modernization theory, although already comprehensive in many respects, has to be enlarged and completed by linking it to insights on social change in general and on political and social modernization in particular. Before we draw some empirical evidence from actual developments in (Dutch) environmental policy, its organisation and implementation (next section), we try to identify the assumptions of the ecological modernization theory. One could list these assumptions in different ways. According to the argument formulated before, we prefer to identify preconditions assumed to appear at the state level, on the market and within civil society and, most importantly, the assumptions about the interrelations between them. By looking from these three perspectives successively, we try to clarify the (hidden) assumptions of the ecological modernization theory and their theoretical and empirical tenability.

The market The ecological modernization theory is based upon a lot of (hidden) assumptions about the situation on 'the market'. We cannot and will not try to enumerate them all, but they mainly deal with (a) the existence of a relative stable situation on the market itself, characterized among other things by the existence of balanced economic networks between the industries themselves, as appears from (common) research and development efforts, from the participation in and the loyalty vis-à-vis the branch organizations, etc.; (b) the existence of a relative stable and balanced relationship between the economic actors in the market and the state; and (c) the existence of a similar relationship with the consumers. While distinguishing an economic, a societal and a policy network, Mol (1995) refers to almost similar categories. The empirical evidence he brought forward on the Dutch chemical industry (Mol, 1995: 355-379) makes clear how different, both in morphology, organization, homogeneity, internal solidarity, modernity, culture etc., the sectors involved are, and how these differences influence the actual presence, the momentum, the pace and the direction of ecological modernization. However, he does not draw conclusions about the tenability of the ecological modernization theory's assumptions in such a variety of situations and contexts.

Another aspect is the actual or possible role of the consumer as an initiator of ecological modernization. It may be clear from the above that ecological modernization theory is suggesting an active, initiating role of the consumer indeed. He or she is assumed to be concerned, well informed, critical and emancipated, challenging entrepreneurs to offer more environmentally sound products, and adapting his or her consuming behavior to the (ecological) demand and supply situation. Spaargaren (1997), inspired by Otnes and others, depicts this role of the consumer in ecological modernization, thereby to some extent correcting the 'productivist' bias of the theory. But at the same time his assumptions and suggestions are seized by analyses made by Bauman, Beck and others, about a consumer highly dependent on technological equipment and on expert systems which he is not able to assess, let alone to handle. In short, they depict a consumer highly dependent on the information (monopolistic organized) industries are willing to give. Therefore, even on a theoretical level, the assumptions of the ecological modernization theory about the role of the consumer seem questionable, for they suggest a powerful consumer, whereas other authors criticize that suggestion. The division of power between industries and consumers on the market is unequally balanced, which will affect the initiating and countervailing role of consumers (see next section).

In short, there is little evidence so far for the development of the power relations on the market to reinforce ecological modernization. The actual organization of many industrial sectors is not favorable for the development of any sensitivity within those sectors about ecological modernization. On the contrary, the privatization of some of the public facilities (see next section) enhances their sensitivity for economic constraints and may therefore even endanger their environmental concern. Even if one assumes the consumer to be concerned and well informed, his or her role and influence seems rather limited.

The state Ecological modernization theory meant a break with some unrealistic expectations about 'the state' and its role in environmental policies. Mol (1995: 46) summarizes statements by Jänicke, Huber and other theorists of ecological modernization when stating that 'bureaucratic environmental policy is criticized for being, among other things, inflexible, economically inefficient and unjust, a brake rather than a motor of technological innovations, unable to monitor and control the billions of material and energy transformations taking place each day, and incapable of stimulating companies to adopt more progressive environmental behavior'. A severe criticism, pertinent in many respects, though caricatural in some others.

But having said that, what is the position and role of the state in ecological modernization theory? As we pointed out earlier, the ecological modernization theory in its 'Huberian' variant is suggesting 'self-regulation' as steering principle, with the role of the state limited to providing favorable conditions for self-regulation. This means (a) a greater emphasis on steering by economic mechanisms and instruments (taxes and subsidies; risk-liability and civil law insurances; responsible consumers challenging entrepreneurs to creativity etc.), and (b) a greater emphasis on decentralized and participatory policy making (area specific policies; target group policies; an emancipated environmental movement and other NGOs participating in decision making etc.).

However, some other indications do give rise to questions about the changing role of the state, e.g., the nation-state as suggested by ecological modernization theory. There is empirical evidence of some nation-states at the same time stimulating environmentally sound technology and investing in the testing of new nuclear weapons (France), or the safe-guarding of a transport of nuclear waste

(Germany). In both cases, neither the policy strategy (from direct regulation to self-regulation) nor the policy organization (from an authoritarian to a participatory policy) was changing along the suggestions of the ecological modernization theory. It is true that both decisions have been contested severely, by Greenpeace, by neighboring nation-states and by an almost global public opinion in the former case, by both locally and nationally organized environmentalists in the latter. Nevertheless it seems to us to be too optimistic and premature to assess, as Beck (1996b) does, the Mururoa case – along with the Brent Spar case – as an example of (successful) environmentally driven sub-politicization on a global level. In fact the authority of nation-states recently (re-)emerged in some other issues and policy domains. The nuclear industry is the example 'par excellence'. Illustrative is also the way different European nation-states have been initiating major works concerning the infrastructure, such as the building of new airports, new highways and new railway tracks – with the 'Chunnel' to be conceived as 'the mother of all infrastructural works'. Though less harming environmentally than the examples mentioned above, no one will dispute the relevance of these decisions and investments for environmental policy and environmental quality. And yet it is hard to see any of the (hidden) assumptions of the ecological modernization theory to be brought in practice. Nation-states on the contrary do operate in these cases in a very classical way. Decision making procedures and styles may differ according to national politico-institutional traditions and differences. In general decision making in these cases is top-down, state-initiated, rather authoritarian than participatory, and with little respect to local protest. We will draw some more conclusions on the role of the state from the (different) styles of actual decision making in environmental policies in the next section.

Civil society The ecological modernization theory assumes a high level of societal and political self-regulation, thereby apparently suggesting balanced power relations between the most important actors. These assumptions are questionable, as we indicated earlier, as far as the market relations are concerned: generally speaking, the consumer is not a real counterpart, let alone a countervailing power to the transnationally organized, quasi-monopolies of the major economic sectors.

And what about the citizen? Is he or she able to oppose successfully to either nation-state or industry decisions with negative impacts on the environment, and to initiate and promote environmentally sound decision making? During the 1970s and 1980s, the development of environmental policy has been accompanied by the opening up of some traditional decision making procedures, such as permit applying, and the creation of some new ones, such as the environmental impact assessment. This renewal of procedures was not only aiming at a better environmental quality, but also an increase in public participation. In many respects this renewal can be regarded as an answer of the nation-state to the protest of the early environmental movement, urging for more public participation in environmental decision making and thereby as a form of political modernization as mentioned before. The actual influence of these forms of participation, however, is very limited as empirical research has made clear (see Valkenburg, 1995). From the late 1980s new forms of participatory decision making developed in environmental policy and related policy domains. As they can be regarded as the result of both political and ecological modernization, giving rise to new policy arrangements, we will draw some conclusions on their actual functioning below.

(New) policy arrangements in environmental policy

In this section we concentrate on the question of how both political and ecological modernization may have influenced the (re)production of traditional (rule-directed) arrangements and new (rule-altering) arrangements.

Since the ecological modernization theory has been launched, it has indeed been of great influence on the strategy formulation, the instrumentation and the organization of environmental policy on different levels: both international organizations (OECD, EU and others) and nation-states (USA, Canada, Germany, the Netherlands and others) have been introducing and implementing major changes in the designing and actual instrumentation and organization of their environmental policies (Weale, 1992). But so far, it is hard to judge whether these changes indicate a fundamental change in the division of roles between the state, the market and civil society, or a more or less opportunistic adaptation of environmental policy to a new, more or less neo-liberal political culture. The changes in organization, designing and implementation moreover seem to be oriented at problem solving or searching for capabilities for problem solving. In other words progress, uniformity and rationality still seem to be guiding principles.

In general, however, in post- and reflexive modernization theory, modern ideas of Progress, Rationality, Uniformity have lost their plausibility. In the risk society progress and rationality will evoke only new problems, which makes reflexivity and multiplurity in governance, policy making and public participation necessary. Theoretically and empirically observed changes presuppose a shift in political modernization from policy arrangements, based on rationality, uniformity and control, into policy arrangements based on reflexivity, multiplurity and self-control. To what extent is this transition taking place? In other words: how reflexive and multiplural are processes of policy making? To answer this question we firstly elaborate on new directions in political modernization in environmental policy and confront these with the core assumptions of the concept of reflexivity, such as self-confrontation, self-control and the radical reinvention of society and politics.

Rule-directed arrangements Firstly, still existing *rule-directed arrangements* based upon neo-corporatism, neo-étatism or neo-liberalism are changing as a result of processes of globalization and reflexive modernization. In the Netherlands for example the agricultural *neo-corporatist policy community* has been weakened by external pressure and internal dissent corollary to the 'crises' of surplus production and agro-environmental pollution. Although there is a change to de-corporatization, this change has not yet resulted in new socio-political arrangements in agriculture (Frouws, 1996). Despite governmental intentions for 'self-regulation', traditional tasks such as implementation, maintenance and control continue to be a task for the central government. Also environmental decision making and implementation, both on a national and international level, went through what can be seen as a process of political modernization since the 1980s.

Illustrative for this renewal of environmental policy is the creation of new consultative bodies in which citizens and NGOs could participate. Some of these consultative bodies, especially on the national level, can be characterized as neo-corporatist: only a limited number of actors, e.g., pressure groups, acknowledged as such by the nation-state, can get entrance to these circuits of decision making. With its recognition as a legitimate partner of the authorities, the environmental movement was allowed to negotiate with the formerly acknowledged pressure groups, e.g., the employers and the unions (the Netherlands, Flanders). Part of the environmental policy field seems indeed to be organized along the lines of (allocative) arrangements known from the socio-economic policy fields, while in other cases (see below) new kinds of negotiating networks are set up, mostly with the (trilateral) participation of public authorities, representatives of private economic actors and the citizen, represented either by the established environmental movement or by newly formed grass roots pressure groups.

During the 1980s the global ideology of the 'New Right' promoted 'the rolling back of the state'. In several Western countries the idea that only individuals can judge what they want and, therefore, the less the state interferes in their lives the more freedom they will have to set their own priorities (Held, 1989) was translated into politics. Deregulation and privatization initiatives were developed to restrict the role of 'the state' in favor of civil society and especially 'the market'. In some sectors the formation and implementation of policy was privatized, which resulted in different institutional arrangements in public utilities. For example, as a result of privatization in the waste water treatment sector three institutional arrangements can be distinguished: the French model of 'out contracting', the Dutch model of 'delegated public management' and the English model of 'privatization'. These 'quasi autonomous non-governmental organizations' (or quangos) are the institutionalization of the handing over of policy formulation and implementation of public utility tasks from 'the state' to private or semi-private organizations. At the same time major merges took place in the energy production sector (natural gas, electricity, especially nuclear power), but also in the waste water treatment sector, in the household and industrial waste treatment industry, and even in the drinking water supply, leading to transnationally (European) organized companies and even to global (quasi-)monopolies. In short, a lot of crucial industrial sectors – with considerable ecological impacts – and some public facilities sectors – crucial for an environmentally sound management of resources – are organized in a way which cannot be regarded as 'balanced networks'.[3] These developments affect and will affect the role of citizens and consumers. On the one hand it is true that industries, over the last years, more often have had to withdraw some products for environmental harm – albeit most for hygienic or health reasons, and industry is more sensitive than ever for its environmental image. However, consumers hardly succeed to really influence production strategies. On the other hand in the public facilities sector, a sector which is highly monopolistic, major merges and privatization processes seem even to have decreased any possible role of the citizen/consumer. In short, the division of power between industries and consumers on the market is not so equally balanced as ecological modernization theory is suggesting. Therefore it is questionable whether the consumer is able to play the initiating and countervailing role the ecological modernization theory is suggesting.

At the same time on both supra- and national level, *neo-étatist arrangements* are developed. Especially in the case of infrastructure neo-étatist arrangements on a national and supra-national level are formed. In the Netherlands the planning of infrastructure is an example of top down policy making. Even new legal arrangements, like the NIMBY (Not In My Backyard) act, were introduced by the Dutch state to accelerate normal procedures. These regulations qualify 'national interests' of a higher order than 'local interests'. Explicit aim is to rule out NIMBY objections in order to accelerate policy making processes. On the supra-national level a combination of neo-étatist and neo-liberal arrangements are developed to govern international economy. According to Hirst & Thompson (1996: 189) governance of international economy is possible at different levels, varying from agreements between the major advanced industrial states (the G7), the creation of international regulatory agencies by a substantial number of states (WTO), to the control over large economic trade blocs (EU, NAFTA).

Rule-altering arrangements Secondly there is an increasing variety of *rule-altering arrangements* like 'policy networks', 'consensual decision making', 'participative democracy' and 'policy-oriented learning' in processes of environmental policy making, inspired by (contextual) theories on politics and policy making. These contextual theories on policy making are a reaction on the inadequacies of rational theories on policy making, which study policy making processes as linear evolving rational processes. A central assumption is that policy making is the result of interactions between public and private actors. An example is the policy network approach. The concept of policy networks refers both to a heuristic analytical concept and a prescriptive organizational policy instrument. Characteristic for the latter, which seems to be becoming predominant in a lot of policy processes, is the optimism about problem solving and steering capacities of policy networks (Van Tatenhove and Leroy, 1995). In environmental policy making processes network managers and facilitators are introduced to resolve policy problems, by trying to organize policy making processes in such a way that consensus and 'win-win-situations' will be the result. The accent in environmental dispute resolving (Glasbergen, 1995) for example is on the process itself, while structural mechanisms of political power building, influencing the definition of problems and issues at hand, are underestimated. In region specific environmental policy in the Netherlands cooperation, negotiation and deliberation between public and private actors play a major role. In these policy networks the state seems to be changing into a labyrinthian configuration of negotiating actors. Policy reality and problems continuously have to be redefined into new settings of negotiation. Instead of grand narratives, like emancipation, equal rights or welfare, local narratives are ruling policy making 'games'.

Also new arrangements arise from changes in modes of participation. According to Valkenburg (1995) 'old' participation models, like the resource mobilization theory, are in processes of reflexive modernization only relevant as far as they are based on new forms of participation. In reflexive modernization participation means that citizens can make their competence object of a mutual learning process. This participation as organized reflexivity is characterized by the equivalency of its participants ('Herrschaftsfreie Dialog'). In environmental policy different models of participation exist side by side. Traditionally the environmental movement and even Green parties were regarded to be intermediary actors mediating between civil society and the state. New arrangements such as co-production show new forms of participation in which not only the environmental movement is a participant. Co-production is a process of collective image building aimed at the development of a shared policy arena of interdependent public and private actors. Characteristic for the process of collective image building is to realize a shared perception among 'stakeholders'. On the basis of criteria such as representativity, authoritativity and diversity, stake-holders are selected (Bekker, 1996). Co-production is a form of contextual 'steering'. The aim is to gain support and to realize shared definition of the situation. Another new form of participation is Beck's constellation of global sub-politics. He describes temporarily rule altering arrangements on a global scale, such as a coalition between governments, NGOs and business around specific policy issues. 'The activity of world corporations and national governments is becoming subject to the pressure of a world public sphere. In this process individual-collective participation in global action networks is striking and decisive; citizens are discovering that the act of purchase can be a direct ballot which they can always use in a political way' (Beck, 1996b: 21).

Conclusions: from uniformity to multiplurity

How reflexive are the arrangements in the examples discussed above? Or in the rhetorical question of Alexander (1996: 134) 'when has sub-parliamentary politics *not* played a primary role in the social life of industrial societies? Were consciousness and social action really focused only on distributive and material issues before the environmental crises of the 1960s?' To understand recent transformations in policy arrangements we make use of the concept of multiplurity.[4]

Translating the concept of multiplurity to processes of policy making, the present situation can be characterized not so much by the transition from policy arrangements of uniformity and control to consensus-building and negotiating, but by the side by side existence of different types of arrangements, both within the nation-state and exceeding it. Politics is both hierarchic and horizontal, and governance is both uni-centric and multi-centric.

State, civil society and market are not separated sub-spheres. These interrelations take place within an interference zone. In this interference zone different logics and rationalities exist and new arrangements will occur. These new arrangements can be on the level of the nation-state system, such as 'policy networks', 'neo-corporatist policy communities' and arenas of co-production as we have indicated above. Central is the existence of a multiplural or labyrinthian society, with many different issues, many different interests, many different loci of power etc. In this context the state is both authoritarian and negotiating.

The ecological modernization theory seems to be too optimistic about the political modernization of nation-states, because its evolutionary scheme presupposes the evolution to reflexive modernization. The kind of political modernization and the changing role of the state as the ecological modernization theory is suggesting may be plausible for some specific sectors and themes of environmental policy, especially as far as the changes in the formulation of both goals and strategies of industrial pollution policy are concerned. But it is unlikely to see the nation-state playing a similar, 'eco-modernist' role in a lot of other cases and policy fields. Instead of a withering away of the state, we may even ask ourselves whether in some policy domains a neo-étatist style of decision making is (re)emerging. Therefore one has to take into account the multiplurity of the state's interests and roles, the variety of specific situations in different policy fields, and different styles of decision making within society. In other words: the interaction between ecological modernization and political modernization results not only in a high degree of political pluralism and participative policy making as political/ecological modernization theory is suggesting. The role of the citizen, of state representatives and managers is also limited by mechanisms and power structures of neo-étatist or neo-corporatist style and character. On the other hand arrangements will exceed the level of the nation-state system, such as interregional coalitions, supra-national coalitions and global sub-politics. As a result of the globalization of economy, culture, politics and environmental problems, responsible policy makers, citizens etc. have to participate in supra-national, national, regional or even supra-national-regional arrangements.

Notes

1. 'A special feature of this politics of the second modernity is that in practice its "globality" does not exclude anyone or anything – not only socially, but also morally or ideologically. It is, in the end, a politics *without opponents or opposing force*, a kind of "enemy less politics"'. (Beck, 1996b: 19)
2. It is remarkable, by the way, that little attention has been payed so far to the similarities and dissimilarities in the way the production factor 'labor' was emerging and becoming institutionalized in the nineteenth and early twentieth century, and the way the factor 'environment' is supposed to do in our time. Although both can rightly be regarded as forces of production from an economic point of view, one has to realize the differences in their sociological functioning (e.g. as sociological divides and political cleavages), and in the socio-historical contexts they appear(ed) within. The latter have to do, among other things, with major changes in the state-market-civil society relations.
3. Although monopoly capitalism in itself is not regarded as a major obstruction for ecological modernization, it does hamper the possible emergence and the actual influence of any 'countervailing power', either from competing industries, from state initiated environmental regulations or from the ecological concern of consumers.
4. According to Kunneman (1996) the central themes of postmodernity are multiplurity and interference. In postmodern thinking interference refers to the frequent influencing of different subsystems, which on the one hand have their own rules and local logic, but on the other hand those rules and logics are valid only in the reciprocal transfer to other orders (Kunneman, 1996: 186).

Bibliography

Alexander, J.C. (1995), *Fin de Siècle Social Theory. Relativism, Reduction, and the Problem of Reason*. London/New York: Verso.

Alexander, J.C. (1996), Critical Reflections on 'Reflexive Modernization'. *Theory, Culture & Society*, 13 (4): 133-138.

Bauman, Z. (1992), *Intimations of Postmodernity*. London and New York: Routledge.

Bauman, Z. (1993), *Postmodern Ethics*. Oxford: Blackwell.

Beck, U. (1992), *Risk Society: Towards a New Modernity*. London: Sage.

Beck, U. (1994), The Reinvention of Politics: Towards a Theory of Reflexive Modernization, in: U. Beck, A. Giddens and S. Lash (eds), *Reflexive Modernization. Politics, Tradition and Aesthetics in the Modern Social Order*. Oxford: Polity Press.

Beck, U. (1996a), Risk Society and the Provident State, in: S. Lash, B. Szerszynski and B. Wynne (eds), *Risk, Environment & Modernity. Towards a New Ecology*. London: Sage, pp. 27-43.

Beck, U. (1996b), World Risk Society as Cosmopolitan Society? Ecological Questions in a Framework of Manufactured Uncertainties. *Theory, Culture and Society*, 13 (4): 1-32.

Bekker, V.J.J.M. (1996), Co-produktie in het milieubeleid: op zoek naar een nieuwe sturingsconceptie, *Bestuurswetenschappen*, 3: 177-194.

Berger P., B. Berger and H. Kellner (1974), *The Homeless Mind: Modernization and Consciousness*. New York: Vintage Books.

Blansch, C.G. le (1996), *Milieuzorg in bedrijven. Overheidssturing in het perspectief van de verinnerlijkingsbeleidslijn*. Amsterdam: Thesis Publishers.

Blowers, A. (1997), Environmental Policy – Ecological Modernization or the Risk Society?. *Urban Studies*, 34 (5/6): 845-871.

Buttel, F.H., A.P. Hawkins and A.G. Power (1990), From Limits to Growth to Global Change – Constraints and Contradictions in the Evolution of Environmental Science and Ideology. *Global Environmental Change*, 1: 57-66.

Dekker, P. (1995), Civil society als partij-ideologie?. *Socialisme en Democratie,* 2: 62-74.

Dryzek, J.S. (1987), *Rational Ecology. Environment and Political Economy*. Oxford/New York: Basil Blackwell.

Dunlap, R.E. and W.R. Catton (1992), Toward an Ecological Sociology: the Development, Current Status and Probable Future of Environmental Sociology. *The Annals of the International Institute of Sociology*, 3: 263-284.

Featherstone, M. (1995), *Undoing Culture*. London: Sage.

Friedman, J. (1995), Global System, Globalization and the Parameters of Modernity, in: M. Featherstone, S. Lash and R. Robertson (eds), *Global Modernities*. London: Sage, pp. 69-90.

Frouws, J. and J. van Tatenhove (1993), Agriculture, Environment and the State. The Development of Agro-environmental Policy-making in the Netherlands. *Sociologia Ruralis*, 2: 220-239.

Frouws, J. (1996), Politieke modernisering van het Groene Front?. *De Sociologische Gids*, 1: 30-45.

Giddens, A. (1985), *The Nation-State and Violence*. Cambridge: Polity Press.

Giddens, A. (1990), *The Consequences of Modernity*. Oxford: Polity Press.

Giddens, A. (1994), Living in a Post-Tradition Society, in: U. Beck, A. Giddens and S. Lash (eds), *Reflexive Modernization. Politics, Tradition and Aesthetics in the Modern Social Order*. Oxford: Polity Press.

Glasbergen, P. (ed.) (1995), *Managing Environmental Disputes; Network Management as an Alternative*. Dordrecht: Kluwer Academic Publishers.

Habermas, J. (1968), *Technik und Wissenschaft als Ideologie*. Frankfurt am Main: Suhrkamp.

Hajer, M.A. (1995), *The Politics of Environmental Discourse: Ecological Modernization and the Policy Process*. Oxford: Clarendon.

Hajer, M.A. (1996), Ecological Modernisation as Cultural Politics, in: S. Lash, B. Szerszynski and B. Wynne (eds), *Risk, Environment and Modernity. Towards a New Ecology*. London/Thousand Oaks: Sage, pp. 246-268.

Hannigan, J.A. (1995), *Environmental Sociology. A Social Constructionist Perspective*. London/New York: Routledge.

Harvey, D. (1989), *The Condition of Postmodernity: An Inquiry into the Origins of Cultural Change*. Oxford: Basil Blackwell.

Held, D. (1989), *Political Theory and the Modern State. Essays on State, Power and Democracy*. Cambridge, Oxford: Polity Press.

Held, D. (1995), *Democracy and the Global Order. From the Modern State to Cosmopolitan Governance*. Cambridge: Polity Press.

Hirst, P. and G. Thompson (1996), *Globalization in Question. The International Economy and the Possibilities of Governance*. Oxford: Polity Press.

Jänicke, M. (1993), Über Ökologische und Politische Modernisierungen. *Zeitschrift für Umweltpolitik & Umweltrecht,* ZfU 2: 159-175.

Kumar, K. (1995), *From Post-industrial to Post-modern Society. New Theories of the Contemporary World*. Oxford/Cambridge: Blackwell.

Kunneman, H. (1996), *Van theemutscultuur naar walkman-ego. Contouren van postmoderne individualiteit*. Amsterdam/Meppel: Boom.

Leroy, P. (1996), *Environmental Science as a Vocation*. Nijmegen: KUN (inaugural lecture).

Luhmann, N. (1989), *Ecological Communication*. Cambridge: Polity Press.

Lyotard, J.F. (1984), *The Postmodern Condition: A Report on Knowledge*. Manchester: Manchester University Press.

Marcuse, H. (1964), *One Dimensional Man. The Ideology of Industrial Society*. London: Sphere Books.

Mol, A.P.J. (1995), *The Refinement of Production. Ecological Modernization Theory and the Chemical Industry*. Utrecht: Van Arkel.

Mommaas, H. (1993), *Moderniteit, vrijetijd en de stad. Sporen van maatschappelijke transformatie en continuïteit*. Utrecht: Van Arkel.

Offe, C. (1986), Nieuwe sociale bewegingen als meta-politieke uitdaging, in: L.J.G. van der Maesen et al. (eds), *Tegenspraken, dilemma's en impasses van de verzorgingsstaat*. Amsterdam: SOMSO staatsdebat, pp. 27-92.

Robertson, R. (1992), *Globalization*. London: Sage.

Robertson, R. (1995), Globalization: Time-Space and Homogeneity-Heterogeneity, in: M. Featherstone, S. Lash and R. Robertson (eds), *Global Modernities*. London: Sage, pp. 25-44.

Sassen, S. (1994), *Cities in a World Economy*. Thousand Oaks: Pine Forge Press.

Spaargaren, G. (1997), *The Ecological Modernization of Production and Consumption – Essays in Environmental Sociology*. Wageningen: Wageningen Agricultural University.

Spaargaren, G. and A.P.J. Mol (1997), Sociology, Environment and Modernity. Ecological Modernization as a Theory of Social Change, in: G. Spaargaren, *The Ecological Modernization of Production and Consumption – Essays in Environmental Sociology*. Wageningen: Wageningen Agricultural University, pp. 63-88.

Spybey, T. (1996), *Globalization and World Society*. Oxford: Polity Press.

Stuurman, S. (1985), *De labyrintische staat. Over politiek, ideologie en moderniteit.* Amsterdam: SUA.

Tatenhove, J.P.M. van (1993), *Milieubeleid onder dak. Beleidsvoeringsprocessen in het Nederlandse milieubeleid in de periode 1970-1990, nader uitgewerkt voor de Gelderse Vallei.* Wageningen: Pudoc.

Tatenhove, J. van and P. Leroy (1995), Beleidsnetwerken: een kritische analyse. *Beleidswetenschap,* 2: 128-145.

Tatenhove, J. van (1996), *Political Modernization and Environmental Policy.* Paper for the International Conference 'Environment, Long-term Governability and Democracy'. France: Abbaye de Fontevraud.

Valkenburg, B. (1995), *Participatie in sociale bewegingen. Een bijdrage aan de theorievorming over participatie, emancipatie en sociale bewegingen.* Utrecht: Van Arkel.

Weale, A. (1992), *The New Politics of Pollution.* Manchester: Manchester University Press.

Wit, Th.W.A. de (1995), De ontluistering van de politiek. Over 'Eén-partij staat Nederland' en postmoderne democratie. *Socialisme en Democratie,* 4: 155-165.

10

Ecological Modernization and Post-Ecologist Politics

Ingolfur Blühdorn

In the context of the widely acknowledged post-Brundtland 'age of the environment', ecological modernization and ecological modernization theory have experienced a comet-like career, and are celebrated as the key to the ecological transformation of late industrial societies. This chapter raises fundamental doubts vis-à-vis the theory of ecological modernization and analyses the corresponding social practices on the basis of the competing *theory of post-ecologist politics* (Blühdorn, 1997). This leads to the thesis that, contrary to its reputation, the practice of ecological modernization must not be seen as a strategy for the preservation or restoration of a certain state of the physical environment. Instead, I will suggest that ecological modernization is a form of societal behavior which responds to the perceived necessity to preserve the myths of modernity. With ecological modernization contemporary societies have found themselves a highly suitable strategy for organizing the peaceful transition to a politics of nature that is largely determined by the principles of power and financial solvency.

Ecological modernization is an unspecific label that is used in several different ways. In the recent literature there are a number of attempts to clarify the meaning of this concept, to spell out different perspectives of society's ecologization, and to assess the adequacy of the ecological modernization approach (e.g. Huber, 1993; Jänicke, 1993; Hajer, 1995, 1996; Mol, 1995, 1996; Christoff, 1996; Spaargaren, 1997). Nevertheless, there is still a lot of uncertainty about what exactly ecological modernization achieves, whether it really takes late modern societies any closer to the desired state of sustainability, and what kind of societal transformation it entails. In this chapter I will attempt to find an answer to some of these questions by breaking the connection between the reformist practice of ecological modernization and the theory of social change that has grown around it. I will suggest that, due to its postmodernist leaning, the *theory of post-ecologist politics* provides a much more suitable theoretical framework for the analysis of environment-induced processes of societal transformation than *ecological modernization theory* can ever hope to offer. After a very brief orientation about the concept of ecological modernization, I will summarize some common – but arguably misguided – points of criticism regarding the suitability of the ecological modernization approach. As judgements about the appropriateness of a particular strategy involves the assessment of the relationship between its means and its ends, section three will undertake an analysis of what actually constitutes the environmental problem. On the basis of this analysis the last two sections are devoted to the post-ecologist reassessment of both the theory as well as the practice of ecological modernization.

Ecological modernization and ecological modernization theory

In the widest sense, the label of ecological modernization covers all reformist environmental efforts since the 1970s which have otherwise also been described as 'environmentalist' (Dobson, 1995). More specifically, the concept is used to refer to an anti-ideological, policy-oriented approach to the environmental problem that superseded both the conservative anti-modernist as well as the leftist-revolutionary approaches typical of the 1970s and early 1980s (Von Prittwitz, 1993a; Hajer, 1995; Mol, 1995; Spaargaren, 1997). The emergence of ecological modernization as a socio-political program of reform indicated the softening of the belief in the incompatibility of high-tech capitalism on the one side and ecological sustainability on the other. As Hajer points out, it signaled the end of 'the sharp antagonistic debates between the state and the environmental movement' (Hajer, 1995: 28f). Given that the structure and achievements of the welfare state seemed to depend on the continuation of the established developmental model, turning away from the path of techno-economic progress no longer appeared as a realistic option. The change of strategy from the 'flight from technology' to 'a technological attack' (Jänicke, 1993: 18) obviously meant 'the acceleration of technological progress', but it was assumed that 'a change in the direction of development' (ibid.) could be achieved at the same time. Particularly since the beginning of the 1990s, so the proponents of this reformist agenda argue, the environment-induced societal transformation can 'no longer be interpreted as mere window-dressing' (Mol, 1995: 2), but the ongoing processes of institutional learning and reorganization amount to a fundamental ecological restructuring of late modern society. In this respect, contemporary ecological modernization is regarded as categorically different from earlier reformist approaches.

Ecological modernization as a theory of social change emerged in the early 1980s. Martin Jänicke and Joseph Huber are widely considered as the founding fathers of this project to devise a theoretical model for late industrial society's attempt to respond to the undesirable side-effects of its own modernization process (Von Prittwitz, 1993b; Mol, 1995; Spaargaren, 1997). In a number of publications Huber has tried to demonstrate the compatibility of economy and ecology, and argued for the necessity to 'ecologize the economy' while 'economising ecology' (Huber, 1982, 1985, 1989, 1991, 1993). Jänicke mainly concentrated on the changing role of democratic institutions, particularly the role of the state (Jänicke, 1978, 1984, 1986, 1988, 1990, 1993). Since the middle of the 1980s he has been arguing for an 'innovative dual structure of the state as a majority legitimated bureaucratic mechanism of intervention and an initiator of processes of negotiation' (Jänicke, 1993: 15). In Jänicke's view, the ecological crisis necessitates and effectively promotes the transition towards 'a more decentralised and consensus oriented kind of politics which focuses the central state on strategic tasks and increasingly devolves the regulation of details to local actors' (ibid.: 24f). A turning point for the development of ecological modernization theory was Beck's concept of reflexive modernization and his discussion of 'the conflict of two modernities' (Beck, 1992, 1995a, 1995b, 1997). For contemporary proponents of ecological modernization theory this element of reflexivity is a major parameter in the social transformation processes they are theorizing (Von Prittwitz, 1993b; Hajer, 1995, 1996; Mol, 1995; Spaargaren, 1997).

An ecologist critique of ecological modernization

As we will argue below, the theory of ecological modernization is probably no longer the cutting edge of green political thought. But whatever we might think about this theory, the practice of ecological modernization is undoubtedly the now dominant response of contemporary European societies to the so-called environmental challenge. As there is no alternative strategy at the horizon, we can safely assume that, in the years to come, ecological modernization will remain the principal approach governments of industrialized countries take to manage their ecological problem. Of course, this does not necessarily mean that ecological modernization is the best possible strategy for achieving our ecological aims. The more radical – but marginal – parts of the eco-movement clearly voice their dissatisfaction with this mode of state policy-making. Also, the fact that most industrialized societies have already been subject to programs of ecological modernization for about one decade does not mean that we have a satisfactory understanding of the kind of societal transformation that is ongoing. On the basis of Maarten Hajer's excellent 1995 book *The Politics of Environmental Discourse: Ecological Modernization and the Policy Process*, whose observations will be supported from the writings of Arthur Mol, Gert Spaargaren and other ecological modernization theorists, we may summarize five central points of criticism regarding the ecological modernization approach.

As Hajer points out, ecological modernization accepts 'the existence of a comprehensive environmental problem' (Hajer, 1995: 28), but frames this problem exclusively in monetary terms and the terms of the natural sciences. On the one hand, this facilitates its incorporation into social, political and economic decision making processes, on the other hand this conceptualization of environmental problems 'explicitly avoids addressing basic social contradictions' (ibid.: 32). To the extent that 'ecological modernization is first and foremost an economy and technology-oriented concept' (Jänicke, 1993: 18), it pre-empts any fundamental ideological conflict and neglects the set of emancipatory concerns which featured prominently in the environmental debate of the 1970s and early 1980s.

Secondly, ecological modernization regards the environmental problem as a 'structural design fault of modernity' (Mol, 1996: 305), or an 'omission in the workings of the institutions of modern society' (Hajer, 1995: 3). Ultimately, the emergence of environmental problems is considered as a management problem which can be solved by means of managerial fine tuning. This fundamentally managerial approach of ecological modernization, Hajer argues, demands 'an almost unprecedented degree of trust in experts and in our political elites' (Hajer, 1995: 11), and it reflects a 'renewed belief in the possibility of (the) mastery and control' of nature (ibid.: 53). Such an approach, however, directly contradicts the ecologist distrust in the expert culture and the demand for democratic involvement at the grass-roots level. Furthermore, the idea of managing, mastering and controlling nature is fundamentally incompatible with the ecologist vision of civilization's genuine reconciliation with nature. Certainly, the proponents of ecological modernization are suggesting that their strategy will 'restore the balance between nature and modern society', and achieve 'a kind of re-embedding' of society into nature (Mol, 1996: 306), but their ideal of an engineered balance evidently differs from the ecologist idea of a reconciliation with nature.

Thirdly, ecological modernization is based on 'the fundamental assumption that economic growth and the resolution of ecological problems can, in principle, be reconciled' (Hajer, 1995: 26). As Arthur Mol confirms, 'the ecological modernization approach diverges from neo-Marxist social theories in that it has little interest in changing the existing relations of production or altering the capitalist mode of production' (Mol, 1995: 41). And Gert Spaargaren seconds, 'there is no principle or theoretical argument making a modern organization of production and consumption and its technology antithetical to sustainability' (Spaargaren, 1997: 16). Whilst the earlier environmental movement was convinced that the principle of growth is unsustainable and has to be replaced by the principle of self-restriction, the proponents of ecological modernization expect that internalizing care for the environment will even initiate new rounds of technical innovation and qualified economic growth. Hajer therefore fears that the strategy of ecological modernization is no more than 'eco-software' which 'will not save the planet if capitalist expansionism remains the name of the game' (Hajer, 1996: 255). Although himself a proponent of this strategy, even Jänicke has suggested that, in the long term, ecological modernization is likely to be unsuccessful as long as its achievements are 'neutralized by high industrial growth' (Jänicke, 1993: 19). In the end, this strategy may enable modern societies to 'push the crisis back into latency' (ibid.: 25), but Jänicke reminds us that unless we address 'eventually also the question of growth' (ibid.: 19), the crisis is unlikely to be 'solved in its causes' (ibid.: 25).

This leads to the more general criticism that ecological modernization theorists believe to be able to solve the ecological crisis within the framework of the existing political, economic and social order. Certainly, there is the demand for 'a reconstruction or rebuilding of some of the central institutions of modern industrial society' (Spaargaren, 1997: 1; see also e.g. Huber, 1993; Jänicke, 1993; Müller-Brandeck-Boquet, 1993; Robert, 1993; Von Prittwitz, 1993b; Zilleßen, 1993; Mol, 1995). Nevertheless, ecological modernization theorists assume that in principle the 'existing political, economic, and social institutions can internalize the care for the environment' (Hajer, 1995: 25), and that 'the environmental issue can be remedied without having to completely redirect the course of social developments' (ibid.: 66). Because of this principle confidence in the existing structures, Hajer describes ecological modernization as 'basically a modernist and technocratic approach to the environment that suggests that there is a techno-institutional fix for the present problems' (ibid.: 32). It was, however, the lack of confidence in techno-institutional fixes, and the argument that the same institutions whose structural inadequacies gave rise to the ecological problem cannot be expected to function successfully as the main instruments of society's ecologization, which made committed ecologists in the 1970s reject merely reformist approaches.

The last and for Hajer probably the most crucial concern vis-à-vis the ecological modernization approach is the emergence of what he calls 'the new environmental conflict' (1995: 8ff). The proponents of ecological modernization are convinced that over the past two decades 'the constant influx of new information about the ecological effects of social practices and institutional arrangements' (Mol, 1995: 394) has established a sufficiently broad foundation for the emergence of an independent ecological rationality that can guide the 'continual redirection of the core institutions' (ibid.). This newly emerging rationality is expected to gradually catch up 'with the long standing dominance of the economic rationality' without implying 'the need for an abolition or abandoning of the economic rationality' (Spaargaren, 1997: 24). Arthur Mol speaks of ecology's 'emancipation from the economic sphere' and the emergence of 'two (increasingly equal) interests and two rationalities' (Mol, 1996: 307). But given the postmodern pluralization and diversification of validity claims which the societal discourse on the environment has to integrate, one may wonder whether such an independent ecological rationality is really available, and whether repairing the structural design fault of modernity might not be substantially more difficult than ecological modernization theorists would want to believe. Hajer argues that although the ecological issue has undoubtedly moved from the periphery into the center of the political debate, and although nobody would seriously deny the existence of an ecological problem, the contemporary eco-debate is still characterized by a 'complex and continuous struggle over the definition and the meaning of the environmental problem itself' (Hajer, 1995: 15). Given the multiple views about what is ecologically necessary and desirable, and their incompatibility with the competing demands for, e.g., economic development and social justice, sufficiently strong foundations for an effective policy of ecological modernization are still not really available.

In particular the first four of these points of criticism reiterate arguments we know from the so-called *Fundi/Realo* controversy (e.g. Wiesenthal, 1993). The common denominator of these basically *fundamentalist* arguments against the *realist* approach of ecological modernization is the deep-seated doubt about the *appropriateness* of this strategy for the ecological problem. If, however, ecologists use the criterion of *appropriateness* to disqualify this reformist approach, they can be expected to support their negative judgement by a proper analysis of the problem, which ecological modernization allegedly fails to address and solve. Yet, when it comes to spelling out what exactly the ecological problem consists of, ecologists can so far only point towards a very poor record of achievement. Against this background there seems to be good cause for remaining sceptical about the ecologist assessment of ecological modernization. Certainly, the point of such scepticism cannot simply be to defend ecological modernization as an *appropriate* strategy in the ecologist sense. It rather seeks to ensure that before we come to any conclusive judgement about the appropriateness of this approach, we first of all have to get a clear understanding of what exactly the ecological problem consists of. It is, after all, not only conceivable that the proponents of ecological modernization have chosen the wrong strategy for resolving the ecological problem, but it could also be possible that the ecologist critics have a wrong understanding of the problem ecological modernization is meant to confront. It goes without saying that with regard to this latter point we should neither rely on the view of ecological modernization theorists, nor on that of their ecologist opponents. What is required is an 'independent' (third party) analysis of the ecological problem.

A post-ecologist analysis of the environmental problem[1]

In the widest sense, environmental problems are phenomena of environmental change. Environmental change, however, is not *eo ipso* problematic. On the contrary, it is an evolutionary normality which raises concern only if it is perceived as environmental degradation. The distinctive criteria of environmental degradation, however, are strikingly underdeveloped. The question: 'When does human-generated environmental change become environmental degradation?' (Goldblatt, 1996: 28) is hardly ever asked in a way that reaches beyond the level of economic reasoning as today's *ultima ratio*.[2] A satisfactory answer has not yet been provided.[3]

In order to distinguish environmental change from environmental degradation we can try to establish empirical criteria in accordance with which certain civilizatory effects or kinds of behavior can be classified as damaging the environment. Ecosystem changes can be analyzed in terms of energy- and substance flows, and so *objective*, scientific definitions might be possible. This is the 'realist' route preferred by the proponents of ecological modernization (e.g. Mol, 1995, 1996; Spaargaren, 1996, 1997), who view nature 'primarily as the material sustenance base to human societies' (Spaargaren, 1996: 8; see also Mol, 1996: 315), and restrict themselves to the 'objective, natural science dimension to environmental problems' (ibid.: 6; see also Spaargaren, 1997: 7f). The obvious drawbacks of this approach are firstly, that it excludes all aspects of nature as a non-material entity, i.e. the 'emotional and sensual experiences, the integrity or intrinsic value of nature, ecological quality, the beauty of the landscape and so on' (Mol, 1996: 315). It should go without saying that such aspects are significant constituents of environmental problems, which cannot simply be bracketed out. Secondly, the sciences may be expected to provide a more or less accurate description of the changes occurring in the physical environment, but they cannot themselves provide normative criteria for the evaluation of the change they measure. In other words, the sciences cannot decide which conditions in or constituents of the natural environment we find worth protecting and why we make a particular choice. Wherever the sciences try to define threshold levels, these remain relative to values which have their origin outside the realm of science.

Alternatively, we can try the *subjective* route of exploring what exactly makes us *concerned* about environmental change. By making the problem-status of environmental change dependent on the emergence of concern, we are accepting the constructionist assumption that environmental problems have a discursive rather than a physical reality.[4] The emphasis is shifted to the question of why we want to protect nature, or certain parts of it, at all. As the spring that drives environmental politics, environmental concern centers around issues like the extinction of species, the finiteness of natural resources, the impairment of human health, the loss of spaces for recreation and inspiration, the destruction of tradition, the responsibility for future generations, etc. The issues are manifold, but in order to facilitate our investigations about the essence of the ecological problem, we may classify environmental concerns into three main categories:

- Concerns about the beauty of nature, its recreational value, the inspiration we gain from it, etc.
- Concerns about personal security, material provision, health issues, and the security of what is generally referred to as the natural foundations of life.
- Concerns about the loss of traditional (moral/religious) values and about human interference with all standards of naturalness.[5]

In the contemporary environmental debate the first category of concerns, i.e. aesthetic issues, play a subordinate role. Firstly, modern environmentalism hopes to increase its credibility by emphasizing its scientific basis and reducing its reliance on evidently subjective criteria. Secondly, no clearly definable catastrophic potential is attached to this dimension of the environmental discourse. Most certainly aesthetic concerns have to be taken into consideration, the really crucial ecological issues, however, are expected to lie elsewhere.

Regarding the second category, the situation seems to be somewhat different. In this context environmentalists discuss the classical triad of problems comprising the finiteness of resources, the growth of the world's population, and the civilizatory emissions into the natural environment. Since the publication of Goldsmith's *Blueprint* we know that the processes of civilizatory development 'are undermining *the very foundations of survival*' (1972: 15; my emphasis). However, this civilizatory process has, at the same time, always been a process of emancipation from nature, which implied that on the one hand, the natural foundations of life were increasingly replaced by artificial surrogates, while on the other hand, the notion of the good life was to an ever larger extent based on 'foundations of life' which had never been natural in the first place. Processes of familiarization and habituation further blurred the distinction between the so-called natural foundations and their civilizatory supplements. As a matter of fact, the experience of naturalness is largely the result of processes of naturalization. It is an established truth that there is no such thing as a state of original naturalness to which civilization could or would even want to revert. Hence, what used to be called the natural foundations of life is more appropriately referred to as *opportunities of life*, which include in a more general sense the conditions and requirements for mere survival as well as a fulfilled life. Undoubtedly, access to and control over these opportunities of life is unevenly distributed. Yet, whilst individuals might find their opportunities of life restricted beyond the bare minimum, there is no indication of the survival of mankind being threatened, as the rhetoric of the natural foundations of life would seem to imply. So the concerns for the so-called natural foundations of life boil down to questions of the acceptable social distribution of opportunities of life. And as long as communities and individuals can develop technological, social and psychological mechanisms which make the results of environmental change, as well as its implications for the social distribution of opportunities of life, *appear natural*, environmental change can undoubtedly continue *ad infinitum* without ever being experienced as problematic.

If neither the aesthetic qualities of nature nor the so-called natural foundations of life appear to be the irreducible essence of our interest in nature and the desire to protect it, we are finally left with the third category of concerns. This category openly addresses what the two others can easily be reduced to: concerns for nature as a normative standard. With regard to aesthetic issues, there could never be any doubt that these are normative issues. In the case of the so-called natural (material) foundations of life, this seemed less evident, yet we could establish that any environmental condition or change only qualifies as an environmental problem, if it does not *appear natural* or acceptable, i.e. if it infringes upon established norms of perception. Ethical concerns, finally, are once again clearly normative concerns.

So our analysis so far has firstly revealed that environmental problems are subjective concerns rather than objective physical realities. Secondly, they are concerns for normative stability. And thirdly, the crucial norm that seems to be worth defending is the norm of *naturalness*. Nature is worthy of protection as an index of norms. We have thus come much closer to a full understanding of the environmental problem, but there remain two questions which we still need to clarify. Firstly, why is it nature that assumes this crucially important role as an index of norms? And secondly, why do we need such an index of norms at all?

Human rationality and the societies which have developed on its basis are evidently full of contradictions and imperfections. There is a wide abyss between social reality and the ideals of reason. On both the individual level as well as the level of society at large, human rationality is fragmentary; on its own it does not make sense. In order to infuse human rationality with meaning, it needs to be tied into – and differentiate itself against – some larger context. Just as a single letter only becomes meaningful when seen within the framework of the complete alphabet, human rationality depends on the framework of absolute reason, i.e. a context that (however invisibly) connects all rational discourses. After the decline of traditional metaphysics, the only context that can fulfil this function, i.e. that can guarantee the meaningfulness of human rationality and life, is nature.[6] Nature is the all-inclusive systematic coherence out of which civilization sought to emancipate itself, but in which it needs to remain embedded. The systematic coherence of nature is the guarantee of the theoretical possibility of the as yet unrealized rational and systematic coherence of society. Nature and naturalness assume the role as absolute and all-embracing norms providing the metaphysical anchorage for all standards of beauty, morality and meaning. At the heart of all concern for nature is therefore the concern for this all-embracing context. Without nature, human rationality invariably becomes contradictory and inconsistent, i.e. it ceases to make sense.[7]

In as much as the systematic coherence of reason, which is guaranteed by the coherence of nature, is primarily essential for the project of modernity and the modern rational subject, concern for nature is ultimately concern for the project of modernity. In a fundamentally modernist fashion, ecologists have sought to hold up nature – the principle of unity – against the postmodernist dissolution of absolute values.[8] To the extent, however, that individuals and communities have learnt to accommodate themselves in conditions of postmodern plurality and fragmentation, the situation has changed, and the ecologist formulation of the environmental problem has become dubious. On-going processes of modernization have fragmented the grand rational coherence. The modernist principle of unity and all-inclusiveness is irretrievably lost. Ecologists have so far believed that their concern – the environmental problem – would remain exempt from this process of pluralization and fragmentation. Yet even the regular re-conceptualization of the ecological problem, from first-generation local problems in the 1970s, via second-generation international problems in the 1980s, to third-generation global problems in the 1990s, did not manage to preserve the ecologist belief in the idea of all-inclusiveness. If, however, the grand coherence which nature symbolized suddenly proves less indispensable than had previously been assumed, the traumatic experience of the loss of nature gradually fades, and the necessity to protect it disappears. In other words, to the extent that the postmodern condition supersedes that of modernity, the essentially modernist concern of ecologists loses its significance. To the extent that we manage to get used to (naturalize) the non-availability of universally valid normative standards, the ecological problem (in the ecologist formulation) simply dissolves.

When we are talking about the dissolution of the environmental problem, two essential clarifications may help to avoid misunderstandings. Firstly, this dissolution does in no way imply that environmental change would come to a halt, that the extinction of species would stop, that the hole in the ozone layer would disappear, or that trends of global warming would reverse. These are physical phenomena which will, of course, continue as before. Yet as we argued above, such phenomena of physical change are not in themselves problematic, but only to the extent that we perceive them as problematic, i.e. their problem status depends on social norms of perception rather than on scientifically measurable empirical data. And provided these norms of perception become sufficiently flexible, the same phenomena which have so far been experienced as catastrophic may assume the status of perfectly acceptable (natural) normality. So, what the dissolution of the (ecologist) environmental problem means is not the disappearance of any physical conditions, but the dissolution of absolute criteria on the basis of which these conditions could be described as unambiguously negative, i.e. problematic for every member of society or even every human being.[9] First and foremost this means that ecologists lose their claim to cognitive or moral superiority, i.e. the basis for their social critique. Ecologists can no longer present themselves as defending the common good against the attack of particular interests. And for the lack of valid ecological criteria they are not just politically but also morally forceless against the societal practice of distributing the opportunities of life according to the principles of power and financial solvency. This radical devaluation of the ecological critique, which indirectly strengthens the politics of social marginalization and exclusion, we may call the transition from an ecologist to a *post-ecologist politics of nature* (Blühdorn, 1997).

Secondly, when talking about the dissolution of the environmental problem, we have to bear in mind that social patterns of perception and norms of evaluation do not change over night. Even though there is plenty of evidence that in contemporary industrialized societies the principle of all-inclusiveness has long been abandoned and superseded by mechanisms of exclusion and marginalization,[10] there is also evidence that the social distribution of opportunities of life along the criteria of power and financial solvency is still widely experienced as barbaric. Certainly, there is broad agreement that the re-installation of the principle of all-inclusiveness, i.e. a re-entry into modernity, is neither philosophically possible nor economically feasible. Nevertheless, the current debate about economic globalization and the resulting 'race to the bottom' in terms of social and environmental standards (e.g. McMichael, 1996) clearly demonstrates that we have not fully embraced the ideas of plurality and exclusion, and therefore continue to have an ecological problem in the ecologist sense. To the extent that contemporary industrialized societies have to be described as late modern rather than postmodern societies, the ecologist discourse continues while the post-ecologist discourse gradually pushes itself to the fore.

Against this background, the politically most crucial and sociologically most interesting question is how society handles this simultaneity of the non-simultaneous. How does late modern society cope with its transition from an ecologist (modernist) to a post-ecologist (postmodernist) politics of nature – a transition which it evidently cannot hold up, but which, from the ecologist perspective, still appears as fundamentally unacceptable? This question represents a kind of post-ecologist reformulation of the environmental problem. It shifts the emphasis from philosophical and moral issues (defending nature as an absolute normative standard) to exclusively strategic-political issues (naturalizing exclusion and pacifying ecologists). With regard to the analysis of the theory and practice of ecological modernization, this post-ecologist reformulation of the ecological problem provides a most interesting point of departure. In the following section we will try to spell out the full implications of this post-ecologist approach.

Reflexivity and self-referentiality: a post-ecologist assessment of ecological modernization theory

From our post-ecologist perspective, we first have to substantiate the claim that ecological modernization theory is largely inadequate as a model for the analysis of contemporary eco-politics. Its most fundamental weakness appears to be what ecological modernization theory shares with its ecologist critics: the complete lack of an analysis into the essence of the environmental problem.

If we follow Arthur Mol, ecological modernization theory is supposed to fulfil a dual function:

> First, ecological modernization is a social theory (...) for analysing the way modern society reacts on and tries to cope with one of its most serious contemporary problems: the ecological crisis. Second, ecological modernization is a normative theory, asserting that processes of environmental reform along the lines of ecological modernization are the best (...) way to conquer the ongoing burdening of modern society's sustenance base. If, and only if, modern society follows the path set out by the ecological modernization theory can the ecological crisis be controlled and eventually solved. (Mol, 1995: 49)

However, with regard to the analytical-descriptive function, one may wonder how we can expect to gain any insights about the ongoing ecological transformation of society, if we base our investigations on a theory that does not undertake any deeper analysis of the problem which is meant to trigger and determine this transformation. With regard to the normative function of this theoretical model, it is even more surprising how ecological modernization theorists can claim to offer the only available solution to the eco-problem, if they have so simplistic an understanding of what this problem actually consists of. The validity of the conceptualization offered above may, of course, be contested, but our analysis has definitely demonstrated that we cannot simply go back to the practice of regarding environmental problems and ecological risks as objective external physical realities which merely need to be internalized into the economic system. Even if we were in principle prepared to accept the assumption that the 'burdening of modern society's sustenance base' is the problem, we would certainly want to know *whose* sustenance base this is meant to be, *what* we may count towards it in terms of its constituents, and *whom or what* we may regard as the burden that is to be removed.

Furthermore, given that the ecological problem remains so strikingly undertheorized, we can hardly expect much of the ecological rationality which according to ecological modernization theorists has now largely emancipated itself from its economic counterpart, thus fulfilling the essential precondition for the subsequent integration of the 'two equally valued domains' (Mol, 1995: 30). If there can be any talk of an independent ecological rationality that is at least vaguely comparable to the economic rationality, we may expect to be told the ecological pendant to the unambiguous economic code of *payment or non-payment*. Any closer investigation, however, immediately reveals that such a code is not easily available.[11] Ecological modernization theory, however, does not even recognize this problem, let alone solve it. With its positivist belief that the environmental problem is of a scientifically measurable physical nature, that we are by and large in agreement about its definition, and that an ecological rationality (our ecological common sense) will tell us how to deal with it most effectively, ecological modernization theory reproduces the full set of modernist beliefs.

Its language of the 'emancipation of ecology', the 'institutional self-reformation' of late modern society, and of the project of 're-embedding' that 'cannot be a reversal of the historical dis-embedding' (Mol, 1995: 29), clearly places ecological modernization theory in the immediate vicinity of early critical theory.[12] Yet, whilst the Frankfurt School was at least aware that the project of enlightenment about the Enlightenment invariably leads into problems of self-referentiality,[13] ecological modernization theorists have no qualms about preaching the ecological self-reformation of the institutions of modern society. Despite its flirtations with the idea of reflexivity, ecological modernization theory undoubtedly remains what its proponents are keen to avoid: 'a belated exponent of the old phase of simple modernization' (Mol, 1995: 393). It fails to take into account that after the abolition of the grand rational coherence any ecological rationality that could guide the institutional transformation of modern society becomes arbitrary in terms of its underlying fundamentalisms. Unsurprisingly, ecological modernization theorists make little effort to spell out or even legitimate the set of values they regard as ecological. Similarly, they become extremely vague when it comes to justifying why exactly the economic rationality should be restricted at all. In this respect, the following quotation can be regarded as paradigmatic:

> *Because of the impossibility of life, and especially meaningful life*, if ecological qualities are not safeguarded, other rationalities (...) should only be allowed to function within the strict boundaries set by ecological rationality. (Mol, 1996: 308; my emphasis)

It is interesting to note that Mol makes reference to what our above analysis of the environmental problem showed to be the crucial point: the category of meaning. But from our above investigations we know that such a statement cannot be sustained. There is no justification for the normative *should*; there is no evidence for the impossibility of life; and even the category of meaning is sufficiently flexible to adapt itself to changing social and cultural circumstances.[14] Unless ecological modernization theorists spell out *whose* life they are talking about, *what* the concept of meaning is meant to signify, and *why* it is required at all, they fail to address some of the most crucial underlying questions. From an eco-politics point of view the confidence in the capability of ecological modernization theory is, of course, fully destroyed if we are told that 'as a typically Western theoretical framework' ecological modernization theory 'is not equipped' to deal with questions related to 'the international distribution of (ecological and other) costs and benefits' (Mol, 1995: 55). As we have seen above, the distribution of – and the participation in – opportunities of life are, more than anything, the central questions an adequate theory of contemporary eco-politics has to deal with. We may therefore legitimately wonder what can be expected from a theory that light-heartedly declares to have 'little attention for aspects of democracy or inequality in connection with environmental reform' (ibid.: 401).

It should go without saying that these deficiencies of ecological modernization theory and the impossibility to establish one independent and unambiguous ecological rationality that could guide society's self-transformation towards sustainability do not mean that modernization processes – which might even be described as ecological improvements – are once and for all impossible. There is plenty of evidence that production and consumption processes can be and are refined in such a way that they come closer to whatever is regarded as ecologically correct and desirable. Efficient mechanisms are available by means of which any monetary value can be placed on whatever is considered as endangered and worthy of protection. These refinement processes, however, do not take the whole of modern society any closer to an objectively definable and fixed state of sustainability, but they are part of a dynamic and multi-dimensional process which involves several competing ideas about what is (ecologically) necessary, and in which both the ecological goal posts as well as the means of achieving them are permanently being reviewed. The crucial point about the concept of reflexivity is that it does not only subject established practices and institutions to reform but, at the same time, also constantly redefines the yardstick of improvement and progress. If reflexivity is taken seriously, it completely replaces the principle of linearity by that of circularity. This total abolition of modern linearity, however, is the crucial point ecological modernization theorists are not prepared to accept. In their fundamentally modernist attempt to internalize the external, they reproduce the modernist dualism (and linearity), and refuse to take account of the fact that nature and ecology are no longer external, but social constructs which are multiple, and can be adapted as required. Against this background, the process of society's so-called ecologization can either be described as a process of constant change without any sense of direction, or – given the firmly established relations of power – as the implementation of very specific interests which have successfully been transfigured as ecological. It is evident that a theory aiming to describe the processes of social transformation in response to the ecological crisis necessarily remains inappropriate, if it does not take into account the transition from the (late) modernist principle of reflexivity to the postmodernist principle of self-referentiality. In particular it remains inappropriate if it does not spell out what kind of values have been installed as ecological, and how their implementation is being organized. Keeping these limitations of ecological modernization theory in mind, we can finally ask what exactly the practice of ecological modernization achieves, and to what extent it may be regarded as appropriate and successful.

Non-convinced ecological communication: a post-ecologist reassessment of the social practice of ecological modernization

At the end of section three we came to the conclusion that despite the ongoing dissolution of the (ecologist) ecological problem, a theory of post-ecologist politics has to confront a problem that we described as the simultaneity of the non-simultaneous. At least from the humanist perspective of ecologism the transition from the ecologist principle of inclusion to the post-ecologist principle of exclusion appeared as barbaric, and the question emerged through what mechanisms modern society can make this transition bearable. The suggestion I now want to make is that this required mechanism is the practice of ecological modernization. With ecological modernization contemporary societies have found themselves a political practice which combines elements of an ecologist and a post-ecologist politics of nature. On the one hand, ecological modernization emulates the conditions of modernity, while on the other hand it does not obstruct the implementation of social exclusion. What ecological modernization preserves or restores is, as we have seen, not any specific state of the physical environment, but the modernist belief in the nature-society dualism, and the illusion of rational progress and control. In a post-ecologist context where binding ecological norms and standards of ecological responsibility are not available, and the social distribution of opportunities of life is largely regulated by the principles of financial and political power, ecological modernization simulates progress towards a more rational/ecological organization of society for the benefit of universal human wealth and welfare. Environmental hyperactivity reassures us that progress and improvement are possible, and reproduces the old Greenpeace belief according to which 'the optimism of action is better than the pessimism of thought'. At the same time, ecological modernization secures the continuation of post-ecologist politics by avoiding a re-ideologization of the environmental debate. Under conditions of fundamental uncertainty and insecurity the re-installation of invariably ideological concepts of nature as guiding principles of social organization remains an attractive option. This could imply chances of development as well as threats; in any case it would mean the abolition of postmodern plurality and the re-entry into a new modernity. In both a positive and a negative sense, ecological modernization is a protective device against such an emergence of new modernities, i.e. a device of securing the continuation of post-ecologist politics.

Ecological modernization promotes and facilitates the continuation of the established socio-economic practices (with adaptations) while at the same time confirming the belief that society is performing the ecological U-turn. Through mechanisms like ecological taxation it delays the exhaustion of whatever society regards as its ecological limitations (specific resources, available waste dump capacities, biodiversity, etc.), and thereby makes time for a three-dimensional process of adaptation: firstly, for the naturalization of new (surrogate) materials, strategies and patterns of behavior replacing what had formerly been perceived as natural; secondly, for the naturalization of the social impact of these new materials, strategies and practices, i.e. increasing social exclusion; and thirdly, for the full emergence of the postmodern individual whose self-understanding and meaning-requirements differ from those of the modernist subject. Once this postmodern individual has fully emerged, this will put an end to the simultaneity of the non-simultaneous, the experience of an ecological problem, and the possibility of ecological criticism. Wherever this threefold transition has not yet been successfully completed, late modern society identifies new imperatives of avoidance, i.e. the need for further ecological modernization.

In the light of these considerations, it would now appear that whilst the *theory* of ecological modernization is as insufficient as the ecologist ideology it sought to replace, the *political practice* of ecological modernization is a highly appropriate strategy for tackling late modern society's environmental problem in its post-ecological reformulation. Its most central characteristic is its attempt to cope with the postmodern situation of fundamental uncertainty and insecurity, which emerges from the postmodern pluralization of (ecological) knowledge and rationality. Borrowing an expression Niklas Luhmann coined, we may describe ecological modernization as the *Ecology of Not-Knowing*. Whether it was intended by Luhmann or not, this title of a 1992 essay can be read in two different ways: firstly, it refers to a particular kind of ecological politics; secondly, it implies the question of how late modern society can cope with being reduced to mere beliefs, i.e. how we accommodate ourselves in a condition where we cannot refer to any reliable normative standards. In his essay, Luhmann does not elaborate on this ambiguity, but it is evident that both of these aspects are highly relevant in our context. As Luhmann points out, the most important requirement in a situation of fundamental uncertainty is the transition from the *communication of persuasion* (*Überzeugung*) to the *communication of agreement* (*Verständigung*):

> This involves a societal code of practice that values discreteness and does not make any effort at all to persuade those who have to get on with one another that they should change their beliefs. The point is not to convince anybody or influence them in any other way. (Luhmann, 1992: 194)

Under conditions of radical uncertainty, so Luhmann points out, late modern societies have to develop a 'culture of non-persuasive communication' (ibid.: 202), which is 'satisfied with *agreement*' (ibid.: 194; my emphasis), and in which 'the only objective is cease-fire' (ibid.). Obviously, this Luhmannite notion of agreement is diametrically opposed to the Habermasian understanding of the concept.[15] Luhmann goes beyond any normative belief in rationality and re-designates the concept of agreement to a condition where the fundamental incompatibility between different opinions is sufficiently invisible, i.e. where communication partners have managed to find formulae which leave sufficient room for individual (contradictory) interpretation. It is immediately evident that the main obstacle to the achievement of communicative cease-fire in Luhmann's sense is any kind of ethics and ideology including all forms of rational fundamentalism. Eco-ethics and ecological fundamentalism strongly base themselves on the power of their supposedly better argument and insist on the possibility – indeed the necessity – to convert society at large. Therefore, they are not 'particularly beneficial to the achievement of agreement' (ibid.: 195), and, as Luhmann points out, they should be introduced into the ecological discourse 'only when it aims to break off communication' (ibid.: 196). But the practice of ecological modernization succeeds in bringing about the agreement ecological fundamentalism fails to achieve.

In conclusion to these considerations, ecological modernization can therefore finally be described as society's *post-ecologist discourse of non-convinced ecological communication*. As we have seen, this discourse renounces ethics and ideology. Its participants are increasingly aware of the fundamental uncertainty on all sides. Given the multiplicity of ecological diagnoses, interests and remedies, they rehearse a culture of knowing about each other's cognitive limitations and political restrictions. Ecological modernization practices a culture of tolerance and respect which seems most adequate in a framework of fundamental uncertainty. In the interest of an adequate assessment of ecological modernization, it is important to note that contrary to what the ecologist criticism seems to imply, ecological modernization has not been purposefully installed in order to serve the interests of only some sectors of society.[16] Instead, it has emerged as a societal practice that is beneficial to all parties involved. What it achieves is the continuation of communication in a context that obstructs Habermasian understanding. Most certainly, Hajer is fully right with his suggestion that 'ecological modernization will set the tone of environmental policy-making in the years to come' (Hajer, 1995: 262). But the reason for this is not, as ecological modernization theorists would like to believe, that it paves the way to ecological sustainability. And it is equally erroneous to argue that this practice will stay with us although it is utterly inappropriate. Ecological modernization is here to stay because it is a highly appropriate remedy for late modern society's problem of making the transition to a post-ecologist politics of nature bearable. As the UN have recently demonstrated in Kyoto, it preserves the belief in rational progress and improvement, and secures social peace whilst the opportunities of life are being redistributed according to the principles of wealth and power.

Notes

1. In this section I am drawing on – and clarifying – ideas from my article 'A Theory of Post-Ecologist Politics' (1997).
2. As a matter of fact, very similar questions *are* being raised in the context of the debate on *sustainability* (for a very helpful overview see Dobson's (1995) analysis and typology of environmental *sustainabilities*). For its lack of philosophical interest and experience, however, this debate conspicuously avoids unpacking the normative implications of sustainability.
3. In the recent literature, particularly studies with a more or less clearly developed constructionist leaning have pointed the direction for the required research (e.g. Yearley, 1991; Hajer, 1995; Hannigan, 1995; Dickens, 1996; Eder, 1996a). As a general trend, however, these studies focus on the social mechanisms by which particular issues

are constructed and installed as focuses of public concern, but they do not provide an answer to the question of why such concern can emerge at all (on the limitations of most social constructionist approaches and suggestions for a refined constructionist research agenda see Blühdorn, 1998).

4. In order to pre-empt any 'realist' criticism of this approach, it might be worth pointing out that considering environmental problems as discursive constructions rather than physical realities does neither amount to denying 'that the world, including its living components, really does exist apart from humanity's perceptions and beliefs about it' (Soulé and Lease, 1995: xv), nor does it dispute that ecological problems have a physical dimension. The criticism that ecological constructionists are trying to talk problems away – see, for example, the wide spread reservations vis-à-vis Niklas Luhmann's view of the ecological issue (Beck, 1988: 165-76; Rucht and Roth, 1992; Metzner, 1993) – in order to 'justify further exploitative tinkering with what little remains' of nature (Soulé and Lease, 1995: xv), rests on the implicit assumption that the discursive reality of social perceptions is *less real* than that of the physical sciences. In a context in which (democratic) policy making evidently responds first and foremost to the vote-securing concerns of the electorate, and in which scientific experts and counter experts continue to generate competing versions of the allegedly objective physical reality, such an assumption is equally ignorant and obsolete. For a more detailed discussion of the constructionist/realist dispute see Blühdorn, 1998.
5. Goldblatt tries a similar distinction. As will be argued below, neither Goldblatt's categories nor the ones suggested above, can really be considered as three irreducible categories. Nevertheless, this distinction reflects the *prima facie* perception of a clear difference between aesthetic, material, and ethical concerns.
6. Habermas's attempt to establish a 'communicative reason' that unfolds in a utopian 'ideal speech situation' may be regarded as the project of finding an alternative post-metaphysical source of meaning (Habermas, 1984, 1987). The issues of nature and ecology, on the other hand, have never featured prominently in Habermas's work (see e.g. Eckersley, 1990, 1992).
7. Hence Bill McKibben's concern that after the end of nature 'there is nothing but us' (1990: 55), and 'we can no longer imagine that we are part of something larger than ourselves' (ibid.: 78).
8. This is not to say that any environmental activist would answer the question for her/his motivation at this abstract level. Obviously a whole series of much more immediate reasons can be given for environmental activism and the necessity to protect nature. Nevertheless, any rational discourse about the question of why we should protect nature ultimately leads to the principle of all-inclusiveness, or the modernist belief in the universality of reason.
9. In as much as such criteria have really never existed, the dissolution of the (ecologist) environmental problem is, of course, no more than the realization that ecologists used to universalize one particular form of rationality, not noticing that they were installing fundamentalisms which are difficult to justify once exposed to social discourse.
10. One may, for example, think of the official farewell to the ideal of full(-time)-employment, the dismantling of the welfare state, and the re-emergence of wide spread poverty even in traditionally rich countries. The ever more evidently uneven social distribution of wealth, and the far-reaching suspension of the individual rights of the poorer strata of society is increasingly regarded as inevitable if not legitimate. Mechanisms restricting access even to essential goods and services to those who can pay for them are widely accepted. Provided the goods in question have been defined as ecologically sensitive, such mechanisms are often even celebrated as so-called eco-taxes.

11. Two possible candidates may be the binary pairs *natural/unnatural* and *sustainable/unsustainable*. In both cases, however, we clearly make reference to values which precede the respective dualism, i.e. which can neither be deduced from the respective distinction, nor measured by its criteria.
12. Early critical theory was, of course, not concerned with environmental problems in today's sense. Nevertheless, its conceptualization of the nature/society relationship, and its investigations into the possibilities of a rational self-critique of society are extremely valuable for contemporary ecological thought. For the relationship between critical theory and ecological thought see e.g. Eckersley, 1992; Dobson, 1993.
13. This problem of the self-referentiality of post-transcendental reason was the point of departure for both main contemporary theoretical currents: on the one hand the attempt to reform and thereby save critical theory (Habermas and Beck), and on the other hand, the post-critical project of systems theory (Niklas Luhmann).
14. As we argued above, meaning emerges from the embeddedness of human rationality into the grand rational coherence which is symbolized by nature. Phrased in more analytical terms, meaning always emerges from *distinction* and *exclusion* – in our case the distinction between the rational and the natural, and the exclusion of the natural from the rational. Obviously, distinction and exclusion can not just be applied in the modernist sense (one single underlying dualism), but just as well in a postmodernist sense, which then, of course, fragments the societal coherence. Nevertheless, distinction and exclusion within society can still function as a source of (obviously no longer all-embracing, metaphysical) meaning.
15. For Habermas the process of reaching agreement implies 'a process of mutually convincing one another in which the actions of participants are coordinated on the basis of *motivation by reasons*' (Habermas, 1984: 392; emphasis omitted in the translation). Habermas, in other words, retains the normative belief in (communicative) reason.
16. In a 1996 article Hajer rehearses the widely shared argument that ecological modernization is not 'the product of the maturation of the social movements', but rather 'the repressive answer to radical environmental discourse' (Hajer, 1996: 254).

Bibliography

Beck, U. (1988), *Gegengifte. Die Organisierte Unverantwortlichkeit.* Frankfurt: Suhrkamp.

Beck, U. (1992), *Risk Society. Towards a New Modernity.* London/Newbury Park/New Delhi: Sage.

Beck, U. (1995a), *Ecological Politics in the Age of Risk.* Cambridge: Polity Press.

Beck, U. (1995b), The Reinvention of Politics: Towards a Theory of Reflexive Modernization, in: U. Beck, A. Giddens and S. Lash, *Reflexive Modernization. Politics, Tradition and Aesthetics in the Modern Social Order*. Cambridge: Polity Press, pp. 1-55.

Beck, U. (1997), *The Reinvention of Politics. Rethinking Modernity in the Global Social Order*. Cambridge: Polity Press.

Beck, U., A. Giddens and S. Lash (1995), *Reflexive Modernization. Politics, Tradition and Aesthetics in the Modern Social Order*. Cambridge: Polity Press.

Blühdorn, I. (1997), A Theory of Post-Ecologist Politics. *Environmental Politics*, 6 (3): 125-147.

Blühdorn, I. (1998), Construction and Deconstruction: Ecological Politics after the End of Nature, in: A. Warhurst (ed.), *Towards an Environmental Research Agenda*. London/Basingstoke: Macmillan.

Böhme, G. (1992), *Natürlich Natur. Über Natur im Zeitalter ihrer technischen Reproduzierbarkeit*. Frankfurt: Suhrkamp.

Christoff, P. (1996), Ecological Modernization, Ecological Modernities. *Environmental Politics*, 5 (3): 476-500.

Dickens, P. (1996), *Reconstructing Nature. Alienation, Emancipation and the Division of Labour*. London/New York: Routledge.

Dobson, A. (1993), Critical Theory and Green Politics, in: A. Dobson and P. Lucardie (eds), *The Politics of Nature. Explorations in Green Political Thought*. London/New York: Routledge, pp. 190-209.

Dobson, A. (1995), *Green Political Thought*. London/New York: Routledge.

Dobson, A. and P. Lucardie (eds) (1993), *The Politics of Nature. Explorations in Green Political Thought*. London/New York: Routledge.

Eckersley, R. (1990), Habermas and Green Political Thought. Two Roads Diverging. *Theory and Society*, 19: 739-776.

Eckersley, R. (1992), *Environmentalism and Political Theory. Towards an Ecocentric Approach*. London: UCL Press.

Eder, K. (1996a), *The Social Construction of Nature*. London: Sage.

Eder, K. (1996b), The Institutionalisation of Environmentalism: Ecological Discourse and the Second Transformation of the Public Sphere, in: S. Lash, B. Szerszynski and B. Wynne (eds), *Risk, Environment and Modernity. Towards a New Ecology*. London/Thousand Oaks/New Delhi: Sage, pp. 203-23.

Giddens, A. (1994), *Beyond Left and Right. The Future of Radical Politics*. Cambridge: Polity Press.

Goldblatt, D. (1996), *Social Theory and the Environment*. Cambridge: Polity Press.

Goldsmith, E. (1972), *A Blueprint for Survival*. London: Tom Stacey.

Habermas, J. (1984), *The Theory of Communicative Action. Volume 1: Reason and the Rationalization of Society* (translated by T. McCarthy). Boston: Beacon Press.

Habermas, J. (1987), *The Theory of Communicative Action. Volume 2: Lifeworld and System: A Critique of Functionalist Reason* (translated by T. McCarthy). Cambridge: Polity Press.

Hajer, M.A. (1995), *The Politics of Environmental Discourse: Ecological Modernization and the Policy Process*. Oxford: Clarendon Press.

Hajer, M.A. (1996), Ecological Modernization as Cultural Politics, in: S. Lash, B. Szerszynski and B. Wynne (eds), *Risk, Environment and Modernity. Towards a New Ecology*. London/Thousand Oaks/New Delhi: Sage, pp. 246-68.

Hannigan, J. (1995), *Environmental Sociology. A Social Constructionist Perspective.* London/New York: Routledge.

Huber, J. (1982), *Die verlorene Unschuld der Ökologie. Neue Technologien und superindustrielle Entwicklung*. Frankfurt: Fischer.

Huber, J. (1985), *Die Regenbogengesellschaft. Ökologie und Sozialpolitik*. Frankfurt: Fischer.

Huber, J. (1989), *Technikbilder. Weltanschauliche Weichenstellungen der Technik- und Umweltpolitik*. Opladen: Westdeutscher Verlag.

Huber, J. (1991), *Umwelt Unternehmen. Weichenstellungen für eine ökologische Marktwirtschaft*. Frankfurt: Fischer.

Huber, J. (1993), Ökologische Modernisierung: Zwischen bürokratischem und zivilgesellschaftlichem Handeln, in: V. von Prittwitz (ed.), *Umweltpolitik als Modernisierungsprozeß. Politikwissenschaftliche Umwelt-forschung und -lehre in der Bundesrepublik*. Opladen: Leske + Budrich, pp. 51-69.

Jagtenberg, T. and D. McKie (1997), *Eco-Impacts and the Greening of Postmodernity.* London/Thousand Oaks/New Delhi: Sage.

Jänicke, M. (ed.)(1978), *Umweltpolitik. Beiträge zur Politologie des Umweltschutzes.* Opladen: Westdeutscher Verlag.

Jänicke, M. (1984), *Umweltpolitische Prävention als ökologische Modernisierung und Strukturpolitik*, Wissenschaftszentrum Berlin (IIUG dp 84-1).

Jänicke, M. (1986), *Staatsversagen. Die Ohnmacht der Politik in der Industriegesellschaft*. Munich: Piper.

Jänicke, M. (1988), Ökologische Modernisierung. Optionen und Restriktionen präventiver Umweltpolitik, in: U.E. Simonis (ed.), *Präventive Umweltpolitik.* Frankfurt/Main: Campus, pp. 12-26.

Jänicke, M. (1990), Erfolgsbedingungen von Umweltpolitik im internationalen Vergleich. *Zeitschrift für Umweltpolitik und Umweltrecht*, No. 13: 213-232.

Jänicke, M. (1993), Ökologische und politische Modernisierung in entwickelten Industriegesellschaften, in: V. von Prittwitz (ed.), *Umweltpolitik als Modernisierungsprozeß. Politikwissenschaftliche Umwelt-forschung und -lehre in der Bundesrepublik.* Opladen: Leske + Budrich, pp. 15-30.

Lash, S., B. Szerszynski and B. Wynne (eds) (1996), *Risk, Environment and Modernity. Towards a New Ecology*. London/Thousand Oaks/New Delhi: Sage.

Luhmann, N. (1992), *Beobachtungen der Moderne*. Opladen: Westdeutscher Verlag.

McKibben, B. (1990), *The End of Nature*. London: Penguin.

McMichael, P. (1996), *Development and Social Change. A Global Perspective.* Thousand Oaks/ London/New Delhi: Pine Forge.

Merchant, C. (1980), *The Death of Nature: Women, Ecology, and the Scientific Revolution*. San Francisco: Harper & Row.

Metzner, A. (1993), *Probleme sozio-ökologischer Systemtheorie. Natur und Gesellschaft in der Soziologie Luhmanns*. Opladen: Westdeutscher Verlag.

Mol, A. (1995), *The Refinement of Production. Ecological Modernization Theory and the Chemical Industry*. Utrecht: Van Arkel.

Mol, A. (1996), Ecological Modernization and Institutional Reflexivity: Environmental Reform in the Late Modern Age. *Environmental Politics*, 5 (2): 302-323.

Müller-Brandeck-Boquet, G. (1993), Von der Fähigkeit des deutschen Föderalismus zur Umweltpolitik, in: V. von Prittwitz (ed.), *Umweltpolitik als Modernisierungsprozeß. Politikwissenschaftliche Umwelt-forschung und -lehre in der Bundesrepublik.* Opladen: Leske + Budrich, pp. 103-112.

Robert, R. (1993), 'Modernisierung der Demokratie. Umweltschutz und Grundgesetz', in: V. von Prittwitz (ed.), *Umweltpolitik als Modernisierungsprozeß. Politikwissenschaftliche Umwelt-forschung und -lehre in der Bundesrepublik.* Opladen: Leske + Budrich, pp. 93-101.

Rucht, D. and R. Roth (1992), Über den Wolken ..., *Forschungsjournal Neue Soziale Bewegungen*, No 2: 22-33.

Soulé, M.E. and Lease (eds) (1995), *Reinventing Nature? Responses to Postmodern Deconstructionism.* Washington, D.C./ Covelo, CA Island Press.

Spaargaren, G. (1996), Ecological Modernization Theory and the Changing Discourse on Environment and Modernity (unpublished script). Paper presented at the Euroconference on 'Environment and Innovation', Vienna, October 23-26.

Spaargaren, G. (1997), *The Ecological Modernization of Production and Consumption. Essays in Environmental Sociology.* Wageningen: Wageningen Agricultural University (dissertation).

Von Prittwitz, V. (ed.) (1993a), *Umweltpolitik als Modernisierungsprozeß. Politikwissenschaftliche Umwelt-forschung und -lehre in der Bundesrepublik.* Opladen: Leske + Budrich.

Von Prittwitz, V. (1993b), Reflexive Modernisierung und öffentliches Handeln, in: V. von Prittwitz (ed.), *Umweltpolitik als Modernisierungsprozeß. Politikwissenschaftliche Umwelt-forschung und -lehre in der Bundesrepublik.* Opladen: Leske + Budrich, pp. 31-49.

Wiesenthal, H. (1993), *Realism in Green Politics. Social Movements and Ecological Reform in Germany* (edited by John Ferris). Manchester: Manchester University Press.

Yearley, S. (1991), *The Green Case. A Sociology of Environmental Issues, Arguments and Politics.* London: Harper Collins.

Zilleßen, H. (1993), Die Modernisierung der Demokratie im Zeichen der Umweltproblematik, in: V. von Prittwitz (ed.), *Umweltpolitik als Modernisierungsprozeß. Politikwissenschaftliche Umwelt-forschung und -lehre in der Bundesrepublik.* Opladen: Leske + Budrich, pp. 81-89.

11

Self-organizing Complexity, Conscious Purpose and 'Sustainable Development'

Ernest Garcia

> Now, all right, let us say we are now paleontologists and we are studying fossil Bread-and-butter-flies and we wonder why they became extinct. The answer is not that they became extinct because their heads were made of sugar. The answer is not that they became extinct because they couldn't find their food. The answer is that they became extinct because they were caught in a dilemma; and the world is made that way, and is not made the linear single-purpose way. (Bateson, 1991: 279)

> Progress and Doom are two sides of the same medal (...) both are articles of superstition, not of faith. (Arendt, 1967: xxix)

> I expanded without time for proper planning, without any pauses to learn from my experiences or my mistakes or my contemporaries, without time for reflection. How then could I have turned out to be anything but a mess? (Rushdie, 1996: 161-62)

'Development', as this word is used in current social thought, connotes a handled process of change. 'Sustainability' refers to the maintenance of social life within the Earth's supporting capacity. 'Sustainable development', then, suggests an intentional and conscious control of the relationship between society and nature. This relationship consists of historical systems embedded in evolving environments. Conscious purposive behavior therefore implies adaptation and learning under conditions of constant change and unpredictability. There are then some conditions – of time and space – which shall be eroded if social interventions in the environment accelerate their pace and globalize their scale.

Connecting the idea of environmental viability to the general conditions of societal learning opens a way to delimit the scene where sociological analysis has a role to play in the debate about social progress and the continuity of life. Along this way, the question is to avoid reductionist, naturalistic lines of reasoning, without dissolving the ecological predicament in a more or less fashionable cultural mood.

Introduction: value assumptions and metaphors

I have never liked the term 'development'. It is too reductionist. It excessively simplifies things. Mechanical engines have been developed. Organisms 'have' a development too, but in a different sense. On the contrary, societies have a history, which is something rather different. In our *historic* epoch, distinctions in this regard are not frequently made, and something which, in politics and sociology, is also designated with the word 'development' seems to be a fundamental matter; on many occasions, it even seems *the* fundamental matter. However, I feel that human collectives do not have a development problem; their problem is to improve life, to get people to live with enough dignity and freedom. At most, what we call development may be a means to achieve this objective, a means which has turned out to be hypostatic. The era of development as a universal objective, with a planetary scope, is already five decades old. During that time, one out of five human beings has reached a material wealth unheard of. On the other hand, inequality and the number of victims of hunger has increased immensely to levels never known before and, in addition, the natural bases for the subsistence of the species have been damaged, maybe irreversibly. The presumed solution to the dilemmas of social evolution led to 'forms of "improvement" that impoverish and dis-empower' (Seabrook, 1993: 250) more and more people in more and more places. We should at least wonder if, after all, the means was not adequate.

I am not very keen on the word 'sustainability' either, as it replaces the eventual commitment with the beauty and diversity of life – and the beauty and diversity of the exchanges between societies and their ecosystems – by an abstraction which vaguely suggests the possibility of ruling these exchanges in a way unknown to us until today. It is acknowledged, at last, that some natural limits do exist. It is now accepted that – by ignoring those limits – we can jeopardize the material supply which permits the expansion of the industrial civilization. Then the question arises: how to sustain all that, both the supply and the expansion? The answer which seems to be succeeding first states that ignorance can be overcome through new 'sciences of the Earth', supported by artificial satellites and computers. The next step is to postulate that the exchange can be controlled even in extreme situations by means of sophisticated 'geo-engineering' formulas and 'eco-sphere management' techniques. Finally, the opportunity to engage in 'a battle to save the planet' is proclaimed. Such a battle, naturally, requires the urgent constitution of suitable major states which will soon demand full power. This looks more and more like a new cold war, inspired – like the previous one – by the principle of 'keeping at bay'. Keeping at bay nature and society, on the more or less blurred boundaries of the carrying capacity of the planet.

The combination of both terms ('sustainable development') is even more unpleasant to me. My feelings are obviously related to the suspicion that in this apparently promising concept the priority lies on the noun and not on the search for a true balance. The years of unstoppable rise of this expression, since the Brundtland Commission consecrated it up to the éclat provoking its current omnipresence at the Rio Summit, coincided with the initial stage of environmentalism as 'the highest state of developmentalism' (Sachs, 1993: 3). I believe this is not just an ideological and political contingency but something inherent to the concept itself, as I will argue in what follows.

In its contemporary use in social sciences, the term 'development' suggests a managed and controlled process of change. It usually includes the determination of the objectives of the process and also of the means considered adequate to achieve them. The word 'sustainability' refers to the maintenance of the economy within the limits of the carrying capacity of the ecosystems on which it depends. The 'sustainable development' notion therefore suggests an intentional and conscious control of the relationship between society and nature.

The attempt to adapt human beings to their environment by means of consciousness can be represented in several ways. However, there are two metaphors whose comparison is especially instructive. The first of them refers to the well-known image of the 'spaceship Earth', proposed by Boulding (1966). The spaceship has a crew and specific amounts of supplies and fuel. The crew member in command can – more or less – manage everything in an efficient manner. Indeed, a spaceship is a mechanism, that is to say, a system of linear relations which can be controlled if its structure and the laws which rule its dynamics are studied in detail. The second metaphor (not so well-known) was proposed by Bateson (1987: 449-50) and refers to one of the situations in which Carroll's Alice was involved: a croquet game in which a flamingo is used as a mallet and a hedgehog as the ball. Bateson described such a situation as a meta-random sequence of events subject to a second order indeterminacy. In this case, conscious control is impossible, and the finalist interventions, id est linear, are similar to trying to fix a clock by randomly nailing a pencil on its mechanism (Commoner, 1978: 40-41).

Both metaphors illustrate, in a sense, the basic difference between mechanics and life. In my opinion, the relationship between society and nature – two complex self-organizing systems – is better described by the second metaphor. The idea of a sustainable development suggests that it is reasonable for somebody to risk his or her neck in a game like the one described above as far as his or her pulse is firm and his or her will is not fickle. In other words, it tries to match the relationship of human beings to the rest of the living creatures in an essentially mechanistic epistemological framework, this being a systemic mistake.

Complexity, change, and social theory

Societies surely possess the basic properties of complex, self-organizing, and reflexive systems. Unlike a simple or mechanical system (for instance, a rocket or the solar system), a complex system (for instance, an organism, an ecosystem, or a city) cannot be described comprehensively by only considering its integrating parts or elements and the linear relationships among them. Self-organizing systems normally evolve into more complex states by increasing the information they contain. This evolution is possible because they are open systems, able to absorb low entropy (free energy and concentrated materials) from their environment or ecosystem, which – on the contrary – they simplify (or degrade). Human beings and societies are also flexible systems capable of learning, or of learning to learn. Societies can condensate nuclei (institutions) which accumulate information and a decision capacity (power) with a view to acting in a conscious, purposeful way.

The fact that living systems can only evolve by increasing the entropy in their environment has been widely known ever since Schrödinger's (1992: 62-75) famous paper on the physical aspect of the cell was first published in 1944. Organizing systems are disorganizing systems too; that is to say, the term 'self-organizing' has no sense unless the system is in close contact and permanent interaction with an environment possessing available order and energy, so that it can manage to 'live' – somehow – at the expense of that environment. The other aspect of self-organizing systems is their high order and mysterious stability, which cannot be explained by the interchange of energy with the environment. Von Foerster (1991) maintained that autonomy, order, and stability can only be tackled if a separation between the flow of 'foods' and the flow of signs is established. This leads up to new distinctions: all living systems are organized in such a way that internal regularities and stabilization are produced; but societies, unlike organisms, organize themselves on the basis of language and individual autonomy. We experience this new dimension of operating coherence as consciousness, as 'our' mind (Maturana and Varela, 1988: 153). Organisms and societies are, then, opposite cases in the series of those meta-systems that are shaped by cell-system aggregation. Different positions in this series can be ordered according to the degree in which the components, in their existence as autonomous units, depend on their participation in the meta-system they compose. Organisms would be meta-systems with scarcely autonomous components, id est, with components having very few or no dimensions of independent existence. On the other hand, human societies would be meta-systems with highly autonomous components, i.e. with components having many dimensions of independent existence (Maturana and Varela, 1988: 132).

Language radically modifies human behavior because who operates in it can describe him- or herself and his or her circumstance, and then reflection and consciousness become possible (Maturana and Varela, 1988: 139-142). This fact is related to some social theory features, for instance: the impossibility to neatly separate objective from subjective dimensions; the fact that the way in which societies build up their own perceptions – and thus their own visions of social change – modifies the conditions in which the change occurs; the need not to consider only cause-effect relationships but also motivations and purposive orientations of the social action, etc.

The above mentioned distinction between second order couplings (organisms) and third order couplings (societies) leads up to some corollaries having a pure sociological content. The organism restricts individual creativity of its integrating units, as these exist for it; the human social system expands the individual creativity of its components, as it exists for these. As a consequence, those societies that stabilize all their members' behavior in a coercive way spoil themselves shifting to the form of organism (Maturana and Varela, 1988: 132).

Following the above sketched perspective leads to realize that only the recovery of random-ness, indeterminacy, plurality of responses, diversity, and history, makes possible a system with a survival value. In an isolated system, survival value depends on all the holes being closed, on the absence of questions; in an open system, on their being open, on the absence of answers (absolute answers obturating questions) (Ibáñez, 1992). 'Competition, exploitation and accumulation – contrary to cooperation, use and storage – have no survival value' (Ibáñez, 1990: 64).[1]

I would like to clarify what exactly is my point now. I do not think that current attempts to create a new science of complexity starting from building models which simulate the behavior of non-linear dynamic systems have a social theory as one of their outcomes. Indeed the constitution of such a general theory is rather doubtful and the present possibilities of applying it to social sciences are minimum (see Lewin, 1994; Waldrop, 1994). Instead of it, I am thinking about empirical features of reflective systems, and some of their negative, impossible implications, which impose limits to all conceivable descriptions of social change (and so to any concept of sustainable development). The dynamics of reflective systems are irreversible or asymmetric as far as time is concerned (societies never go back to a previous state). Then, they have a history, that is, the activity of the elements constrains the number of possible future states although they do not determine them. The future states are unpredictable in an inherent way, and there is a continuous emergence of novelties as a consequence of the indeterminist dynamics which combines gradual organizational changes with changes (which may be intense) due to disturbances. Finally, as information grows in the systems, the possibility of describing them by means of theories (id est a few basic laws) is reduced and, therefore, their comprehensive description is longer and longer (and may be exclusive to each society and non-applicable to others). As a consequence, theories of social change, rather than instruments for the accumulative production of facts or for the control of future events, are means through which the reflexive agents become more aware of their circumstances (Bohman, 1991: 234). The sustainable development debate cannot escape from these limits. The very notion of 'unsustainability' can provide a starting point in laying the foundations of this point of view.

Several meanings of 'unsustainability'

Concern about a development being sustainable has arisen because, in one way or another, people perceive the current trends of industrial civilization as unsustainable. However, 'unsustainability' means different things. This section considers four different meanings of 'unsustainability' (hence, of sustainability as well), from the nearest to the mechanistic paradigm to the more 'holistic' one. They are to some extent alternative, but my proposal is to consider them as being complementary, as different ways of referring to the same set of facts from different perspectives.

1. The meaning which is nearest to the mechanistic paradigm refers to the *exceeding of the carrying capacity.* Unsustainability, then, is the tendency to collapse caused by surpassing limits. This meaning is implicit in the usual environmentalist advice that nothing can indefinitely grow in a finite environment. In this context, sustainability implies that the physical scale of the social system, namely the totality of human bodies and their associated artifacts, and the metabolic flow of energy and materials needed to reproduce them, should be maintained under the natural capacity for supplying resources. Resources are sources of free energy and concentrated materials, and sinks for bound energy and sparse matter.

This point of view can be seen in the well known reports on the limits to growth (see Meadows et al., 1992). In this last book we find, on the one hand, calculations related to the carrying capacity. The models worked up to accomplish these calculations are highly sophisticated, of course, and they illuminate substantial aspects of the human predicament. They do not need sociological inputs to be built. Population, capital growth and natural environment data are all that is required. On the other hand, moral and political recommendations are proposed, in an openly normative way. They are like commandments: develop visions!, build networks!, tell the truth!, learn!, love! (see Meadows et al., 1992: Chapter 8).

Malthus, whose legacy has been uncomfortable for so many sociologists, is the inescapable reference here. Reasons for discomfort are partly ideological. Malthus was a gloomy thinker, indeed. His work is full of symptoms of that cultural climate which has been described as a 'Western philosophy love affair with death' (Benhabib, 1996: 135). However, there are theoretical reasons too. A careful re-reading would possibly uncover forgotten dimensions of his work. But his recognized inheritance, alive in sociological tradition, has in essence only two aspects. First of all, calculations about the relationship between population growth and food production. Secondly, moral advises. Little room for social theory between these two pieces. Present-day followers of Malthusian tradition, including the most enlightened ones, derive socio-political rules from biological models.[2]

2. Unsustainability as the outcome of a catastrophic *imbalance in the co-evolutive process*. If one of the species 'at stake' receives a much too high energy subsidy, then that species imposes upon the ecosystem a radical simplification, a drastic bio-diversity loss. This has happened ever since the human species developed its special ability to oxide the 'necrosphere' (Margalef, 1991: 250) and to appropriate primary photosynthetic production on a large scale. Then, sustainability implies leaving enough 'room and food' for the creatures remaining.

This meaning is implicit in an article, widely quoted, by Vitousek and colleagues (1986). Its starting point is remembering that we need to take into account that human beings are not the only terrestrial consumers of solar energy captured through photosynthesis. It includes a calculation according to which almost 40% of the potential net primary production[3] over land is directly used, co-opted or lost due to human activities. It is directly used by the human beings and their domestic animals as food, fibers, or wood; it is co-opted in order to use the land for farming or grass for the cattle; it is lost as a result of urbanization, clearing up of forests, desertification, or over-exploitation. The population and economic growth demands an even greater appropriation of the products of photosynthesis.

The above mentioned article falls upon the debate about carrying capacity for human beings, of course.[4] But the point here is that it does not only warn about human beings, but also about the effects on other species (Vitousek et al., 1986: 368), outlining in its conclusions the possibility of 'extinctions that could cause a greater reduction in organic diversity than occurred at the Cretaceous-Tertiary boundary 65 million years ago' (Vitousek et al., 1986: 372). The subsequent loss of options for humanity would entail unpredictability and indeterminacy both in specific signs and in the time of appearance. The limits depending on the drastic reduction in the biological diversity which will take place – if the space is too crowded and the food scarce for the other species – are previous to those of the carrying capacity calculated only for human beings and 'their animals and vegetables'.

The co-evolutive approach allows a special kind of integrated social-natural theory, in which knowledge, organization, technology and values co-evolve with nature (see Norgaard, 1994: 27). However, Norgaard's view seems to imply a common framework of concepts for cultural and biological evolution: '...the well-being of people can improve to the extent that their ways of knowing, social organization, and technologies select for an evolutionary course of the biosphere which complements their values' (Norgaard, 1994: 39). Today this common framework does not exist. May be it cannot exist, as far as linguistic couplings and organic couplings are of a different order. In this sense, the co-evolutive approach can reduce sociology to a rather misty expanded ecology.[5]

3. Unsustainability as acceleration of *entropic degradation* as a consequence of too large or too intense production processes. This meaning is implicit in the proposition that nothing lasts forever, that no process can indefinitely endure in a finite environment. This perspective stresses the fact that industrial civilization has been possible thanks to an uncommon mineral bonanza not easy to repeat. It perceives, as fundamental limits, the extreme uncertainty surrounding the replacement of fossil fuels with more plentiful and less polluting new energy sources as well as the fact that materials cannot be indefinitely recycled. Then, sustainability tends to identify with conservation.

Georgescu-Roegen is the conspicuous representative of this perspective. Despite the fact that it is grounded in thermodynamic laws, it only can be developed by focusing on the social world. The point was introduced by this author stating that it would be absurd to think that the economic process exists only for producing waste; instead, the true product of this process is the *enjoyment of life*: '... the entropy reversal as seen in every line of production bears the indelible hallmark of purposive activity. And the way this activity is planned and performed certainly *depends upon the cultural matrix of the society in question*' (Georgescu-Roegen, 1971: 18-19, emphasis added). Georgescu-Roegen's 'sociological' contributions were rather scarce (1971: 292-316; 1977; 1984). This notwithstanding, it is easy to see that the 'cultural matrix' is not a mere corollary of the physical analysis in the debate on sustainability which can be derived from his approach.

4. Unsustainability as a *blockade of the societal learning devices*, as a consequence of an excessive acceleration and a much too high connectance. If the debate on the environmental crisis is more than a melancholic contemplative exercise it is because human beings are supposed to be able to learn and therefore modify their behavior by causes other than direct physical constriction. Now then, conscious learning has some conditions. Two of these conditions, which are very important, are time and availability of error margins. Learning requires time to select viable adaptations. It also requires places untouched by the effects of error, from where it can be corrected. Both conditions emanate from the basic fact that error is unavoidable. If the system accelerates too much, decision centers begin to make bigger and more frequent mistakes. If the system globalizes too much, if all of its elements are strongly connected, mistakes spread everywhere, and there will not be any alternative spaces open to eventually successful essays. In acceptance of this, sustainability is in maintaining flexibility, avoiding boundless acceleration and globalization in the system.[6]

Some features of the environmental impact of man-made chemicals are strongly related to the pace of their introduction and the time scale of their length in the natural world. It is the case, for instance, of hormone-disrupting chemicals. Many of them resist normal breakdown and accumulate in the body. 'This pattern of chronic hormone exposure is unprecedented in our evolutionary experience, and adapting to this new hazard is a matter of millennia not decades' (C. Hugues, cited in Colborn et al., 1997: 81-82).

Similar concerns are often expressed when the possible effects of biotechnologies are estimated. For instance, one of these effects is the destruction of centuries-old, proved and viable local traditions of agricultural knowledge (Shiva, 1995). They are suddenly replaced by worldwide experiments whose effects, in evolution and history, in biology and culture, are mostly unpredictable. In these conditions, some worrying questions arise: how many experiments? at what pace? how global should they be?

The fourth of the above discussed meanings clearly allows for many relevant sociological contributions, because societal learning is always embedded in culture, organizations, and institutions. It is not so clearly the case for the other three. For instance, the 'limits-to-growth' meaning of sustainability requires two main elements to be established. On the one hand, calculations about carrying capacity (about maximum or optimum physical scale). On the other hand, ethical – perhaps religious – rules. No (or very little) social theory is needed. The 'co-evolutive balance' approach seems to require a unified framework for biological and cultural evolution, and then the basic distinction between societies and populations of organisms fades away. The analysis starting from the entropy law needs to take in the enjoyment of life – its unavoidable counterpart. Social construction of needs and other cultural processes should then be introduced. This is to say, if sustainability analyses carried out on the basis of the approaches which are closest to the mechanistic paradigm were able to solve all the problems in the environmental debate, then social theory would not have a distinct role in that debate .

I don't think this is the case. Substituting mono-cultures of timber for genuine forests could be perceived as being sustainable under the first approach, but not so under the second one. Combining intensive recycling with vast nature reserves could be sustainable from the point of view of biological balance, but unsustainable from the point of view of entropic degradation. Accelerated replacement of entropy by information could make some parts of the world sustainable under the general conditions of the third approach, but does not fulfil the learning requirements of the last. Through the following pages these shifts are developed in a more detailed way.

Beyond reductionism

The 'limits-to-growth' approach depends on the idea of sustaining or carrying capacity. The sustainability condition of the social evolution would then be that the physical dimension of the economy (including the 'sociomass', as Boulding (1995: 29) called the total mass of human beings and their associated artifacts, as well as the periodical throughput of energy and matter required for its reproduction) remained within the carrying capacity of the planet. For this perspective, the sustainable development notion is meaningful, although – once the maximum (or optimum) physical scale is reached – reference could only be made to qualitative aspects (procuring more service from constant or decreasing energy and material inputs). In this context, the distinction made between growth as 'something getting bigger' and development as 'something getting better' (Boulding, 1995: 30) is relevant.

We will now take a respite. The idea that a material system can undergo important qualitative changes without altering its physical dimension seems rather bizarre to me. It may be backed, but it is certainly less obvious than what some of its defenders argue. As defended by many biologists, it may be true that the total mass of the biosphere has not been substantially altered during its evolutionary course, but to me it does not seem so easy to say that about any particular species. In general, in the living world, the notion of development seems to be extremely close to the ideas of growth and form, that is, to quantitative and qualitative variations (Tyler Bonner, 1992: xviii). I do think this requires further analysis with a view to discarding what cannot be maintained in the proposals of 'zero-growth' or 'steady state', due to an excessively idealistic vision. I shall not focus on this for the time being, though.

Apart from what has been said, there are two reasons for which the carrying capacity approach seems insufficient in my opinion. Firstly, the maximum scale cannot be determined (there is no way we can know when the flamingo's neck will bump into the hedgehog nor the direction which the latter will follow). Secondly, this approach does not resolve the problem of human needs (or in other words, although it permits to deal with sustainability to a certain extent, it does not say anything about society). Table 1 shows such an appreciation.

The left column is a combination of the sustainable development operational rules proposed by Kerry Turner (1988: 12-20) and Herman Daly (1991: 256).[7] The column on the right indicates what happens with such rules if applied in the only context where they are meaningful, as they are anthropocentric, id est a society where people believe it is worth living (why should it be sustained otherwise?). If we were to delete the terms 'desirable', 'needs', or 'equity', then both columns would be equivalent. However, we cannot rule them out, which implies the inevitable disappearance of the straightforward advantages of the environmental sustainability rules.

Maybe the core of the subject could be explained by stating that sustainability has to do not only with ecology but also with ethics, aesthetics, and politics, or – optionally – with a single word: mind, that annoying form of complexity.

Whoever may find so much concision extreme, may now be able to pay attention to a somehow more extensive argument. If we ask about the limits of the relationship between humanity and nature, we would start with the biological level, in terms of population and carrying capacity of the environment (as in numerous approaches which associate environmental problems to the population boom). The most serious defenders of this point of view do know that any calculations taking into account only the number of human beings have no meaning, and then they tackle the issue by saying that the impact caused by a human group on the environment results from three factors. The first one is the number of people, the second one is the amount of resources consumed by the average individual, and the third one is the environmental destruction rate caused by the technologies which supply us with commodities. That is to say, Impact = Population x Wealth x Technology (Ehrlich and Ehrlich, 1993: 52).

Table 1 *Notions of sustainability*

Environmental sustainability	Environmental and social sustainability
1. Renewable resources harvesting rates must be equal or inferior to the capacity of natural regeneration of the ecosystems, and waste emissions must remain within the natural assimilative capacity.	1. The desirable level of exploitation of the renewable resources is equal or inferior to the capacity of regeneration/assimilation of the ecosystems, provided this permits the satisfaction considered sufficient of the needs and an acceptable equity level.
2. The exploitation of non-renewable resources must be as slow as possible, preferably using renewable substitutes and exhausting the most abundant non-renewable resources before their more scarce substitutes; the emission of pollutants must remain within the natural assimilative capacity.	2. Adoption of a desirable pace for the exhaustion of non-renewable natural resources, that is, the slowest pace compatible with a sufficient level of satisfaction of the needs of the human beings and an acceptable equity level in their distribution.
3. Technological change should increase the service derived from each natural resource unit consumed and should promote the substitution of non-renewable resources by renewable ones.	3. The technological change needs to increase the service derived from each natural resource unit consumed and to promote the substitution of non-renewable resources by renewable ones, within the framework of acceptable levels of consumption and equity.
4. The physical scale of economy must remain within the carrying capacity of the ecosphere.	4. The physical scale of economy must remain well under the carrying capacity of the ecosphere, in order to provide the necessary flexibility for the unpredictable social evolution.

As a matter of fact, it is not very practical to be aware of this if wealth and technology are afterwards considered as mere external variables. However, if both factors are internalized, the result is a debate which never comes to satisfactory conclusions (Tabah, 1995: 75-76) about the relative influence of population growth, technology, and consumption on the environmental deterioration. As a consequence of different technologies and levels of wealth, the inter-individual variability in energy consumption is so great that it is radically different from any other species. Hunters/harvesters need from 2,500 to 3,000 calories a day, north-American urbanites need at present 200,000 (Catton, in Jensen, 1995: 135). Should we conclude that they belong to different species?

In fact, depending on different hypotheses about technology and consumption, the estimations made on the carrying capacity of the Earth range from 1-2 (Pimentel et al., 1994) to 50 (Revelle, 1974) billion people, this being a conclusion which – as has been said – 'inspires little confidence' (Clarke, 1995: 42).

There is still a relevant human peculiarity. As long as non-renewable resources are extensively exploited, the question to be answered is not 'how much population?', but 'how much population and for how long?'. Additional annoying difficulties join the game, linked to the duration of a process which changes.

To sum up, if the population variable is analyzed in an isolated way, we will not get anywhere. And if we only introduce in the analysis a small portion of the real connections, then the apparent empirical advantage of the carrying capacity concept vanishes (Livi-Bacci, 1997: 224). This result has been acknowledged by many experts, who frequently say that the environmental problems are related to the growth of the human population in an indirect way and through various intermediate factors of a social, economic, and political nature.

The immediate step is to admit that the limits of the society/nature relationship depends on technological change (which may find a way to use materials or energy sources not exploited previously, that is, make them into resources). In human societies, biology is not independent from technology. A consequence of the latter – the proliferation of artifacts irrevocably associated to the human body – entails a considerable implication: some people can have exo-somatic extensions enormously bigger than others. Social inequality is now included in the analysis, or – in other words – there is no way to detach biology from technology nor sociology, nor politics. On the other hand, in the framework of the debate on conscious purpose and sustainability, the arguments about the carrying capacity are only interesting if – unlike any other species – human beings are described as being able to find, in their environment, a 'feeding' source and not exploit it until exhaustion. That is, if they are able to alter their structure of needs for reasons other than the existence or lack of means to meet these needs. We have now added the culture (values and myths also), without eliminating the previous levels. This is enough for now. I have just stopped at one particular point in my route (population) and I have gone through others very quickly, but it does not really matter where we start, as none of these interrelations can be avoided.

In Table 2 the column in the middle refers to the different mind/nature social mediations which should be taken into account. They are demographic, technologic, economic, organizational, political, cultural and symbolic. They are only *relatively dependent* on each other. *Dependent* because the starting point, wherever it can be located, is connected by multiple yarns to all the remaining levels. For instance, the notion of 'limit' that appears in the population-resources debate, built upon feelings of excess and guilt, belongs to the vast off-spring of the *hubris/némesis* dialectics, a mythical narrative which is omnipresent in Western culture (see Zoja, 1993). *Relatively* because each level is autonomous and can not be reduced to any other level. There is no way of deducing the symbolic world of *némesis* from any concrete historical experience of overloading natural environments. There is no way of forecasting the technical novelties which a social group will be able to produce by studying the endowment of natural resources within that group's reach.

Table 2 *Mind-nature social mediations*

	Nature		
S	compatible with biosphere	carrying capacity	tends to be unlimited
o	solar energy	energy supply	nuclear fusion
c	limited	substitution of production factors	very high
i	egalitarian	social structure	polarized
e	pluriversalist	political order	globalizing
t	orientated to sufficiency	structure of needs	orientated to affluence
y	harmony myths	constituting narratives	dominance myths
		Mind	

Source: Garcia, 1995b: 51

Each term from the columns on the left and right side represents a pole or extreme position in the autonomous field of problems arising at the corresponding level. For instance, the energetic alternatives to the fossil fuels issue has its own entity, it is not a mere corollary of positions chosen at the other levels, whatever they can be. The grouping of terms on the right or left side is significant only as a limit trend. From a theoretical point of view, a lot of different groupings could be conceived. In practice there are many 'bridges' and 'crossed paths'. For example, there are people who consider that population growth should be stopped, but not so economic growth. It is possible to imagine a solar *and* global age. Or anything else.

Sometimes a sharp edged alternative between eco-centric and anthropocentric visions is set up. This alternative is to some extent positional. It has something to do with the possibility of circulating through that complex semantic field. When the attention is centered upon human appropriation of photosynthesis, the adoption of an eco-centric point of view is not so difficult. To the contrary, if social perception of risk is focused on, such an association hardly could be spontaneous. Because of these positional shifts, anthropocentrism and eco-centrism are equivocal poles for the environmental dilemma. Neither outside nature nor completely inside it there is a place for civilization.[8] Civilization is rather eccentric. As much as social theory, of course.

The semantic field of 'sustainable development'

How does all this relate to sustainable development? Thinking about this also means taking into account specific forms of the above mentioned interrelations (Garcia, 1995a). Table 3 (p. 243) summarizes the issues at stake.

This table is in itself a drastic simplification. In despite of it, possible combinations of the varied items it contains would generate hundreds of meanings of 'sustainable development'. Sure, making sense of many of them would be a task doomed to failure. A fair amount, however, would conserve a reasonable look. Here they have been reduced to three, striving after minimum losses.

1. Sustainable development as sustained growth. Expansion of production and consumption should then be maintained, together with the consolidation of affluent culture and lifestyles. Reduction of inequality tends to be delayed until a wealthier future (only the growth permits redistribution). Global scale dependence tends to be strengthened. Technological innovation should assure solutions to eventual scarcities or pollution crises.
2. Sustainable development as qualitative improvement without increase in the physical scale, that is to say, as evolution of a homeostatic, steady-state or zero-growth economy. In most of its versions, state intervention should guarantee the generalized satisfaction of basic needs in a context of global interdependence. Transition to a solar age would drastically reduce the consumption of non-renewable resources, thus allowing many inter-resource substitutions.
3. Sustainability is always uncertain, submitted to the permanent necessity of adapting in random conditions. In order to gain flexibility, societies should liberate themselves from considering development as an objective instead of a means (in some versions, development is seen as being the cause of both poverty and environmental degradation). More integration of the economy into natural cycles should permit a fair satisfaction of basic needs. Sufficiency as a cultural regulator, and egalitarian community institutions interconnected by middle intensity relations complete the picture.

Columns on the table are ideal constructions, like sort of attractors in the abstract space of possibilities. Some current proposals coincide almost completely with one of these constructions, but most of them do not. There is nothing surprising about it: the different levels are only relatively dependent on each other; then, there are lots of cross-links between them.

Is there any way of reducing all that to a clear-cut programmatic rule? I do not think this is the case. Let us consider, for instance, the proposal which identifies sustainable development with using energy and raw materials in a more efficient way. The usual discourse says that sustainable development should improve the ecological compatibility, equity, and the satisfaction of needs. Nevertheless, the term 'improvement' refers to some assessment criterion which ultimately can only be the people's feeling that the society or the world in which they live is better than it was the previous year or whenever, and different groups can have different valuations, and so on. Fair enough, let us assume now that the historically dominant needs structure is such that people only feel improvement if their material wealth increases, as seems to be happening with most of the billion people who benefit from industrial civilization. Then sustainable development is an impossible dream and can only be summoned in the magical world of *abracadabra*.

The magic word already exists and – as it is the case with magic words – it belongs to a different language: *prodequisus!* (productivity, equity, sustainability) (Altvater, 1994: 220). Or, as the politicians – who should not utter abracadabras if they want to keep their serious appearance – are used to saying, the 'solution' is the simultaneous promotion of economic growth, social justice, and the protection of the environment. In this context, understanding that a reduced energy and materials consumption is an improvement could mean – at least by now – eco-efficiency, the new motto by Schmidheiny (1992: 62-63) and his greenish executives. But it only makes sense in the short term (Von Weizsäcker et al., 1997: 357), because an indefinite path of increasingly immaterial economy is impossible. For instance, reducing by half the energy and materials per unit of output is difficult, but it seems plausible (and adequate, of course). However, a 'sustainable' growth on that basis would have to reduce also the material intensity in a sustainable way: a fifth, a tenth, a hundredth? I do not need to go on.

To sum up: the approach which argues that growth means increasing and development means improving will only be meaningful in the long term if the current idea of 'improvement' changes, that is, if we accept that a different cultural notion of 'meeting the needs' would be required. Basically it is a historic-cultural question. To the greater efficiency, we need to add an idea of sufficiency, because without it there is no way to round up the figures.

The story would not be very different if sufficiency, or population control, or solar power, or whatever, had been taken as reference points instead of efficiency. The analysis always stumble over a bundle of social mediations. There lies the task for a critical environmental social theory.[9]

Entropy, acceleration and instability

Let us comment on the last problem for the approach of the limits to growth (development without growth, in its latest versions). Despite all the reservations about it in previous pages, many of its warnings are – in my opinion – basically reasonable and realistic. However, there will always be people willing to remember that the Earth is not an isolated system but a closed system which exchanges energy, although not materials, with the external world; so that an abundant energy source could permit a reiterated recycling of materials and the development of costly engineering systems for the terrestrial confinement of material waste and the expulsion of energy waste (heat) into outer space.[10] Or in other words, there will always be people willing to renew – now under the sustainable development flag – the old idea that for a 'technological species' there is no limit to the economic growth related with the carrying capacity of the Earth.

In order to theoretically discuss the last point of view, we must appeal to a more refined and deeper version of the limits dilemma, like the one proposed and defended by Georgescu-Roegen. I cannot go into detail here, so I should just summarize it in the two following laws: no material process can last indefinitely in a finite environment, and no material can be recycled in an indefinite way.[11] The thesis derived from this (social life can only be maintained by reducing the capacity of the environment to sustain it in the long term) depends on the Entropy Law. From this perspective, the notion of sustainable development is essentially contradictory. It belongs to the same class as other useless old myths (the perpetual mobile or the immortal organism, for instance).

Table 3 *Semantic field of the debate on development and sustainability*

	Sustainable growth	Development without growth	Bioeconomic conservationism
Population	limited by the opportunities of the output growth	limited by the carrying capacity of the biosphere	limited by the carrying capacity of the biosphere
Energy supply	replacement of fossil fuels by more abundant and concentrated sources (nuclear fusion?)	replacement of fossil fuels by renewable energies, solar era	more participation of renewable energies, high uncertainty about substitution
Technology	high replaceability of natural resources by man-made capital	low replaceability of natural resources by capital; high replaceability of non-renewable by renewable	heterogeneity and, therefore, limited replaceability of natural resources by natural resources
Economy	monetary calculation of environmental externalities, integrated accounting	accountancy of the natural heritage in physical magnitudes, optimum scale calculations	political ecology, valuation of resources through social conflict
Social structure	only the growth permits some redistribution and reduction of poverty	equity by redistribution (centralized)	different forms of communitarian egalitarism
Political order	globalizing, liberal	globalizing, technocratic, centralizing	'pluriversalist', community-based, decentralizing
Structure of needs	affluence, mass consumption culture, regulation by momentary calming of the anxiety	sufficiency, austerity, moral regulation	sufficiency, rejection of extravagance, aesthetic and political regulation
Constituting narratives	myths of dominance, hubris (Prometheus, Faust, 'multiply yourselves and rule the world!')	myths of limits, (Icarus, Cassandra, aurea mediocritas)	myths of harmony (Goddess, Mother Earth, Prakriti)
Sustainability criterion	weak sustainability (intergenerational transmission of a constant or growing amount of natural 'capital' plus man-made capital)	strong sustainability (intergenerational transmission of a constant amount of natural resources)	quasi-sustainability as slowing down and deglobalization (parsimonious use of the resources with a view not to speeding up the unavoidable entropic degradation)
Vision of sustainable development	sustainable development as a new ('environmentally aware') expansive phase of the present industrial era	sustainable development as a new historic era of qualitative improvement without increasing the physical dimension	sustainable development as a self-contradictory concept (similar to the perpetual mobile and the immortal organism)

In my account, the bio-economic approach encompasses a pretty precise generalization of the material conditions of all social changes. The fact that society, like any other living system, can only evolve by increasing the entropy in its environment is now widely known. If the disorder introduced in the environment is very big, the system may enter a new adaptation form, consuming more energy (but also increasing the degradation of the environment even further). When this is applied to the contemporary debate on development and sustainability, we need to remember – at least – that there is no guarantee that an alternative power source will be found more abundant and less polluting than fossil fuel. In addition, we need to keep in mind that materials also run out irrevocably.[12] The conclusion drawn by this Romanian economist and mathematician is most reasonable: the only valid law in tackling the environmental crisis is conservation (and so it makes sense speaking of 'bio-economic conservationism').

The problem with the analysis starting from the second law is that it does not allow the making of sophisticated calculations nor brilliant formal models. On the basis of the entropy law, there is no technical way to decide whether a particular production rate is sustainable or not. Georgescu-Roegen was aware of that. He said, in a rather abrupt way, that the entropy law 'does not help an economist to say what precisely will happen tomorrow, next year, or a few years hence'. We only can be sure – he added – that 'any use of the natural resources for the satisfaction of non-vital needs means a smaller quantity of life in the future' (Georgescu-Roegen, 1971: 19-21). Doubtless, the application of this criterion on specific and immediate contexts inevitably leads to an undefended direct exposure to abstruse and intricated moral dilemmas.

There is, however, an aspect related with all this which permits us to take a step further. A system depending on growing energy inputs becomes more unstable. Anybody could now adduce that this situation is creative. As it is indeed, but nothing allows us to say that frantically emerging novelties are really what we are interested in from the particular perspective of the human species. We can now approach the subject from a somehow different point of view formulated by the astro-physician Peter Kafka (1993: 346), who has suggested that the environmental crisis is basically a matter of speed and globalization. This point of view assumes that access to low entropy sources and sinks for high entropy waste are essential prerequisites for social life. It accepts that unsustainability has to do with entropic degradation as a destructive consequence of current practices in using raw materials. That is to say, it does not deny the conclusions of bio-economic analysis; it simply adds a dose of space-time concretion to this analysis (although it does not transform it into a *calculable*). More necessary conditions have to be taken into account: 'The time-scales and the degree of diversity in the process of trial and error are decisive for the probability of "success"' (Kafka, 1993: 346). And so, a new notion of 'limits' arise. According to this new notion, a system turns unsustainable if (a) it accelerates excessively, having no time to select the most feasible adaptations; if (b) it is too global, that is, it is unable to fail in some of its parts surviving in others, and so it risks everything at the turn of a card. A conscious control or management device in such a context is doomed to act tentatively, to make bigger and more frequent mistakes. If – in addition – the control center is connected to the most remote parts of the system and if it even has a powerful technology (Bateson, 1987: 440) capable of intensely or deeply altering the ecosystem, then we have all the ingredients for a major disaster.

Let us consider acceleration. A society becomes unsustainable when it has more and more new options at shorter and shorter intervals. When – for instance – it introduces thousands of new chemical substances into the natural environment every year, or when the same begins to be done with thousands of genetically modified organisms. This is not exactly the same excess as that in the physical scale, and it is not even equivalent to entropic increasing, although the similarity here is important. We are instead dealing with an essential failure of the information system, with a powerful error-amplifying device.

The preceding comment suggests that sustainability would be, if anything, slowing down, parsimony (or in cruder terms, less development). A seemingly paradoxical conclusion, as far as many reports on the environment indicate that there is little time before the environmental balance is damaged dramatically. In other words, the reports describe a situation which apparently demands urgent and quick actions. The answer to the dilemma is that there is not such an urgency to do things, but to leave them undone, or – as Walter Benjamin put in his metaphor – to press the emergency brakes.[13] There are two conceivable social answers to the over-heating of the Earth, for instance. The first one is to drastically reduce energy consumption. The second one is putting into orbit gigantic mirrors to reflect a part of the incoming solar radiation, inserting aluminium balls in the stratosphere (for critical remarks, see Rosenberg and Scott, 1994: 59-60), injecting carbon dioxide in the oceans and geological containers (Dessus and Claverie, 1995), or to keep it frozen in hyper-fridges (Fritsch et al., 1994: 108-121). Only the first answer (non-action route) allows for a reasonable hope in at least not generating more problems than solutions.

In contemporary sociology, it is constantly said that we are living a time of intense and fast changes. But there is a lot of confusion in this perception. Instead, we should say that this era prevents social change, as changing takes time for detecting and correcting the mistakes. As pointed out by philosophy of science, the mind can discover error, but not truth. An excessively accelerated system loses this quality and becomes rigid, unable to develop the flexibility which is needed for selecting viable novelties.

Naturally, the previous observation is also anthropocentric. Something will always happen. Something will also happen to an accelerated system, even many things. The thing is we were talking about intentional and conscious answers, and if the speed is too fast, answers can be intentional (and in fact they probably are because they are human) but – let us put it this way – they are less and less 'conscious'.

Globalization and diversity

I would like to propose a reflection about a shrewd comment on his own astronaut metaphor which Boulding made years after he first formulated it. We quote: 'The most worrying thing about the Earth is that there seems to be no way of preventing it from becoming one world. If there is only one world, then if anything goes wrong, everything goes wrong. Also by the generalized Murphy's Law, every system has some positive probability, however low, of irretrievable catastrophe' (Boulding, 1993: 312-313).

Boulding said this keeping the analogy, that is, applied to an 'Earth-Machine'. But this is even more worrying if we take into account that the Earth is not a machine. He felt this as well but did not draw all the consequences:

> Perhaps the greatest weakness of the metaphor is that the spaceship presumably has a clear destination and a mission to accomplish. It is essentially a planned economy. The evolutionary process, however, is not a significant planned economy any more than an ecosystem. The biological ecosystem is not even a community, in spite of the fact that biologists sometimes call it that, it is the wildest example of free private enterprise and does not even have a mayor. (Boulding, 1993: 313)

It is striking that – in our time, when nobody dares defend the planning of the economy and when it is a commonplace that planning replaces random by error – there are so many people willing to support an idea of sustainability which implies the planning of nature, let alone, a global planning. After all, plans made for society – despite not having been very bright until now – can be conceived in theory. The mind, the system which has turned reflexive, can aim at a certain level of self-regulation (provided not too many new options are imposed at the same time). But it cannot aim at regulating life, upon which it is dependant, since the part cannot entail the whole.

In nature, reduction of diversity takes place in accelerated renewal processes, which does not fit too well with the parsimony co-substantial to sustainability. On the contrary, a very large diversity is found in systems which preserve little energy for the changes and which – despite their stable outlook – are highly vulnerable to external impulses. Therefore an intermediate level of diversity seems appropriate in front of evolutive errors. The idea arises that globalization of economy, centralization of power, and cultural unification, are inherently anti-environmental and consequently not recommendable as suitable structures for environmental performance. In line with this, the 1970s environmentalism did defend decentralization to a great extent. Nowadays, the climate is changing – excuse the joke! – and the tendency seems to be rather different. Some arguments are usually accepted without much discussion, for instance: '...many environmental problems transcend the local level, and some of the more intractable ones are global in scope. Institutions of scale appropriate to deal with such issues are necessary' (Dryzek, 1992: 37). Even Arne Naess (1995: 404) is riding the wave, and defends the adoption of a new motto: 'Think globally, act globally'.

The environmental motto used to go: 'think globally, act locally'. Today, we frequently hear about 'global government' or 'global control'. But both government and control are relative to action, not to thought. In the really existing globalism, the usual trend is to invert – and pervert – the old motto, transforming it into 'local thought and global action'.

Somehow, this discussion has always been confusing. The expression 'think globally' referred mainly to a certain capacity to see beyond immediate interests, to consider oneself a part of a whole with the people of the future, and with the rest of the living beings. I will mention here a passage by Naess himself, as an example. When he wanted to illustrate what global thought actually meant, he wrote down an answer given by a member from an indigenous community to somebody who asked about his opposition to the construction of a dam in the land which traditionally had belonged to his people: 'This place is part of myself' (Naess, 1995: 404). Note he said 'this place' and not 'this world', which – by the way – is very reasonable, as semi-gods capable of having the world in their heads are rarely found.

The advice to act locally was also quite sensible, if we take into account the possibilities of the average people to whom it was addressed. Of course, one could allege that all actions affect the Earth and therefore everybody always acts globally. Although this statement cannot be sustained (the systemic principle that everything is related to everything else must not be interpreted too literally, as in the world only a part of the possible relationships actually become effective), we can admit it in the discussion. Then some individuals are more global than others: they hold more power or they have longer arms; the relations are asymmetric.[14] They are asymmetric in their material dimension: some populations depend almost exclusively on local resources and therefore degrade the local environment, whereas others are nearly exclusively based on external resources, causing degradation everywhere. They are also asymmetric as to information: there are no Latin American networks advising, for instance, on how to deal with the Canadian forests (Gudynas, 1993: 173). As a result, the truly existing globalization tends to discriminate the perspectives and interests of those who have less power (this was, for instance, the general feeling of the women who attended the Rio Summit) (Venkateswaran, 1995: 219).

The so-called 'global environmental change' is – to a great extent – an ideological construction, at least as far as the problems and the institutions which deal with them are concerned.

The way in which problems are presented sometimes results from an 'arbitrary pattern of global labeling' (Buttel and Taylor, 1994: 237). A simple look at the usual lists is enough to realize that reality is much less schematic than suggested by the summary 'environmental problems are global'. Climatic change has to do with 'global commons' – greenhouse gas sinks – but in its predictable effects there is a great geographical diversity, and even the appraised temperature rise is just an average value. And of course, if we were to consider the causes instead of the effects, the diversification is even more obvious. Indeed, large oil or automobile companies have a global reach, but this globality has very little to do with the diffuse responsibility patterns which are normally associated with the sentence 'this problem is global'. The same could be applied to the hole in the ozone layer. In fact, if any progress has been made at all in this field, it is because there were just a few large producers of CFCs. The most pugnacious global element in the bio-diversity debate is the bio-tech companies' wish to have free access to the genetic reserves all around the world. Similarly, there is not a lot of globality, except for the fact that they spread everywhere, in industrial pollution, desertification, soil erosion, scarcity and loss of quality of fresh water, etc. The 'globalizing' presentation of the environmental crisis has more to do with other things: it permits to concentrate on the symptoms instead of the causes (which favors the apparent consensus and makes the political 'management' less conflicting); it permits to simplify and gives a somehow controllable outlook to the agenda (something convenient for all the participants in the 'global environmental complex', from governments to news super-agencies, or different international organisms, corporations and large NGOs). All this involves a shifting of the gravity center which is not only ideological: in Rio, a meeting was held on the climatic change but not on the automotive industry; desertization was discussed but not beef farm-factories, etc.

Similar things could be said about the suitable institutions. The management of commons (even if they are global, as the atmospheric carbon dioxide sink) leads to a debate on hierarchy and mutual support. Theoretically, this management could be approached from a center with separate power or by means of a set of reciprocity rules (like those which permit the sustainable use of resources in numerous small communities which elude the tragedy of the commons through a communal management, without having to delegate on any eco-Leviathan). I do not think there is a clear way-out to this dilemma (and basically, I do not think there is just one way-out). In any case, it may be worth remembering this diversity of options (Keohane and Ostrom, 1995: 21) in order to try to move the debate away from the polarization 'world state or uncooperative anarchy', which excludes many intermediate possibilities for a more or less institutionalized cooperation.

The results of taking into account these dimensions of the subject would blend some of the sociological attempts which, by accepting the ideological framework of globalism, reduce the social content of the environmental crisis to a conflict in which the global capitalist system (made up with transnational companies, the transnational capitalist class, and the consumption culture-ideology) is threatened by an enemy who, being weaker though not less ambitious, turns into a 'global environmentalist system' (Sklair, 1994: 207) (with the corresponding elements: transnational environmentalist organizations, transnational environmentalist elites, and an environmentalist culture-ideology). The excessive schematism of this giants-fight should be sufficient by now, at least, to moderate its ambitions.

Ending remarks

In the previous comments, there is plenty of political and cultural critique. I do not think this polemic impulse should be repressed, not because a more abstract and 'neutral' formulation is not possible. In fact, all which has been previously said could be summarized by saying that a system depending on growing energy inputs tends to extreme instability forms and, if it only contains one evolutive line then the instability will be fatal. However, if the analysis outlined in the first part of this paper is right, the specific system I am talking about is a very particular one which cannot detach its energy consumption from its ghosts. What we could call the 'emerging paradigm of sustainable development' – greenish version of the end of history – seems to be prisoner to the basic myths of the European patriarchy, of a pact with the devil in return of knowledge and dominance over the world. A transaction, as everybody knows, whose payment cannot be deferred in a 'sustainable' way.

The beneficial effects that could perhaps be derived from following the old environmentalist advice (scale down, slow down, democratize, decentralize (Roszak, 1993: 312)) are undoubtedly excluded, despite its promises, from the new program which announces the sustainability of development by means of more energy, more megatechnics, and more concentration of power. Combining social betterment and the continuity of life is not exactly a problem, it is rather a dilemma proposed by a sphinx. Let us put it in a different way: it is the re-establishment of history.

Notes

1. The Spanish sociologist Jesús Ibáñez did a creative profound work on this line, mainly on epistemology, research techniques and sociology of consumption. In the last years of his life, he had an interest in environmental matters and he proposed a sociological framework defined by three characteristics: a) it is *ecological*, i.e. because of the fact that every open system exists in an ecosystem, the survival unit (as well as the object of knowledge) is not the system but the circuit between system and ecosystem; b) it is *global*, with this word meaning 'a system cannot survive if all its related ecosystems do not survive'; c) it is *complex*, i.e. it goes from elements to the whole, it integrates randomness (dis-order, noise), it informs of singularities, and it includes the observer into observations – since the object is a product of a subject's objectifying action (Ibáñez, 1992).
2. See, for instance, Hardin (1993). There is, in Hardin's book, an important exception to his overall reductionist drive. Although commons must be managed, different schemes to manage them (market, state, community) can be socially established. His theory has no answer to the query about the best system of distribution (Hardin, 1993: 218-220).
3. The net primary production is the amount of energy which remains after having subtracted the respiration of primary producers (mainly plants) from the total amount of energy (most of it being solar) which is fixed biologically, and it 'provides the basis for maintenance, growth, and reproduction of all heterotrophies' (Vitousek et al., 1986: 368).
4. Although the authors display the usual reservation and clarify that the information they supply cannot be used directly for the calculation of the long-term carrying capacity for human beings (as it also depends on consumption and technology (Vitousek et al., 1986: 372-373)), it is obvious that they provide a powerful argument in favor of the less expansive forecasts and therefore in favor of population control: a society with a low consumption and not a very sophisticated technology but very dense would not be sustainable either.
5. However, it also has to be said that other aspects of Norgaard's book hint at combining several approaches in 'a co-evolving patchwork quilt of discursive communities' (Norgaard, 1994: 172).
6. In my opinion, the best formulation of the theoretical bases of this point of view can be found in Kafka (1993). With other grounds, many members of the so called 'culturalist school' work on lines which refer to the learning approach.
7. There is a remarkable difference between both sources: Daly proposes a rule for the use of nonrenewable resources according to which their consumption should be established by the pace set by the development of renewable substitutes. On the limit, only a strictly solar civilization would be sustainable, that is, a civilization using exclusively solar power and biospheric materials. I think this is too extreme a proposal. There has never been (and there will probably never be) a solar civilization in this sense. Civilization is near synonymous to nonrenewable resource consumption. Therefore, a parsimony rule – like the one proposed in the table – seems more plausible. A second difference is that Daly insists more on a rule with global scope, according to which sustainability means to remain within the carrying capacity of the biosphere. Simplifications embodied in operational rules for environmental sustainability are surely the outcome of a

conscious purpose in their proposers' mind, motivated by polemic and pedagogic reasons, and cannot be found in other arguments by the same authors. In fact, I totally agree with the latest Daly's programme: 'the most pressing need is to stop the exponential expansion of this subsystem boundary under the current regime of economic imperialism – but without falling prey to the seductions of ecological reductionism' (Daly, 1996: 12).

8. Some of the texts of Hannah Arendt contain valuable insights into this question. She turned down the idea of finding a place totally outside the natural world (1977: 280). On the other hand, by stating that freedom depends on a complex articulated society, on such an artificial social construction as 'a republic with a constitution' (see Young-Bruehl, 1993: 511), she rejected the nostalgia for the return to an imaginary 'natural community'. The criterion that comes off, that civilization implies to overcome the naturalism and also to impose limits to the hubris, constitutes a quite fruitful formulation of the social conditions of sustainability.
9. You may have probably noticed that – so far – the argumentation of this chapter flows totally separate from the thesis which states that there is no problem in getting wealthier because then there will be more resources for the cleaning of the environment (post-industrialism) and a greater social willingness to do so (post-materialism). It is obvious that this is only meaningful, and only partially meaningful, if the portion of the product devoted to make up for the environmental damage is small. If such a portion increases a lot (let us play the numbers game again: up to what? 10%, 50%, 90%?) then the assumed wealth would be undermined and canceled. This is more or less what Jim O'Connor (1991) noted in his thesis on the second contradiction or the growing cost of supplying the natural conditions of production.
10. One of the attempts in this regard – probably the most ambitious – is Fritsch et al., 1994 (see Chapter 3 and 4 especially).
11. Georgescu-Roegen's classic, *The Entropy Law and the Economic Process* (1971), is a compulsory reference. The controversy caused with what he called 'the energetic dogma', that is, the belief that no materials scarcity will impose limits provided there is abundant energy, was developed in different works (see Georgescu-Roegen, 1975, 1979, 1982a, 1982b).
12. Georgescu-Roegen's conviction that 'matter matters too' led him to the formulation of an entropy law for materials (the so-called 'fourth law of thermodynamics'), according to which matter, the same as energy, exists in available and unavailable qualities, and it is also subject to irreversible degradation (1993: 197). Many critics of this idea have disregarded its economic, anthropomorphic context, and they have perceived it, at best, as a blurred extension of the second law. Well, perhaps it is not a good physical law, but it is sensible to take it into consideration as a good economic law, as it may be known by anybody familiar with already existing experiences of recycling.
13. Riechmann (1991) is a contribution to the current debate on the environmental crisis – more concretely, to its political dimensions – inspired by Benjamin's metaphor.
14. Giddens, for instance, supports that 'at present, everyday actions of the individuals have global consequences'. He describes worldwide trade as the main indicator of interconnectance, illustrating this idea in the following way: 'My decision to buy a particular clothing item, for instance, or a specific type of food, has many global implications' (Giddens, 1995: 57-58). But then, it makes sense to say that consumers are 'more global' when they are 'more affluent' (in a double sense: they have more money to buy things and a greater diversity of options).

Bibliography

Altvater, E. (1994), *El Precio del Bienestar: Expolio del Medio Ambiente y Nuevo Desorden Mundial* Translated into Spanish by M. Ardid. València: Edicions Alfons el Magnànim.

Arendt, H. (1967), *The Origins of Totalitarianism.* London: George Allen and Unwin.

Arendt, H. (1977), *Between Past and Future. Eight Exercises in Political Thought.* New York: Penguin.

Bateson, G. (1987), *Steps to an Ecology of Mind.* London: Jason Aronson.

Bateson, G. (1991), *A Sacred Unity: Further Steps to an Ecology of Mind.* New York: A Cornelia & Michael Bessie Books/HarperCollins.

Benhabib, S. (1996), *The Reluctant Modernism of Hannah Arendt.* London: Sage.

Bohman, J. (1991), *New Philosophy of Social Science: Problems of Indeterminacy.* Cambridge: Polity Press.

Boulding, K.E. (1966), The Economics of the Coming Spaceship Earth, in: H. Jarrett (ed.), *Environmental Quality in a Growing Economy.* Baltimore: Johns Hopkins University Press, pp. 3-15.

Boulding, K.E. (1993), Spaceship Earth Revisited, in: H.E. Daly and K.N. Townsend (eds), *Valuing the Earth: Economics, Ecology, Ethics.* Cambridge, Mass.: MIT Press, pp. 311-313.

Boulding, K.E. (1995), The Limits to Societal Growth, in: E. Boulding and K.E. Boulding, *The Future: Images and Processes.* London: Sage, pp. 26-39.

Buttel, F. and P. Taylor (1994), Environmental Sociology and Global Environmental Change, in: M. Redclift and T. Benton (eds), *Social Theory and the Global Environment.* London: Routledge, pp. 228-256.

Clarke, J.I. (1995), The Interrelationship of Population and Environment, in: D.E. Cooper and J.A. Palmer (eds), *Just Environments: Intergenerational, International and Interspecies Issues.* London: Routledge, pp. 34-46.

Colborn, T., D. Dumanoski and J. Peterson Myers (1997), *Our Stolen Future. Are We Threatening Our Fertility, Intelligence, and Survival? – A Scientific Detective Story.* London: Abacus.

Commoner, B. (1978), *El Círculo que se Cierra.* Translated into Spanish by J. Ferrer. Barcelona: Plaza & Janés.

Daly, H.E. (1991), *Steady-state Economics. Second edition with new essays.* Washington: Island Press.

Daly, H.E. (1996), *Beyond Growth: The Economics of Sustainable Development.* Boston: Beacon.

Dessus, B. and M. Claverie (1995), ¿Hay que Almacenar el Dióxido de Carbono en el Fondo de los Océanos?, *Mundo Científico*, n. 159, July-August: 640-645.

Dryzek, J.S. (1992), Ecology and Discursive Democracy: Beyond Liberal Capitalism and the Administrative State. *Capitalism, Nature, Socialism,* 3 (10): 18-43.

Ehrlich, P.R. and A.H. Ehrlich (1993), *La Explosión Demográfica: El Principal Problema Ecológico.* Translated into Spanish by C. Batlle. Barcelona: Salvat.

Fritsch, B., S. Schmidheiny and W. Seifritz (1994), *Towards an Ecologically Sustainable Growth Society: Physical Foundations, Economic Transitions, and Political Constraints.* Berlin: Springer.

Garcia, E. (1995a), *El Trampolí Fàustic: Ciència, Mite i Poder en el Desenvolupament Sostenible.* Alzira: Germania.

Garcia, E. (1995b), Notas sobre 'Desarrollo Sustentable' y Propósito Consciente. *Ecología Política*, n. 10: 45-59.

Georgescu-Roegen, N. (1971), *The Entropy Law and the Economic Process.* Cambridge, Mass.: Harvard University Press.

Georgescu-Roegen, N. (1975), Energy and Economic Myths. *Southern Economic Journal,* XLI: 347-381.

Georgescu-Roegen, N. (1977), Inequality, Limits and Growth from a Bioeconomic Viewpoint. *Review of Social Economy,* XXXV (December): 361-375.

Georgescu-Roegen, N. (1979), Energy and Matter in Mankind's Technological Circuit, in: P.M. Nemetz, *Energy Policy: The Global Challenge.* Toronto: Butterworth, pp. 107-127.

Georgescu-Roegen, N. (1982a), The Crisis of Resources: Its Nature and its Unfolding, in: G.A. Daneke, *Energy, Economics, and the Environment: Toward a Comprehensive Perspective.* Lexington, Mass.: Lexington Books, pp. 9-24.

Georgescu-Roegen, N. (1982b), Energetic Dogma, Energetic Economics, and Viable Technologies, in: J.R. Moroney, *Advances in the Economics of Energy and Resources,* Vol. 4. Greenwich, Conn.: JAI Press, pp. 1-39.

Georgescu-Roegen, N. (1984), Feasible Recipes Versus Viable Technologies. *Atlantic Economic Journal,* XII (1): 21-31.

Georgescu-Roegen, N. (1993), Thermodynamics and We, the Humans, in: *Entropy and Bioeconomics. First International Conference of the EABS. Proceedings.* Edited by the European Association for Bioeconomic Studies. Milano: Nagard, pp. 184-201.

Giddens, A. (1995), Living in a Post-traditional Society, in: U. Beck, A. Giddens and S. Lash, *Reflexive Modernization: Politics, Tradition and Aesthetics in the Modern Social Order.* Cambridge: Polity Press, pp. 56-110.

Gudynas, E. (1993), The Fallacy of Ecomessianism: Observations from Latin America, in: W. Sachs, *Ecology: A New Arena of Political Conflict.* London: Zed Books, pp. 170-179.

Hardin, G. (1993), *Living within Limits: Ecology, Economics, and Population Taboos.* New York: Oxford University Press.

Ibáñez, J. (1990), Hacia una Etica de la (Eco)responsabilidad, *Contrarios,* n. 4: 56-69.

Ibáñez, J. (1992), El Paradigma Ecológico en Sociología. Paper presented to the Workshop 'La Sociología Frente a la Crisis Ecológica'. València: UIMP.

Jensen, D. (1995), *Listening to the Land: Conversations about Nature, Culture and Eros.* San Francisco: Sierra Club Books.

Kafka, P. (1993), Conditions of Creation: The Invisible Hand and the Global Acceleration Crisis, in: *Entropy and Bioeconomics. First International Conference of the EABS. Proceedings.* Edited by the European Association for Bioeconomic Studies. Milano: Nagard, pp. 344-369.

Keohane, R.O. and E. Ostrom (eds) (1995), *Local Commons and Global Interdependence: Heterogeneity and Cooperation in Two Domains.* London: Sage.

Lewin, R. (1994), *La Complexité: Une Théorie de la Vie au Bord du Chaos.* Translated into French by B. Loubières. Paris: InterÉditions.

Livi-Bacci, M. (1997), *A Concise History of World Population* (Second Edition). Oxford: Blackwell.

Margalef, R. (1991), *Teoría de los Sistemas Ecológicos.* Barcelona: Publicacions de la Universitat de Barcelona.

Maturana, H. and F. Varela (1988), *El Arbol del Conocimiento: Las Bases Biológicas del Entendimiento Humano.* Santiago de Chile: Editorial Universitaria.

Meadows, D.H., D. Meadows and J. Randers (1992), *Beyond the Limits: Global Collapse or a Sustainable Future.* London: Earthscan.

Naess, A. (1995), The Third World, Wilderness, and Deep Ecology, in: G. Sessions, *Deep Ecology for the Twenty-first Century.* Boston: Shambhala, pp. 397-408.

Norgaard, R.B. (1994), *Development Betrayed: The End of Progress and a Coevolutionary Revisioning of the Future.* London: Routledge.

O'Connor, J. (1991), The Second Contradiction of Capitalism: Causes and Consequences, in: *Conference Papers,* CES/CNS Pamphlet 1: 1-10.

Pimentel, D., R. Harman, M. Pacenza, J. Pecarsky and M. Pimentel (1994), Natural Resources and an Optimum Human Population, *Population and Environment,* 15 (5): 347-369.

Revelle, R. (1974), Food and Population. *Scientific American,* 231: 161-170.

Riechmann, J. (1991), *¿Problemas con los frenos de emergencia? Movimientos ecologistas y partidos verdes en Alemania, Holanda y Francia.* Madrid: Revolución.

Rosenberg, N. J. and M.J. Scott (1994), Implications of Policies to Prevent Climate Change for Future Food Security. *Global Environmental Change: Human and Policy Dimensions,* 4 (1): 49-62.

Roszak, T. (1993), *The Voice of the Earth: An Exploration of Ecopsychology.* London: Bantam.

Rushdie, S. (1996), *The Moor's Last Sigh.* London: Vintage.

Sachs, W. (1993), Global Ecology and the Shadow of 'Development', in: W. Sachs, *Global Ecology: A New Arena of Political Conflict.* London: Zed Books, pp. 3-22.

Schmidheiny, S. (con el Consejo Empresarial para el Desarrollo Sostenible) (1992), *Cambiando el Rumbo: Una Perspectiva Global del Empresariado para el Desarrollo y el Medio Ambiente.* México: Fondo de Cultura Económica.

Schrödinger, E. (1992), *What is Life?: The Physical Aspect of the Living Cell (With Mind and Matter & Autobiographical Sketches).* Cambridge: Cambridge University Press.

Seabrook, J. (1993), *Victims of Development: Resistance and Alternatives.* London: Verso.

Shiva, V. (1995), *Monocultures of the Mind. Perspectives on Biodiversity and Biotechnology.* London: Zed Books.

Sklair, L. (1994), Global Sociology and Global Environmental Change, in: M. Redclift and T. Benton, *Social Theory and the Global Environment.* London: Routledge, pp. 205-228.

Tabah, L. (1995), Population Prospects with Special Reference to the Environment, in: D.E. Cooper and J.A. Palmer, *Just Environments: Intergenerational, International and Interspecies Issues.* London: Routledge, pp. 72-89.

Turner, R.K. (1988), *Sustainable Environmental Management: Principles and Practice.* London: Belhaven.

Tyler Bonner, J. (1992), Introduction, in: D'Arcy W. Thompson, *On Growth and Form.* Cambridge: Cambridge University Press, pp. xiv-xxii.

Venkateswaran, S. (1995), *Environment, Development and the Gender Gap.* New Delhi: Sage.

Vitousek, P.M., P.R. Ehrlich, A.H. Ehrlich, and P.A. Matson (1986), Human Appropriation of the Products of Photosynthesis. *BioScience* 34 (6): 368-374.

Von Foerster, H. (1991), Sobre Sistemas Autoorganizadores y sus Ambientes, in: *Las Semillas de la Cibernética.* Translated into Spanish by M. Pakman. Barcelona: Gedisa, pp. 39-55.

Von Weizsäcker, E.U., Hunter L. Lovins and A.B. Lovins (1997), *Factor 4. Duplicar el Bienestar con la Mitad de los Recursos Naturales. Informe al Club de Roma.* Translated into Spanish by Adan Kovacsics. Barcelona: Galaxia Gutenberg/Círculo de Lectores.

Waldrop, M.M. (1994), *Complexity. The Emerging Science at the Edge of Order and Chaos.* London: Penguin.

Young-Bruehl, E. (1993), *Hannah Arendt.* Translated into Spanish by M. Lloris. Valéncia: Edicions Alfons el Magnànim.

Zoja, L. (1993), *Crescita e Colpa.* Milano: Anabasi Spa.

Index

agency (role of human), 50
attitude-behavior paradigm, 59
biodiversity agreement, 160
Blueprint for Survival, 41, 215
Brent Spar, 200
Brundtland report, 45, 230
carrying capacity, 233, 236, 237, 239
cassava connection, 155
cell-tissue society, 42
Chernobyl, 3, 61, 62, 108, 165
Chicago School, 5
citizen-consumers, 8, 25
class concepts and the environment, 31, 32
classical sociological theory, 17 ff
cleaner technology (transfer of), 158; *see also* technology
climate change (policies), 158 ff
co-evolution (theory of), 235 ff
constructionism/constructivism (social), *see* realism versus constructivism
consumption, 57, 199
  domestic, 35, 58
  critique of, 159
co-production, 204
counterproductivity (theory), 7, 42, 43
developing countries (role of), 30, 111, 229 ff
discourse coalitions, 141, 166
disembedding mechanism, 194
dominant western worldview, 25
ecocentrism versus anthropocentrism, 240
ecological modernization (theory), 4, 5–7, 12, 13, 19, 28, 29, 30, 41–71, 135 ff, 163–185
  and the critique, 30, 32, 51, 196, 209–228
ecological switch-over, 48
economic growth, 152, 212
entropy law, 235, 242
environmental degradation (concepts of), 26, 214–215
environmental economics, 43, 235 ff
environmental justice; *see* inequalities
environmental movement, 25, 26, 30, 35, 73, 140, 169, 196, 202, 213
environmental problems (definitions and perceptions of), 28, 152, 153, 154, 157; *see also* sustainability (concepts of)
environmental sociology (North American variant of), 11, 12, 18, 19, 24 ff
equity and environment; *see* inequalities
European Union (EU), 122, 136, 138, 140, 155, 201, 203
exemptionalism/exeptionalism; *see* human exceptionalism paradigm
export processing zones, 133
Friends of the Earth International (FOEI), 141; *see also* environmental movement
free riders, 112, 157
General Agreement on Tariffs and Trade (GATT), 138, 153

global commons, 160
global environmental management, 155, 160, 161
Greatest Permissible Pollution (GPP), 172, 173
greener production (attitutes towards), 175 ff
greening of industry, 178
Greenpeace, 141, 169, 200
Group of Seven (G7), 155, 203
High Consequence Risks (HCR), 62, 131, 144
human ecology, 4–5, 21, 22, 34, 35
Human Exemptionalist Paradigm (HEP), 4, 10, 11, 19, 23, 24, 110
industry (role of), 52, 154, 172 ff
industrial ecology, 9
industrial society (theory), 49
industrialism versus capitalism, 48, 51
inequalities (and environment), 31, 35, 107, 110, 133, 157, 159
institutional learning, 166
institutionalism, 130
Intergovernmental Panel on Climate Change (IPCC), 7, 14, 135
International Monetary Fund (IMF), 138, 155
International Standard Organization (ISO), 145
Kyoto, 223
liberalization (of trade and markets), 157
lifestyle (concept of), 60, 64
Limits to Growth, 233, 236
Lisbon Group, 121, 124, 145
Marxism (neo-), 20, 21, 23, 42, 82, 89 ff,
modernization theories, 189
  simple modernization, 28–29
Multi National Enterprises (MNE), 137
nation state (role of), 2, 139 ff
neo-corporatism, 190
New Ecological Paradigm (NEP), 4, 10, 18; *see also* Human Exemptionalist Paradigm
New Industrializing Countries (NIC), 137
New Social Movements, 189
NIMBY, 203
Non-Governmental Organizations (NGO's), 202
nuclear energy, 108
ontological security, 76
Organization for Economic Cooperation and Development (OECD), 136–138, 201
patent rights (in nature), 160
policy network approach, 203
political modernization, 13, 187–208
pollution charges, 171, 175 ff
Pollution Prevention Pays/Pollutor Pays Principle (PPP), 6–7, 172–173, 181
pollution quotas, 160
populationism, 237–238
postmodernism/post-modernity, 4, 7 ff, 129, 192, 216 ff
post-traditional politics, 63, 192
power (concepts of), 111 ff; *see also* inequalities
precautionary principle, 159
Probabilistic Risk Assessment (PRA), 80
property rights, 113
Rational Actor Paradigm (RAP), 75, 77, 78–81
rationality (concepts of), 6, 22, 48, 52–56, 130, 195, 213, 216
realism (neo) and international relations, 123, 129
realism versus constructivism, 9, 12, 19, 27 ff, 63, 82, 103–119, 156, 214, 224
reflexive modernization (theory), 6, 19, 28–29, 90, 121, 127, 189–190, 211, 221
regimes (international environmental policy), 64, 140, 142

Rio Earth summit, 141, 160, 248
risk society (theory), 3, 12, 19, 28–29, 31, 73–101, 190
double risk societies, 13, 164 ff
Santa Barbara oil spring, 115
science and technology (role of), 10, 29, 51, 63, 134, 191, 214
SHELL, 138
social Darwinism, 21, 24
social facts, 20
social theory (role of), 152, 161
social versus physical factors, 104
society-nature relationships, 231
sociology of consumption, 8, 57, 197
Socio-Material-Collective-Systems (SMCS), 59
sovereignty, 157
state (role of the), 134, 187, 191, 195–199, 210
state-market relationships, 177, 188
state-society-relations, 33
sub-politics, 29, 64, 134, 139,144, 192, 200, 204
sustainability (concepts of), 55, 152, 157, 229–253
sustainable development, 141
system theory, 232 ff
systems of provision, 57
technology (role of), 43, 49, 52, 60, 105, 158
Temporarily Permissible Pollution (TPP), 172
Third World countries, 111; *see* developing countries
Toxic Release Inventory (TRI), 110
treadmill of production, 25
UNCED, 160; *see* Rio Earth summit
Union Carbide, 138
United Nations (UN), 122, 161
universal principles, 141
value change, 25
World Bank, 31, 138, 155, 158
World Commission on Environment and Development (WCED), *see* Brundtland report
World Health Organization (WHO), 153
World Trade Organization (WTO), 122, 138, 153, 155, 203
World Wide Fund for Nature (WWF), 141, 169